Advanced Applications of Remote Sensing for Agricultural Crops

Advanced Applications of Remote Sensing for Agricultural Crops

Edited by Callen Elliott

Larsen & Keller
www.larsen-keller.com

Advanced Applications of Remote Sensing for Agricultural Crops
Edited by Callen Elliott
ISBN: 979-8-88836-078-1 (Hardback)

Larsen & Keller
Published by Larsen and Keller Education,
5 Penn Plaza,
19th Floor,
New York, NY 10001, USA

Cataloging-in-Publication Data

Advanced applications of remote sensing for agricultural crops / edited by Callen Elliott.
p. cm.
Includes bibliographical references and index.
ISBN 979-8-88836-078-1
1. Agriculture--Remote sensing. 2. Crops--Remote sensing. 3. Plants--Remote sensing.
4. Vegetation monitoring. I. Elliott, Callen.
S494.5.R4 A38 2023
577.82--dc23

For more information regarding Larsen and Keller Education and its products, please visit the publisher's website www.larsen-keller.com

Table of Contents

Preface

The main aim of this book is to educate learners and enhance their research focus by presenting diverse topics covering this vast field. This is an advanced book which compiles significant studies by distinguished experts in the area of analysis. This book addresses successive solutions to the challenges arising in the area of application, along with it; the book provides scope for future developments.

Remote sensing is the technology used for acquiring information about an object or any phenomenon without making any physical contact with the object. There are several applications of remote sensing in the field of agriculture. It plays a crucial role in observing and monitoring the health of crops. It also helps in identifying the crops affected by conditions such as bad weather and pests. This information is essential for making timely crop management decisions. It is used to determine the properties of soil including soil texture, soil organic matter, soil pH level, and moisture content. Some other major applications of remote sensing in agriculture include predicting weather conditions, monitoring water conditions, precision farming, observing air quality, and monitoring and predicting climate changes. This book outlines the advanced applications of remote sensing for agricultural crops. A number of latest researches have been included to keep the readers updated with the latest concepts in this area of study. The book aims to serve as a resource guide for students and experts alike and contribute to the growth of research on this topic.

It was a great honour to edit this book, though there were challenges, as it involved a lot of communication and networking between me and the editorial team. However, the end result was this all-inclusive book covering diverse themes in the field.

Finally, it is important to acknowledge the efforts of the contributors for their excellent chapters, through which a wide variety of issues have been addressed. I would also like to thank my colleagues for their valuable feedback during the making of this book.

Editor

1

Satellite and Proximal Sensing to Estimate the Yield and Quality of Table Grapes

Evangelos Anastasiou [1,*], Athanasios Balafoutis [1,2], Nikoleta Darra [1], Vasileios Psiroukis [1], Aikaterini Biniari [3], George Xanthopoulos [1] and Spyros Fountas [1]

1 Department of Natural Resources Management & Agricultural Engineering, Agricultural University of Athens, Iera Odos 75, 11855 Athens, Greece; a.balafoutis@certh.gr (A.B.); nicoletadarra@aua.gr (N.D.); vassilispsiroukis@gmail.com (V.P.); xanthopoulos@aua.gr (G.X.); sfountas@aua.gr (S.F.)

2 Institute for Bio-Economy & Agri-Technology, Centre of Research & Technology Hellas, Dimarchou Georgiadou 118, 38221 Volos, Greece

3 Faculty of Crop Science, Agricultural University of Athens, Iera Odos 75, 11855 Athens, Greece; kbiniari@aua.gr

* Correspondence: evangelos_anastasiou@aua.gr

Abstract: Table grapes are a crop with high nutritional value that need to be monitored often to achieve high yield and quality. Non-destructive methods, such as satellite and proximal sensing, are widely used to estimate crop yield and quality characteristics, and spectral vegetation indices (SVIs) are commonly used to present site specific information. The aim of this study was the assessment of SVIs derived from satellite and proximal sensing at different growth stages of table grapes from veraison to harvest. The study took place in a commercial table grape vineyard (*Vitis vinifera* cv. Thompson Seedless) during three successive cultivation years (2015–2017). The Normalized Difference Vegetation Index (NDVI) and Green Normalized Difference Vegetation Index (GNDVI) were calculated by employing satellite imagery (Landsat 8) and proximal sensing (Crop Circle ACS 470) to assess the yield and quality characteristics of table grapes. The SVIs exhibited different degrees of correlations with different measurement dates and sensing methods. Satellite-based GNDVI at harvest presented higher correlations with crop quality characteristics ($r = 0.522$ for berry diameter, $r = 0.537$ for pH, $r = 0.629$ for berry deformation) compared with NDVI. Proximal-based GNDVI at the middle of veraison presented higher correlations compared with NDVI ($r = -0.682$ for berry diameter, $r = -0.565$ for berry deformation). Proximal sensing proved to be more accurate in terms of table grape yield and quality characteristics compared to satellite sensing.

Keywords: spectral vegetation index; precision viticulture; remote sensing; table grapes; crop yield and quality estimation

---

1. Introduction

Table grapes have been included in the human diet since ancient times. The global production of table grapes reached 22.7 million tons in 2017 [1]. Table grapes can be defined according to their berry colour (green, blue or red cultivars), the existence of seeds in berries (in seeded or seedless cultivars) and the berry size (small, medium and large sizes). Seedless table grape varieties are the most popular nowadays as they are more consumer friendly [2]. Table grapes are affected strongly in terms of quantity and quality characteristics by agronomical practices and field conditions. The impacts of pre-harvest activities influence the major quality variables of table grapes—total soluble solids, acidity and berry diameter—which are strongly connected with storability and consequently, with shelf life, which has an impact on the final table grape price [3–6]. As a consequence, there is a need to sufficiently and effectively assess table grape production at the field level in order to arrange better postharvest operations (storage, packing, transportation to the final markets) during the growing season. This assessment of final yield and quality at the field level during the growing season would be very valuable to the farmers and remote sensing could provide such services.

Remote sensing has proved to be game changing for the agricultural sector, as it is one of the backbones of precision agriculture. Remote sensing estimates the properties of a plant through non-destructive processes in a fast and accurate way [7]. Remote sensing has a wide range of applications in agriculture, including crop growth monitoring, crop yield and quality estimation, identification of irrigation needs, as well as biotic and abiotic damage (e.g., pests infestations, disease infections, hail damage, flood and drought damage) [8].

Spectral vegetation indices (SVIs) are mathematical quantitative combinations of the absorption and scattering rates of plants in different bands of the electromagnetic spectrum and are used to measure the properties of a crop. SVIs provide a simple yet elegant method for measuring plant responses throughout the season, exploiting the basic differences between soil and plant spectra, and are often calculated as a type of relationship between reflected light in the visible and near infrared (NIR) wavelengths. Healthy green plants have relatively low reflectance and transmittance in the visible regions of the spectrum (high absorbance of light for photosynthesis). Their reflectance and transmittance, however, are usually high in the near infrared (NIR) region. The most well-known SVI is the Normalized Difference Vegetation Index (NDVI) [9]. NDVI has been correlated with crop parameters, such as wet biomass, leaf area index, plant height and grain yield [10]. Based on the NDVI, more indices have been generated that present equal or better performances in the estimation of crop-related parameters and are based on the same or different bands of the electromagnetic spectrum, such as the Normalized Difference Red Edge Index (NDRE) [11], the Advanced Normalised Vegetation Index (ANVI) [12], the Green Normalised Difference Vegetation Index (GNDVI) [13], the Green Red Vegetation Index (GRVI) [14], the Damage Sensitive Spectral Index (DSSI) [15], the Optimized Soil-Adjusted Vegetation Index (OSAVI) [16], the Ratio Vegetation Index (RVI) [17], the Simple Ratio (SR) [18] and the Enhanced Vegetation Index (EVI) [19]. Despite the usefulness of the SVIs in assessing the crop growth and yield parameters, the platform that is used to acquire the different bands is very important and can be divided into satellite, aerial and proximal platforms, based on the distance to the assessed crop [20,21]. Consequently, satellite and aerial remote sensing can provide regular spatially- and temporally-explicit information across large areas when compared to proximal sensing which is mainly used for small scale applications with higher resolution [22].

After the first satellites were launched and effectively provided usable data, the need for higher spatial resolution satellite imaging systems with quicker revisit cycles became obvious. Nowadays, satellites with sub-meter resolution like GeoEye and WorldView, and only 1 day revisit time can be used in agriculture [23]. In addition, there are freely available satellite imagery with 20 m resolution and 5 days revisit time, like the one that is provided by the Sentinel satellites [24].

Satellite images from the SPOT satellites were used to zone the South African viticultural terroirs [25]. Landsat imagery was used to locate vineyards in Spain [26]. Satellite data from Landsat 8, MODIS and GOES were used to provide daily field scale evapotranspiration estimates over two vineyards [27]. High-resolution multispectral satellite imagery from IKONOS was used to map the leaf area of wine grape vineyards, providing plant growth models and decision support for irrigation and canopy management from the resulted maps [28]. Kandylakis and Karantzalos (2016) used high resolution satellite imagery and found that satellite data have the potential to describe wine grape quality parameters [29]. Borgogno-Mondino et al. (2018) found that the NDVI derived from Landsat 8 imagery was highly correlated with NDVI derived from aerial imagery at a vineyard scale when assessing vine vigor to produce prescription maps [30]. However, some studies indicated that the spatial resolutions of medium resolution satellite imagery are not sufficient for assessing vineyards due to the narrow vine spacing; this problem is more intensive in vineyards with large heterogeneity, and satellite data of higher resolution can produce comparable results with aerial platforms [31,32]. At the same time, temporal resolution and cloud cover that can occur at the time the satellite passes are also limitations that should be taken into consideration [33]. Due to the latter, there is limited research on the application of medium resolution satellite imagery in estimating wine and table grape yield and quality characteristics.

In comparison to remote sensing, proximal sensing is based on the usage of ground-based moving vehicles carrying various types of sensors that are suitable for continuous measurements of soil or canopy parameters [31]. The advantages of proximal sensors advantages are (1) their high resolution imagery; (2) their total independence from external parameters and limitations (e.g., cloud cover); (3) their suitability for small fields; and (4) their simple application (i.e., mounting the sensor on the tractor). Cerovic et al. (2008) presented a study on a non-destructive optical method that allowed both flavonol and anthocyanin contents of intact berry skin to be measured. The data was acquired with three optical sensor devices (a Dualex FLAV, a Dualex ANTH and a prototype Multiplex) that were used for the screening of grape chlorophyll fluorescence [34]. Llorens et al. (2011) compared Ultrasonic and Light Detection and Ranging (LIDAR) sensors with the traditional manual and destructive canopy measurement procedure in vineyards, and both of the sensors performed well. The authors concluded that the ultrasonic sensor is an appropriate tool for the determination of the average canopy characteristics, while the LIDAR sensor provides more accurate and detailed information about the canopy [35]. A mobile terrestrial laser scanner was used by del-Moral-Martinez et al. (2016) to map the Leaf Area Index (LAI) of a vineyard, and then the impacts of different scanning methods (on-the-go or discontinuous systematic sampling) on the reliability of the resulting maps were examined, with an analysis of correlation between maps. It was found that the terrestrial sensor can be used discontinuously in specific sampling sections separated by up to 15 m along the rows, although this method naturally reduces the amount of field data acquired [36]. Gatti et al. (2016) tested the performance of a terrestrial multi-sensor (MECS-VINE) in terms of reliability and degree of correlation with several canopy growth and yield parameters in grapevines. The results showed high correlation between the Canopy Index and any canopy parameter at any date, especially canopy gaps and leaf layer number, as well as a good correlation between cluster and berry weight, suggesting that the sensor is also potentially able to make accurate yield estimations [37]. On the contrary, Kazmierski et al. (2011) and Fountas et al. (2014) found SVIs to have limited performance in an assessment of wine grape yield and quality parameters through proximal sensing [38,39].

While many studies have been conducted on wine grapes using satellite and proximal remote sensing, the application of these methods in table grapes is considered relatively new. In particular, the application of freely available satellite imagery on table grapes is very limited, as is the comparison of this method with proximal sensing. Table grapes differ from wine grapes mainly due to the use of different trellis systems which have increased exposure to sunlight and thus, could have different impacts on the spectral measurements. Therefore, the main aim of this study was to assess the use of freely available satellite and proximal remote sensing methods on the production of table grapes and the correlation with yield and quality.

2. Materials and Methods

2.1. Field Site

The study was conducted at a commercial table grape vineyard located in Southern Greece during the 2015, 2016 and 2017 cultivation years (Figure 1). The vineyard was planted in 2006 with *Vitis vinifera* L. cv. Thompson seedless (37°54.532′ N, 22°44.798′ E, Corinth, Greece) at 1.4 ha. The variety was grafted onto 1103 Paulsen rootstock. The field exhibited variation in soil texture with two different soil types (sandy clay loam and clay loam). Table grapes were trained to a double cross arm trellis system and spaced at 1.8 m × 2.6 m. The vineyard received numerous vineyard operations (canopy management) to adjust the more vigorous vegetation. In addition, plant growth regulators and fertilizers were excessively used to reach commercial standards regarding the berry diameter and sugar content along with high yield. The vineyard was irrigated with approximately 2400 mm/ha, while 16 spraying applications of foliar fertilizers, plant protection products and crop growth regulators and 4 canopy management practices were applied on an annual basis. A regular grid of 36 cells (298–404 m^2 per cell) was set up to facilitate field sampling in order to assess crop vigour, yield and grape quality covering the total area, as shown in Figure 2. This methodology was also used by Tagarakis et al. (2013) and Farid et al. (2016) [40,41].

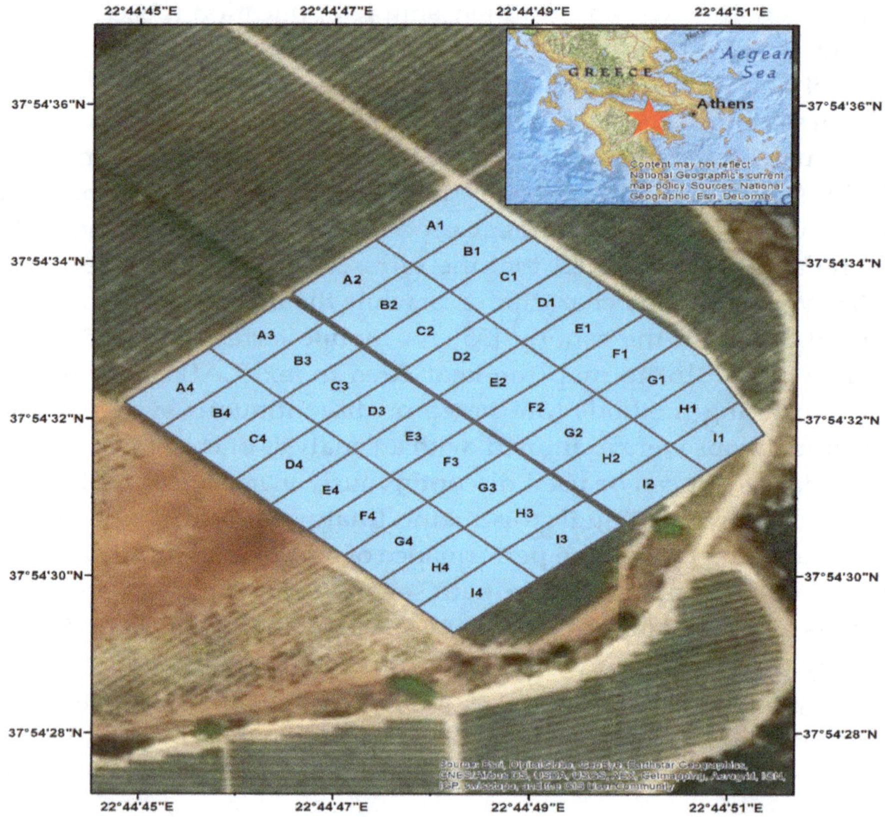

Figure 1. Satellite image of the table grapes in the commercial vineyard.

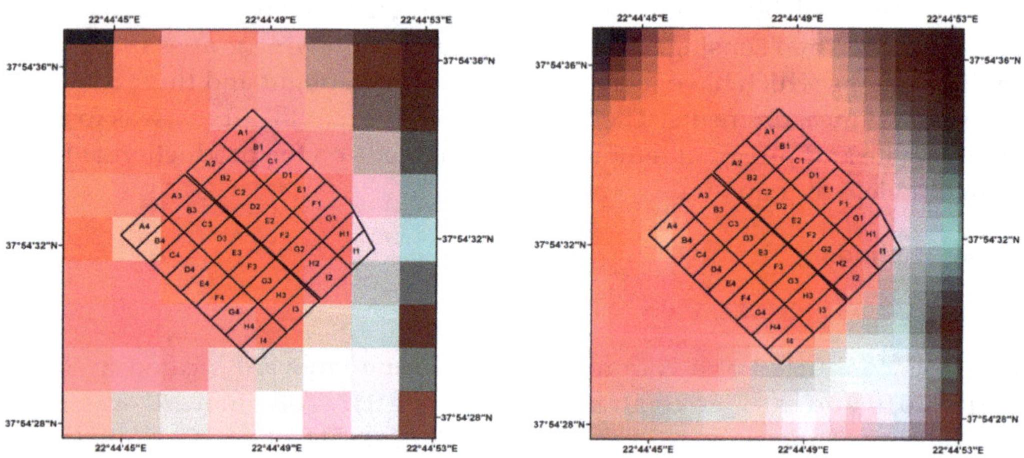

(a) (b)

Figure 2. Landsat 8 false color images with (**a**) 30 m and (**b**) 10 m resolution.

2.2. Remote Sensing Measurements

Crop vigour at 3 different developmental stages of the berry, namely (i) veraison (SV), (ii) mid of veraison (MV) and (iii) technological maturity (H), was assessed by measuring the Normalized Difference Vegetation Index (NDVI and the Green Normalized Difference Vegetation Index (GNDVI) (Table 1). The selection of these specific development stages was due to the fact that during these stages, there is a rapid change in berry composition, such as sugar accumulation, which is reflected to final yield and quality [42]; several studies have found high correlations between SVIs and grape yield and quality characteristics at these developmental stages [20,23,39].

Table 1. The spectral vegetation indices that were used in this study.

Spectral Vegetation Index	Equation	Bibliography
Normalized Difference Vegetation Index	$NDVI = \frac{NIR-RED}{NIR+RED}$	[43]
Green Normalized Difference Vegetation Index	$GNDVI = \frac{NIR-GREEN}{NIR+GREEN}$	[13]

A Crop Circle proximal canopy sensor (ACS-470, Holland Scientific Inc., Lincoln, NE, USA) with the sensor located at a height of 1.5 m from the soil surface and 1.2 m horizontally from the vines was used to scan the side canopy area in order to assess crop vigour from proximal sensing, while Landsat 8 satellite imagery was used to assess it through satellite remote sensing. Three lenses with different band absorptions were used to calculate the above SVIs with the Crop Circle canopy sensor (550 nm—GREEN, 670 nm—RED and 760 nm—NIR), while Band 3 (GREEN), Band 4 (RED) and BAND 5 (NIR) from Landsat 8 were utilized for the calculation of the same SVIs. The reason for choosing the NIR wavelength was its high reflection in healthy leaves due to its relationship with many leaf structural features [44]. The red wavelength region was chosen because it presents strong absorption peaks for assessing the chlorophyll content [45], and some researchers have found that higher absorption in the green wavelength region increases the efficiency of plant photosynthetic activity compared to the red wavelength region [46]. Specifically, atmospherically corrected Landsat-8 satellite images (L2 products) were downloaded from the EarthExplorer, which is the satellite imagery data hub of the United States Geological Survey (USGS) agency (https://earthexplorer.usgs.gov/). The satellite imagery was resampled at a 10 m resolution by bilinear resampling in order to address the cell grid scale of the experimental site and to match the resolution of the freely available Sentinel-2A imagery (Figure 2). According to Zhang et al. (2018) similar results can be obtained between Sentinel-2A and Landsat-8 surface reflectance values by employing this method [47]. The measurement dates of the proximal canopy sensing and satellite sensing can be seen in Table 2.

Table 2. Dates of acquisition of satellite and proximal sensing data.

Year	Satellite Imagery Dates	Proximal Sensing Dates
2015	16 July 2015 1 August 2015 2 September 2015	18 July 2015 30 July 2015 1 September 2015
2016	18 July 2016 3 August 2016 19 August 2016	16 July 2016 2 August 2016 17 August 2016
2017	14 July 2017 30 July 2017 15 August 2017	9 July 2017 26 July 2017 16 August 2017

2.3. *Table Grape Measurements*

The table grapes were hand harvested during 2–3 September 2015, 21–22 August 2016 and 16–17 August 2017. The actual yield was estimated during the harvest period by measuring the total number of bins per cell and multiplying it with the average bin weight of the harvested table grapes. Fifty berries were randomly taken in each vineyard cell at harvest by sampling five berries from one cluster per grapevine from a total of ten grapevines to assess berry and must parameters [48]. Berry diameters were estimated using the image analysis software, ImageJ 1.46r (Research Services Branch, NIMH, Bethesda, MD, USA). Specifically, each berry from the berry sample was placed in a custom made berry holder in an upright position and photographed with a digital camera (Konica Minolta Dimage Z6 (6.0 Mpixel), Minolta Co. Ltd., Osaka, Japan). The retrieved digital image was imported into ImageJ to calculate the Feret's diameter of every berry, where the mean value for every berry sample lot was calculated. The individual berry fresh weight was determined for each sample of 50 berries per vineyard cell using a digital scale. The initial sample of 50 berries per cell was split in half to measure the deformation force in each of the 25 berries (deformation distance was taken as 1 mm based on Bourne (2002) [49]), and the rest of the 25 berries were used to measure the maximum force to detach the berry's peduncle. The texture analysis was carried out in a Texture Analyser TA-XT2i (SMS, Surrey, UK). Upon completion of the non-destructive measurement, berry juice was extracted. The extracted juice was used to measure the total soluble solids (°Brix) with a digital refractometer (SR400), the total titratable acidity with a Fruit Acidity Meter GMK-708 (G-won Hitech Co., Seoul, Korea) and the pH with a pH meter AD8000 (Adwa Hungary Kft., Szeged, Hungary). The ripeness index is important for the table grape market and was calculated by dividing the total soluble solids (TSS) with the total titratable acidity (TTA), as presented in Equation (1) [50].

$$\mathrm{RI} = \frac{\mathrm{TSS}}{\mathrm{TTA}} \tag{1}$$

2.4. *Map Construction and Statistical Analysis*

The maps for vegetation images, yield and quality parameters for the three years of study were produced using ArcGIS 10.2 software (ESRI Inc., Redlands, CA, USA). For the assessment of the performance of satellite and proximal remote sensing on the estimation of table grape yield and quality, a statistical analysis was executed, including descriptive statistics, Pearson's correlation and a regression model. The regression model analysis was performed only for the SVIs that presented the highest correlation for each approach and at a certain crop stage. The statistical analysis was conducted with statistical software (Statgraphics 16, StatPoint Technologies Inc., Warrenton, VA, USA).

3. Results

3.1. *Descriptive Statistics*

3.1.1. Table Grape Yield and Quality Characteristics

The table grape parameters presented different degrees of variation, as shown in Figure 3. Specifically, pH, total soluble solids and mean berry diameter presented coefficients of variance (CV) equal or lower than 10%. However, yield, berry detachment, total titratable acidity, berry deformation and the ripeness index presented higher CVs, with the highest values being given by yield and grape detachment (32% and 33%, respectively) (Table 3).

Table 3. Descriptive statistics of the yield and quality characteristics of table grapes for the three years.

Table Grape Parameters	Min	Max	Mean	SD	CV (%)
Yield (kg/ha)	4518	38432	20612	6575	32%
Berry Detachment (N)	4.265	14.373	8.332	2.752	33%
pH	2.90	3.93	3.49	0.29	8%
Total Soluble Solids (TSS, °Brix)	15.7	23.2	18.6	1.3	7%
Total Titratable Acidity (TTA, %)	0.10	0.39	0.22	0.05	21%
Berry Diameter (mm)	19	28	23	2.215	10%
Berry Deformation (N/mm)	1.131	3.227	1.934	0.501	26%
TSS/TTA	45	209	87	21	24%

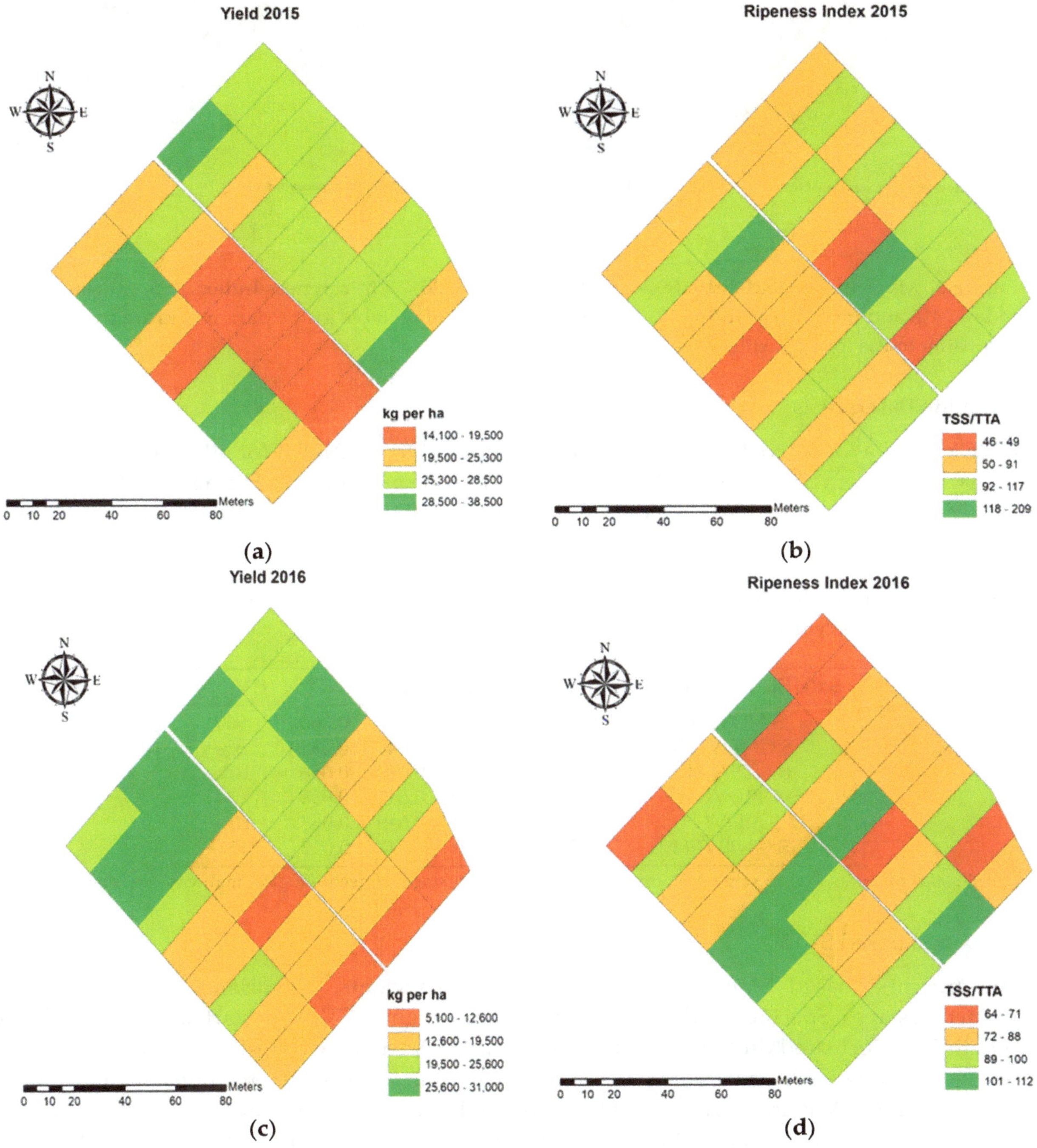

Figure 3. *Cont.*

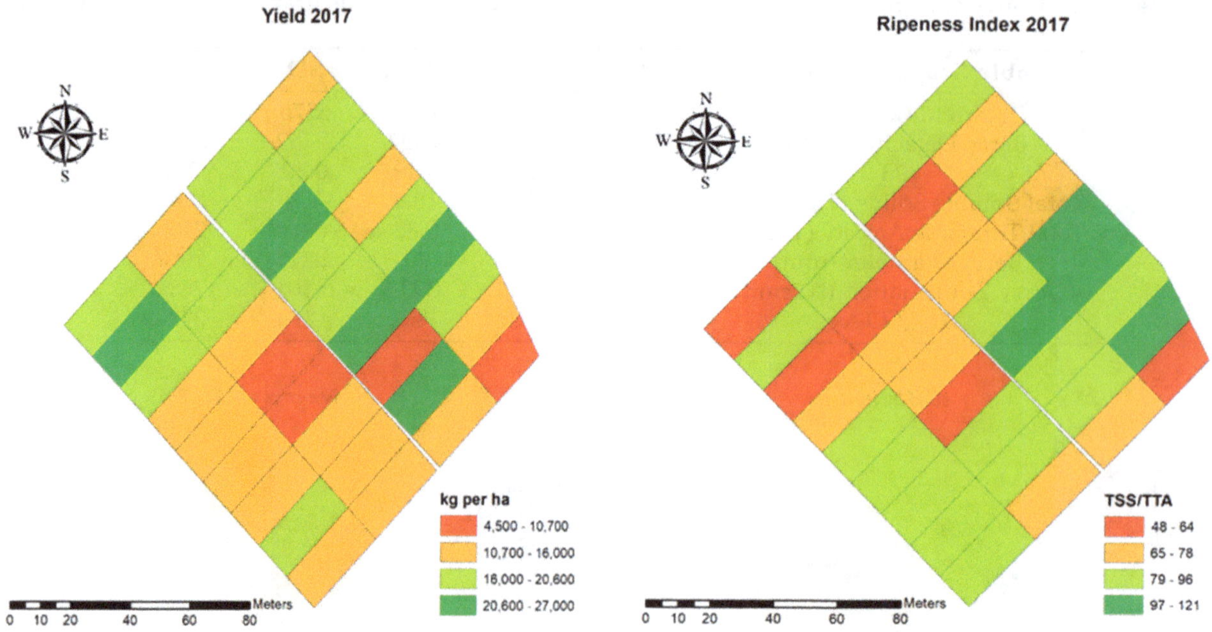

(e) (f)

Figure 3. Maps of (**a**) yield 2015, (**c**) yield 2016, (**e**) yield 2017, (**b**) ripeness Index 2015, (**d**) ripeness Index 2016 and (**f**) ripeness Index 2017. The value ranges of all table grape parameters are presented according to the four quartiles.

3.1.2. Satellite Based SVIs

GNDVI had a lower coefficient of variance compared to the NDVI for all different crop stages, while the CVs for both SVIs increased between the SV crop stage and the H stage (Table 4). The SVIs had higher ranges of values near harvest, indicating that the top canopy's vine leaves had higher photosynthetic rates near harvest [51,52].

Table 4. Descriptive statistics of the satellite based spectral vegetation indices (SVI) at different crop stages for the three years.

Satellite Sensing SVIs	Min	Max	Mean	SD	CV (%)
S NDVI SV	0.505	0.749	0.658	0.060	9%
S GNDVI SV	0.544	0.701	0.648	0.032	5%
S NDVI MV	0.509	0.769	0.649	0.066	10%
S GNDVI MV	0.544	0.710	0.641	0.038	6%
S NDVI H	0.493	0.762	0.649	0.068	10%
S GNDVI H	0.525	0.720	0.643	0.043	7%

SV: start of veraison crop stage, MV: middle of veraison crop stage, H: technological maturity crop stage.

3.1.3. Proximal Based SVIs

The proximal-based SVIs presented different coefficients of variance during the three developmental stages. Specifically, the NDVI presented the highest CV at the start of veraison, while the CVs of both SVIs decreased during the next crop stages. NDVI had the fastest drop in CV compared to GNDVI, with a value of 3% during the MV and H crop stages. Moreover, the highest values of SVIs occurred during the MV crop stage. This was due to the fact that that leaves that are found in the center of vine canopy are considered fully developed and present the highest photosynthetic activity compared to the others [51,52] (Table 5).

Table 5. Descriptive statistics of the proximal-based spectral vegetation indices at different crop stages for the three years.

Proximal Sensing SVIs	Min	Max	Mean	SD	CV (%)
P NDVI SV	0.187	0.924	0.631	0.166	26%
P GNDVI SV	0.336	0.691	0.573	0.091	16%
P NDVI MV	0.717	0.824	0.793	0.022	3%
P GNDVI MV	0.339	0.718	0.615	0.068	11%
P NDVI H	0.623	0.836	0.775	0.027	3%
P GNDVI H	0.390	0.771	0.615	0.059	10%

SV: start of veraison crop stage, MV: middle of veraison crop stage, H: technological maturity crop stage.

3.2. *Pearson's Correlation*

3.2.1. Satellite SVIs x Table Grapes Characteristics

Yield did not present correlations with NDVI nor GNDVI at any crop stage. Berry detachment correlated positively with both SVIs and had the highest correlation during the H crop stage. Moreover, GNDVI presented the highest correlation ($r = 0.536$, $p < 0.01$) among the two different SVIs with the berry detachment parameter. The same pattern of correlations was presented with the pH parameter, with GNDVI again showing the highest correlation ($r = 0.537$, $p < 0.01$). Both SVIs were negatively correlated with the total soluble solids, with the highest correlation during the MV stage ($r = -0.362$ and $r = -0.373$ for NDVI and GNDVI, respectively). There was no correlation between the TTA and the SVIs that were derived from satellite remote sensing. The berry diameter was correlated with both SVIs only at H stage, with GNDVI presenting the highest correlation ($r = 0.522$, $p < 0.01$). Berry deformation was correlated in all three crop stages with both SVIs, while GNDVI at the H crop stage presented the highest correlation with $r = 0.629$ ($p < 0.01$). The TSS/TTA ratio did not present any correlation with the SVIs at any crop stage. In conclusion, at harvest, GNDVI presented significant correlations at the $p < 0.01$ level with berry detachment, pH, total soluble solids, berry diameter and berry deformation, indicating that it can be used to assess the multiple crop yield and quality parameters at harvest (Table 6).

Table 6. Pearson's correlation matrix between the satellite-based spectral vegetation indices and the table grape yield and quality characteristics at different crop stages.

Satellite Sensing SVIs	Yield	Berry Detachment	pH	TSS	TTA	Berry Diameter	Berry Deformation	TSS/TTA
S NDVI SV	−0.016	0.140	0.155	−0.307 **	−0.043	0.131	0.222 *	−0.011
S GNDVI SV	0.083	0.153	0.161	−0.281 *	−0.056	0.172	0.244 *	−0.013
S NDVI MV	−0.136	0.182	0.254 **	−0.362 **	0.020	0.128	0.278 **	−0.085
S GNDVI MV	−0.069	0.225 *	0.306 **	−0.373 **	0.033	0.181	0.338 **	−0.097
S NDVI H	0.048	0.459 **	0.471 **	−0.335 **	0.008	0.424 **	0.551 **	−0.022
S GNDVI H	0.150	0.536 **	0.537 **	−0.322 **	0.006	0.522 **	0.629 **	−0.009

** Correlation is significant at the 0.01 level; * correlation is significant at the 0.05 level.

3.2.2. Proximal SVIs x Table Grapes Characteristics

Both SVIs presented correlations with table grape parameters at the SV and MV crop stages, while there were no correlations at harvest. Berry detachment presented a negative correlation with GNDVI at the SV and MV crop stages and positive with NDVI at the MV crop stage. The same pattern of correlations was presented with the pH parameter. In addition, the SVIs were correlated with TSS during the SV and MV crop stages with the highest correlations being presented in MV for NDVI ($r = -0.255$) and SV for GNDVI ($r = 0.497$). On the contrary, there was no correlation between TTA and the SVIs at all crop stages. The berry diameter was correlated with NDVI (SV and MV crop stages) and GNDVI (SV and MV crop stages) presenting the highest correlation with the latter at MV stage

($r = 0.682$, $p < 0.01$). Berry deformation presented the highest correlation with the GNDVI at MV crop stage. The TSS/TTA ratio presented a significant correlation only with NDVI at SV ($r = 0.208$, $p < 0.05$) (Table 7). Finally, GNDVI at the MV crop stage presented significant correlations at the $p < 0.01$ level with berry detachment, pH, total soluble solids, berry diameter and berry deformation, indicating that it could be used to assess multiple crop yield and quality parameters a few weeks before harvest.

Table 7. Pearson's correlation matrix between the proximal-based spectral vegetation indices and the table grape yield and quality characteristics at different crop stages.

Proximal Sensing SVIs	Yield	Berry Detachment	pH	TSS	TTA	Berry Diameter	Berry Deformation	TSS/TTA
P NDVI SV	0.259 **	−0.037	−0.163	0.241 *	−0.173	0.014	−0.009	0.208 *
P GNDVI SV	0.085	−0.400 **	−0.627 **	0.497 **	−0.066	−0.395 **	−0.444 **	0.160
P NDVI MV	0.218 *	0.488 **	0.524 **	−0.255 **	−0.041	0.465 **	0.549 **	0.037
P GNDVI MV	−0.423 **	−0.572 **	−0.493 **	0.321 **	0.169	−0.682 **	−0.565 **	−0.121
P NDVI H	−0.115	−0.108	−0.146	0.102	0.002	−0.196 *	−0.156	0.014
P GNDVI H	−0.077	0.108	0.153	−0.039	−0.086	0.110	0.039	0.142

** Correlation is significant at the 0.01 level; * correlation is significant at the 0.05 level.

3.3. Regression Analysis

GNDVI at harvest presented the highest correlations when compared with the NDVI for the different crop stages. For this reason, a linear regression analysis was performed using this SVI in order to evaluate its performance in assessing the yield and quality characteristics of table grapes for the highly correlated table grape characteristics. Accordingly, the assessment of proximal-sensed GNDVI at the MV crop stage to estimate the yield and quality characteristics of table grapes was performed through linear regression models only for the characteristics that had significant Pearson's correlations. The results of this analysis are presented in the sub-sections below.

3.3.1. Satellite GNDVI H x Table Grape Characteristics

The regression models between satellite-derived GNDVI and table grape yield characteristics presented different degrees of accuracy. Specifically, the best fitted model was for the estimation of berry diameter (adjusted R^2 = 88%), while the other models' accuracies were between 28% and 83%. The linear regression model that was developed for TSS had a coefficient of determination with adjusted R^2 of 28%, while the berry diameter model had an adjusted R^2 of 88%. Regarding the produced yield of table grapes, the model had a coefficient of determination equal to 33% (Table 8 and Figure 4).

Table 8. Regression model of correlated crop parameters with satellite-derived GNDVI at the harvest crop stage.

Regression Model	Year	Adjusted R^2	RMSE
Yield = 50389 − 37530 × S GNDVI H	2015		
Yield = −18,473 + 64276 × S GNDVI H	2016	33%	5382 kg/ha
Yield = 5379 + 16583 × S GNDVI H	2017		
Berry Detachment = −2.14 + 20.95 × S GNDVI H	2015		
Berry Detachment = 13.86 − 13.77 × S GNDVI H	2016	83%	1.13 N
Berry Detachment = 0.24 + 11.52 × S GNDVI H	2017		
pH = 4.31 − 0.79 × S GNDVI H	2015		
pH = 3.61 − 0.79 × S GNDVI H	2016	83%	0.12
pH = 4.03 − 0.79 × S GNDVI H	2017		
TSS = 6.12 + 18.27 × S GNDVI H	2015		
TSS = 25.15 − 9.55 × S GNDVI H	2016	28%	1.10 °Brix
TSS = 16.71 + 2.34 × S GNDVI H	2017		

Table 8. *Cont.*

Berry Diameter = 24.7 + 1.5 × S GNDVI H	2015		
Berry Diameter = 20 + 1.5 × S GNDVI H	2016	88%	0.8 mm
Berry Diameter = 21 + 1.5 × S GNDVI H	2017		
Berry Deformation = 0.21 + 3.44 × S GNDVI H	2015		
Berry Deformation = −0.58 + 3.44 × S GNDVI H	2016	77%	0.24 N/mm
Berry Deformation = −0.3 + 3.44 × S GNDVI H	2017		

The values of adjusted R^2 and RMSE refer to all three years of this study.

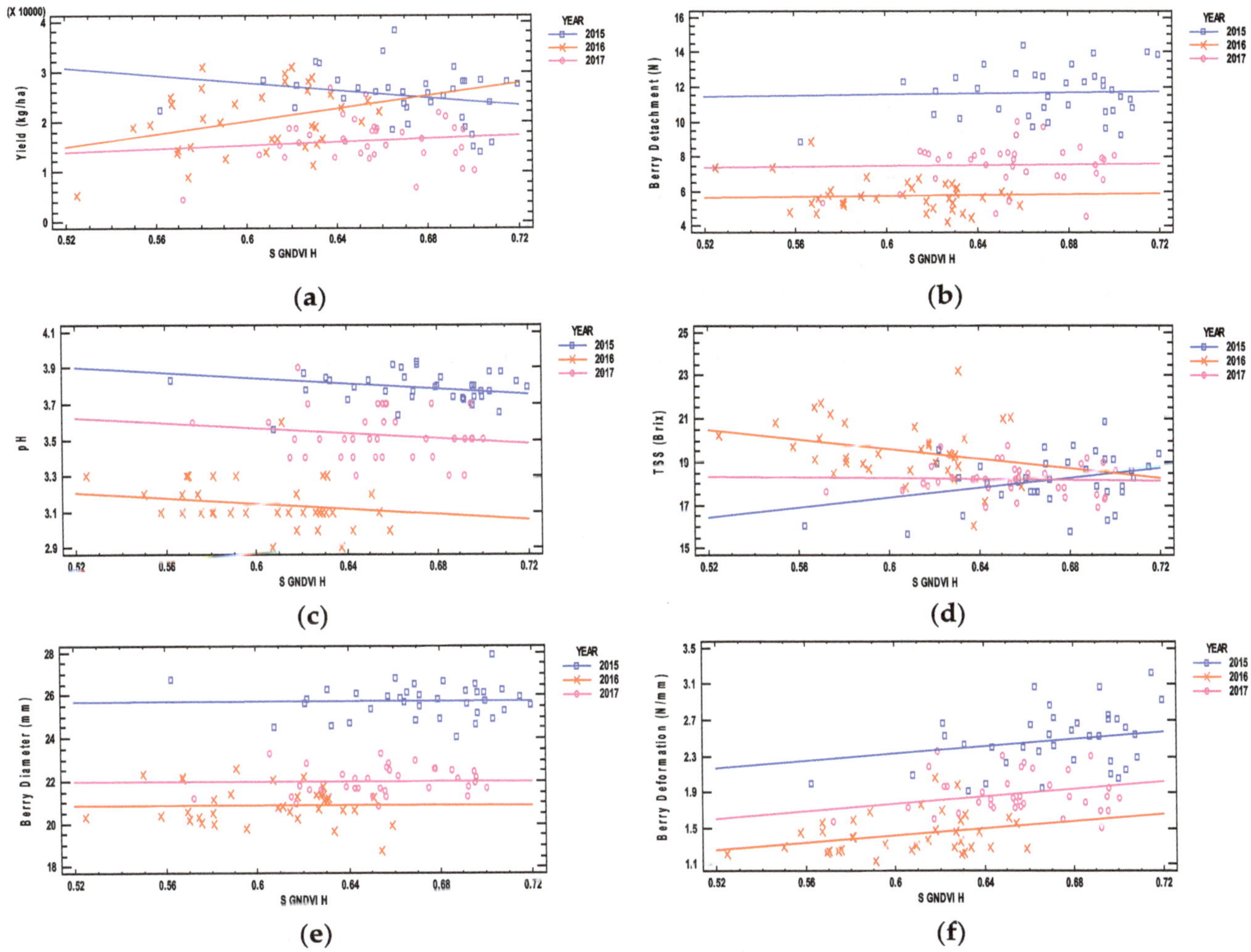

Figure 4. Scatterplots of S GNDVI H with (**a**) yield, (**b**) berry detachment, (**c**) pH, (**d**) total soluble solids, (**e**) berry diameter, and (**f**) derry deformation.

3.3.2. Proximal GRVI MV x Table Grape Characteristics

The linear regression models of table grape yield and quality characteristics with proximal sensed GNDVI at the MV crop stage presented coefficients of determination ranging between 26% and 89%. The berry diameter regression model had the highest coefficient of determination with an adjusted R^2 of 89% and the lowest was for estimating TSS (adjusted R^2 = 26%). The yield estimation model had a coefficient of determination of 31% (Table 9 and Figure 5).

Table 9. Regression model of correlated crop parameters with proximal-derived GRVI at the middle of veraison crop stage.

Regression Model	Year	Adjusted R^2	RMSE
Yield = 34965 − 17742 × P GNDVI MV	2015		
Yield = 31970 − 17742 × P GNDVI MV	2016	31%	5450 kg/ha
Yield = 27654 - 17742 × P GNDVI MV	2017		
Berry Detachment = 10.68 + 1.78 × P GNDVI MV	2015		
Berry Detachment = 4.61 + 1.78 × P GNDVI MV	2016	81%	1.21 N
Berry Detachment = 6.41 + 1.78 × P GNDVI MV	2017		
pH = 3.64 + 0.27 × P GNDVI MV	2015		
pH = 2.97 + 0.27 × P GNDVI MV	2016	83%	0.12
pH = 3.34 + 0.27 × P GNDVI MV	2017		
TSS = 15.78 + 4.31 × P GNDVI MV	2015		
TSS = 16.74 + 4.31 × P GNDVI MV	2016	26%	1.12 °Brix
TSS = 15.43 + 4.31 × P GNDVI MV	2017		
Berry Diameter = 27.4 − 3 × P GNDVI MV	2015		
Berry Diameter = 22.8 − 3 × P GNDVI MV	2016	89%	0.7 mm
Berry Diameter = 23.9 − 3 × P GNDVI MV	2017		
Berry Deformation = 2.66 − 0.33 × P GNDVI MV	2015		
Berry Deformation = 1.65 − 0.33 × P GNDVI MV	2016	72%	0.21 N/mm
Berry Deformation = 2.1 − 0.33 × P GNDVI MV	2017		

The values of adjusted R^2 and RMSE refer to all three years of this study.

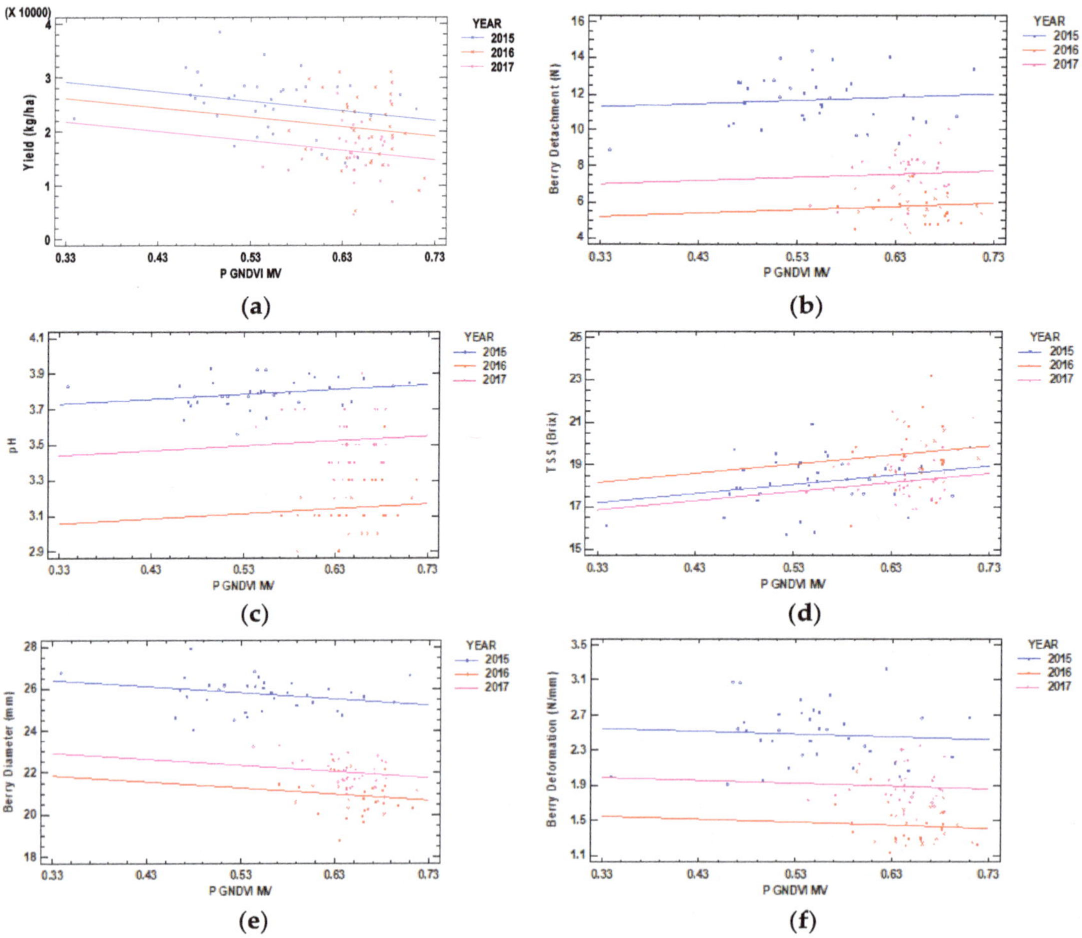

Figure 5. Scatterplots of P GNDVI MV with (**a**) yield, (**b**) berry detachment, (**c**) pH, (**d**) total soluble solids, (**e**) berry diameter, and (**f**) berry deformation.

4. Discussion

All the yield and quality characteristics of the table grapes presented high CV values except for the berry diameter and total soluble solids. It is worth mentioning that the market value of Thompson seedless table grapes is based on the total soluble solids and berry diameter. Due to this, many agricultural operations, like the spraying of crop growth regulators are focused on producing homogenized products in accordance with to the aforementioned parameters. The SVIs presented correlations with specific yield and quality characteristics. Particularly, they were only correlated with yield, berry detachment, berry diameter, berry deformation, pH and total soluble solids. Similar results have also been found by other researchers [39,53], while there have been no other studies presenting any correlations of SVIs with berry deformation and berry detachment. However, the correlation of these two parameters with other yield and quality characteristics is also in accordance with other studies on mechanical properties of berries [54–56]. In detail, the satellite-based SVIs presented positive correlations with all of the aforementioned characteristics, except total soluble solids, during all developmental stages, while there were correlations between proximal-sensed SVIs with the same variables for the different crop stages. This is explained by the numerous operations (i.e., trimming) that took place during vine growth and affected the side canopy of the vine leaves. This study is aligned with Fountas et al. (2014) who stated that many field operations in vineyards, like in this vineyard, may affect the correlations of SVIs with the yield and quality parameters of grapes [39].

The highest correlations between the SVIs and the table grape yield and quality characteristics were presented at different crop stages for each of the two remote sensing methods. Specifically, the SVIs that were derived through proximal sensing had the highest correlations with table grape yield characteristics during the middle of veraison, while the satellite-derived SVIs presented their highest correlations during harvest. The latter is in accordance with Sun et al. (2017), who found that the best crop stage for estimating wine grape yield from satellite-derived data is before harvest, while Garcia-Estevez et al. (2017) found that the highest correlation of NDVI derived from proximal sensing with yield parameters of wine grapes was at veraison [57,58]. Thus, higher resolution data, such as proximal sensing data, can provide earlier crop yield and quality estimations compared to medium resolution remote sensing data. Landsat 8, with its small spatial resolution of 30 m, is a good and cheap approach, but the new satellites, such as Sentinel 2, that provide freely available data of higher resolution, are expected to facilitate this procedure. Moreover, this is explained by the physiology of the vines. Specifically, the vine leaves are developed as the vine shoots grow, resulting in more mature leaves being found near the trunk. This means that the more photosynthetically active leaves are found near the trunk at the beginning of vine development, while as the vine grows to senescence, these leaves are found at the edge of the shoot [51,52]. Consequently, the leaves that are at the side canopy, which are measured with proximal sensing, have higher values of SVIs earlier in the growing season compared to the satellite-based SVIs that measure the photosynthetic activity of the top canopy leaves. This is an indication that the use of proximal sensing can present earlier and more accurate estimations of important yield and quality characteristics without being affected by soil background effects and lower resolution, which is the case for the satellite SVIs.

Furthermore, it was found that the satellite-based GNDVI at harvest provided comparable results and, in some cases, better results for the estimation of table grape yield and quality characteristics when compared to proximal-based SVIs. This finding comes in accordance with Tattaris et al. (2016) study, in which the satellite imagery provided better results on estimating yield when compared with aerial and proximal remote sensing [59]. This was also presented by Yang et al. (2013) who found that the use of high resolution satellite imagery can provide comparable results with aerial data of higher resolution [60].

Moreover, according to this study, there are higher correlations between satellite-derived GNDVI and proximal-derived GNDVI with table grape yield characteristics when compared with NDVI. Hall and Wilson (2013) found that SVIs, like the EVI (Enhanced Vegetation Index), perform better than NDVI in estimating wine grape yield characteristics, thus supporting the findings of this study [61].

Moreover, the saturation of NDVI at high values had lower correlations with crop parameters, like the grapevine biomass and crop coefficient, resulting in lower estimations of crop yield characteristics and crop evapotranspiration [62,63]. Consequently, there is need to assess the performance of different SVIs in order to use the most appropriate method according to the variable being measured.

5. Conclusions

In this study, an assessment of different remote sensing methods (satellite and proximal) for the estimation of crop yield characteristics was conducted during three crop growing seasons (2015, 2016 and 2017) on a commercial table grape vineyard cultivated with Thompson Seedless variety grapes. Two different spectral vegetation indices were calculated based on the spectral information that was derived from the different methods. The statistical analysis indicated that proximal sensing resulted in higher accuracy in estimating crop yield characteristics when compared with satellite sensing-derived estimations. Proximal sensing provided higher correlations earlier in the growing season than the satellite sensing approach, suggesting that the first method can be used as a tool for on time scheduling of table grape yield and quality estimation.

This type of study is valuable, as non-destructive methods for crop quantitative and qualitative characteristics are increasing due to the benefits they provide to the end users (farmers, crop consultants, public bodies, etc.). Further research is needed to assess new spectral vegetation indices and sampling resolutions using different sensing methods (e.g., comparison of Unmanned Aerial Vehicle-based SVIs with proximal and satellite sensing SVIs).

Author Contributions: E.A., A.B. (Athanasios Balafoutis) and S.F. conceived and designed the experiments; E.A., V.P. and A.B. (Athanasios Balafoutis) performed the experiments; E.A. and N.D. analyzed the data; K.B., G.X. and S.F. contributed their experience in the analysis and presentation of data.

Acknowledgments: We would like to thank Dimitris Theodorou and Pegasus COOP for their collaboration in conducting our field experiments in their field.

References

1. Fresh Deciduous Fruit: World Markets and Trade (Apples, Grapes, & Pears). Available online: https://apps.fas.usda.gov/psdonline/circulars/fruit.pdf (accessed on 29 April 2018).
2. Strik, B.C. Growing Table Grapes. 32. Available online: http://smallfarms.oregonstate.edu/sites/default/files/publications/growing_table_grapes_ec1639_may_2011.pdf (accessed on 29 April 2018).
3. Rather, J.A.; Wani, S.H.; Haribhushan, A.; Bhat, Z.A. Influence of girdling, thinning and GA3 on fruit quality and shelf life of grape (*Vitis vinifera*) cv. perlette. *Elixir Agric.* **2011**, *41*, 5731–5735.
4. Sen, F.; Oksar, R.; Kesgin, M. Effects of Shading and Covering on 'Sultana Seedless' Grape Quality and Storability. *J. Agric. Sci. Technol.* **2016**, *18*, 245–254.
5. Tehrani, M.M.; Kamgar-Haghighi, A.A.; Razzaghi, F.; Sepaskhah, A.R.; Zand-Parsa, S.; Eshghi, S. Physiological and yield responses of rainfed grapevine under different supplemental irrigation regimes in Fars province, Iran. *Sci. Hortic.* **2016**, *202*, 133–141. [CrossRef]
6. Hussein, M.A.; Abd-elall, E.H. Attempts to Improve Berry Quality of flame seedless Grapevines. *Egypt. J. Hortic.* **2017**, *44*, 235–244. [CrossRef]
7. Moran, M.S.; Inoue, Y.; Barnes, E.M. Opportunities and limitations for image-based remote sensing in precision crop management. *Remote Sens. Environ.* **1997**, *61*, 319–346. [CrossRef]
8. Mondal, P.; Basu, M. Adoption of precision agriculture technologies in India and in some developing countries: Scope, present status and strategies. *Prog. Nat. Sci.* **2009**, *19*, 659–666. [CrossRef]
9. Panda, S.S.; Ames, D.P.; Panigrahi, S. Application of Vegetation Indices for Agricultural Crop Yield Prediction Using Neural Network Techniques. *Remote Sens.* **2010**, *2*, 673–696. [CrossRef]
10. Thenkabail, P.S.; Smith, R.B.; De Pauw, E. Hyperspectral Vegetation Indices and Their Relationships with Agricultural Crop Characteristics. *Remote Sens. Environ.* **2000**, *71*, 158–182. [CrossRef]

11. Gitelson, A.; Merzlyakb, M.N. Quantitative estimation of chlorophyll-u using reflectance spectra: Experiments with autumn chestnut and maple leaves. *J. Photochem. Photobiol. B Biol.* **1994**, *22*, 247–252. [CrossRef]
12. Pena-Barragan, J.M.; Lopez-Granados, F.; Jurado-Exposito, M.; Garcia-Torres, L. Spectral discrimination of Ridolfia segetum and sunflower as affected by phenological stage. *Weed Res.* **2006**, *46*, 10–21. [CrossRef]
13. Chang, J.; Clay, D.E.; Dalsted, K.; Clay, S.; O'Neill, M. Corn (*Zea mays* L.) Yield Prediction Using Multispectral and Multidate Reflectance. *Agron. J.* **2003**, *95*, 1447–1453. [CrossRef]
14. Ranjitha, G.; Srinivasan, M.R.; Rajesh, A. Detection and Estimation of Damage Caused by Thrips *Thrips tabaci* (Lind) of Cotton Using Hyperspectral Radiometer. *Agrotechnology* **2014**, *3*. [CrossRef]
15. Genc, H.; Genc, L.; Turhan, H.; Smith, S.E.; Nation, J.L. Vegetation indices as indicators of damage by the sunn pest (Hemiptera: Scutelleridae) to field grown wheat. *Afr. J. Biotechnol.* **2008**, *7*, 173–180.
16. Jia, L.; Yu, Z.; Li, F.; Gnyp, M.; Koppe, W.; Bareth, G.; Miao, Y.; Chen, X.; Zhang, F. Nitrogen Status Estimation of Winter Wheat by Using an IKONOS Satellite Image in the North China Plain. In *Computer and Computing Technologies in Agriculture V*; Li, D., Chen, Y., Eds.; Springer: Berlin/Heidelberg, Germany, 2012; Volume 369, pp. 174–184. ISBN 978-3-642-27277-6.
17. Li, F.; Gnyp, M.L.; Jia, L.; Miao, Y.; Yu, Z.; Koppe, W.; Bareth, G.; Chen, X.; Zhang, F. Estimating N status of winter wheat using a handheld spectrometer in the North China Plain. *Field Crops Res.* **2008**, *106*, 77–85. [CrossRef]
18. Glaser, J.; Casas, J.; Copenhaver, K.; Mueller, S. Development of a broad landscape monitoring system using hyperspectral imagery to detect pest infestation. In Proceedings of the IEEE2009 First Workshop on Hyperspectral Image and Signal Processing: Evolution in Remote Sensing, Grenoble, France, 26–28 August 2009; pp. 1–4.
19. Huete, A.; Didan, K.; Miura, T.; Rodriguez, E.; Gao, X.; Ferreira, L. Overview of the radiometric and biophysical performance of the MODIS vegetation indices. *Remote Sens. Environ.* **2002**, *83*, 195–213. [CrossRef]
20. Hall, A.; Lamb, D.W.; Holzapfel, B.P.; Louis, J.P. Within-season temporal variation in correlations between vineyard canopy and winegrape composition and yield. *Precis. Agric.* **2011**, *12*, 103–117. [CrossRef]
21. Baluja, J.; Diago, M.P.; Goovaerts, P.; Tardaguila, J. Assessment of the spatial variability of anthocyanins in grapes using a fluorescence sensor: Relationships with vine vigour and yield. *Precis. Agric.* **2012**, *13*, 457–472. [CrossRef]
22. Joshi, N.; Baumann, M.; Ehammer, A.; Fensholt, R.; Grogan, K.; Hostert, P.; Jepsen, M.R.; Kuemmerle, T.; Meyfroidt, P.; Mitchard, E.T.A.; et al. A Review of the Application of Optical and Radar Remote Sensing Data Fusion to Land Use Mapping and Monitoring. *Remote Sens.* **2016**, *8*, 70. [CrossRef]
23. Hall, A.; Lamb, D.W.; Holzapfel, B.; Louis, J. Optical remote sensing applications in viticulture—A review. *Aust. J. Grape Wine Res.* **2002**, *8*, 36–47. [CrossRef]
24. Gascon, F.; Bouzinac, C.; Thépaut, O.; Jung, M.; Francesconi, B.; Louis, J.; Lonjou, V.; Lafrance, B.; Massera, S.; Gaudel-Vacaresse, A.; et al. Copernicus Sentinel-2A Calibration and Products Validation Status. *Remote Sens.* **2017**, *9*, 584. [CrossRef]
25. Vaudour, E.; Carey, V.A.; Gilliot, J.M. Digital zoning of South African viticultural terroirs using bootstrapped decision trees on morphometric data and multitemporal SPOT images. *Remote Sens. Environ.* **2010**, *114*, 2940–2950. [CrossRef]
26. Rodríguez, J.R.; Miranda, D.; Álvarez, C.J. Application of satellite images to locate and inventory vineyards in the designation of origin "Bierzo" in Spain. *Trans. ASABE* **2006**, *49*, 277–290. [CrossRef]
27. Semmens, K.A.; Anderson, M.C.; Kustas, W.P.; Gao, F.; Alfieri, J.G.; McKee, L.; Prueger, J.H.; Hain, C.R.; Cammalleri, C.; Yang, Y.; et al. Monitoring daily evapotranspiration over two California vineyards using Landsat 8 in a multi-sensor data fusion approach. *Remote Sens. Environ.* **2016**, *185*, 155–170. [CrossRef]
28. Johnson, L.; Roczen, D.; Youkhana, S.; Nemani, R.; Bosch, D. Mapping vineyard leaf area with multispectral satellite imagery. *Comput. Electron. Agric.* **2003**, *38*, 33–44. [CrossRef]
29. Kandylakis, Z.; Karantzalos, K. Precision viticulture from multitemporal, multispectral very high resolution satellite data. *ISPRS Int. Arch. Photogramm. Remote Sens. Spat. Inf. Sci.* **2016**, *41*, 919–925. [CrossRef]
30. Borgogno-Mondino, E.; Lessio, A.; Tarricone, L.; Novello, V.; de Palma, L. A comparison between multispectral aerial and satellite imagery in precision viticulture. *Precis. Agric.* **2018**, *19*, 195–217. [CrossRef]

31. Matese, A.; Gennaro, S.F.D. Technology in Precision Viticulture: A State of the Art Review. Available online: https://www.dovepress.com/technology-in-precision-viticulture-a-state-of-the-art-review-peer-reviewed-fulltext-article-IJWR (accessed on 29 April 2018).
32. Erena, M.; Montesinos, S.; Portillo, D.; Alvarez, J.; Marin, C.; Fernandez, L.; Henarejos, J.M.; Ruiz, L.A. Configuration and specifications of an unmanned aerial vehicle for precision agriculture. *ISPRS Int. Arch. Photogramm. Remote Sens. Spat. Inf. Sci.* **2016**, *XLI-B1*, 809–816. [CrossRef]
33. Marçal, A.; Gonçalves, J.; Cunha, M. Analysis of the temporal signature of vineyards in Portugal using VEGETATION. In Proceedings of the 26th EARSeL Symposium, New Developments and Challenges in Remote Sensing, Warsaw, Poland, 29 May–2 June 2006; Millpress: Rotterdam, The Netherlands, 2007; pp. 377–384.
34. Cerovic, Z.G.; Moise, N.; Agati, G.; Latouche, G.; Ben Ghozlen, N.; Meyer, S. New portable optical sensors for the assessment of winegrape phenolic maturity based on berry fluorescence. *J. Food Compos. Anal.* **2008**, *21*, 650–654. [CrossRef]
35. Llorens, J.; Gil, E.; Llop, J.; Escolà, A. Ultrasonic and LIDAR Sensors for Electronic Canopy Characterization in Vineyards: Advances to Improve Pesticide Application Methods. *Sensors* **2011**, *11*, 2177–2194. [CrossRef] [PubMed]
36. Del-Moral-Martínez, I.; Rosell-Polo, J.R.; Company, J.; Sanz, R.; Escolà, A.; Masip, J.; Martínez-Casasnovas, J.A.; Arnó, J. Mapping Vineyard Leaf Area Using Mobile Terrestrial Laser Scanners: Should Rows be Scanned On-the-Go or Discontinuously Sampled? *Sensors* **2016**, *16*, 119. [CrossRef] [PubMed]
37. Gatti, M.; Dosso, P.; Maurino, M.; Merli, M.C.; Bernizzoni, F.; José Pirez, F.; Platè, B.; Bertuzzi, G.C.; Poni, S. MECS-VINE®: A New Proximal Sensor for Segmented Mapping of Vigor and Yield Parameters on Vineyard Rows. *Sensors* **2016**, *16*, 2009. [CrossRef] [PubMed]
38. Kazmierski, M.; Glémas, P.; Rousseau, J.; Tisseyre, B. Temporal stability of within-field patterns of NDVI in non irrigated Mediterranean vineyards. *OENO One* **2011**, *45*, 61. [CrossRef]
39. Fountas, S.; Anastasiou, E.; Balafoutis, A. The influence of vine variety and vineyard management on the effectiveness of canopy sensors to predict winegrape yield and quality. In Proceedings of the International Conference of Agricultural Engineering, Zurich, Switzerland, 6–10 July 2014; p. 8.
40. Tagarakis, A.; Liakos, V.; Fountas, S.; Koundouras, S.; Gemtos, T.A. Management zones delineation using fuzzy clustering techniques in grapevines. *Precis. Agric.* **2013**, *14*, 18–39. [CrossRef]
41. Farid, H.U.; Bakhsh, A.; Ahmad, N.; Ahmad, A.; Mahmood-Khan, Z. Delineating site-specific management zones for precision agriculture. *J. Agric. Sci.* **2016**, *154*, 273–286. [CrossRef]
42. Sonnekus, N. Development and Change that Occurs in Table Grape Berry Composition during Growth. Ph.D. Thesis, Stellenosch University, Stellenbosch, South Africa, 2015.
43. Huete, A. A comparison of vegetation indices over a global set of TM images for EOS-MODIS. *Remote Sens. Environ.* **1997**, *59*, 440–451. [CrossRef]
44. Vescovo, L.; Wohlfahrt, G.; Balzarolo, M.; Pilloni, S.; Sottocornola, M.; Rodeghiero, M.; Gianelle, D. New spectral vegetation indices based on the near-infrared shoulder wavelengths for remote detection of grassland phytomass. *Int. J. Remote Sens.* **2012**, *33*, 2178–2195. [CrossRef] [PubMed]
45. Wu, C.; Niu, Z.; Tang, Q.; Huang, W. Estimating chlorophyll content from hyperspectral vegetation indices: Modeling and validation. *Agric. For. Meteorol.* **2008**, *148*, 1230–1241. [CrossRef]
46. Terashima, I.; Fujita, T.; Inoue, T.; Chow, W.S.; Oguchi, R. Green Light Drives Leaf Photosynthesis More Efficiently than Red Light in Strong White Light: Revisiting the Enigmatic Question of Why Leaves are Green. *Plant Cell Physiol.* **2009**, *50*, 684–697. [CrossRef] [PubMed]
47. Zhang, H.K.; Roy, D.P.; Yan, L.; Li, Z.; Huang, H.; Vermote, E.; Skakun, S.; Roger, J.-C. Characterization of Sentinel-2A and Landsat-8 top of atmosphere, surface, and nadir BRDF adjusted reflectance and NDVI differences. *Remote Sens. Environ.* **2018**. [CrossRef]
48. Neto, F.J.D.; Tecchio, M.A.; Junior, A.P.; Vedoato, B.T.F.; Lima, G.P.P.; Roberto, S.R. Effect of ABA on colour of berries, anthocyanin accumulation and total phenolic compounds of 'Rubi' table grape (*Vitis vinifera*). *Aust. J. Crop Sci.* **2017**, *11*, 199–205. [CrossRef]
49. Bourne, M.C. *Food Texture and Viscosity: Concept and Measurement*, 2nd ed.; Academic Press: London, UK, 2002; ISBN 0-12-119062-5.

50. Muñoz-Robredo, P.; Robledo, P.; Manríquez, D.; Molina, R.; Defilippi, B.G. Characterization of Sugars and Organic Acids in Commercial Varieties of Table Grapes. *Chil. J. Agric. Res.* **2011**, *71*, 452–458. [CrossRef]
51. Escalona, J.M.; Flexas, J.; Bota, J.; Medrano, H. Distribution of leaf photosynthesis and transpiration within grapevine canopies under different drought conditions. *Vitis* **2003**, *42*, 57–64.
52. Bertamini, M.; Nedunchezhian, N. Photosynthetic functioning of individual grapevine leaves (*Vitis vinifera* L. cv. *Pinot noir*) during ontogeny in the field. *Vitis* **2003**, *42*, 13–17.
53. Martinez-Casasnovas, J.A.; Agelet-Fernandez, J.; Arno, J.; Ramos, M.C. Analysis of vineyard differential management zones and relation to vine development, grape maturity and quality. *Span. J. Agric. Res.* **2012**, *10*, 326. [CrossRef]
54. Esgici, R.; Özdemir, G.; Pekitkan, G.; Eliçin, K.; Öztürk, F.; Sessiz, A. Engineering Properties of the Şire Grape (*Vitis vinifera* L. cv.). Scientific Papers. *Ser. B Horticult.* **2017**, *LXI*, 195–204, Print ISSN 2285-5653.
55. Giacosa, S.; Torchio, F.; Rio Segade, S.; Giust, M.; Tomasi, D.; Gerbi, V.; Rolle, L. Selection of a Mechanical Property for Flesh Firmness of Table Grapes in Accordance with an OIV Ampelographic Descriptor. *Am. J. Enol. Vitic.* **2014**, *65*, 206–214. [CrossRef]
56. Fernandes, J.C.; Cobb, F.; Tracana, S.; Costa, G.J.; Valente, I.; Goulao, L.F.; Amâncio, S. Relating Water Deficiency to Berry Texture, Skin Cell Wall Composition, and Expression of Remodeling Genes in Two *Vitis vinifera* L. Varieties. *J. Agric. Food Chem.* **2015**, *63*, 3951–3961. [CrossRef] [PubMed]
57. Sun, L.; Gao, F.; Anderson, M.; Kustas, W.; Alsina, M.; Sanchez, L.; Sams, B.; McKee, L.; Dulaney, W.; White, W.; et al. Daily Mapping of 30 m LAI and NDVI for Grape Yield Prediction in California Vineyards. *Remote Sens.* **2017**, *9*, 317. [CrossRef]
58. García-Estévez, I.; Quijada-Morín, N.; Rivas-Gonzalo, J.C.; Martínez-Fernández, J.; Sánchez, N.; Herrero-Jiménez, C.M.; Escribano-Bailón, M.T. Relationship between hyperspectral indices, agronomic parameters and phenolic composition of *Vitis vinifera* cv. Tempranillo grapes: Hyperspectral indices, agronomic parameters and phenolic composition of *V. vinifera*. *J. Sci. Food Agric.* **2017**, *97*, 4066–4074. [CrossRef] [PubMed]
59. Tattaris, M.; Reynolds, M.P.; Chapman, S.C. A Direct Comparison of Remote Sensing Approaches for High-Throughput Phenotyping in Plant Breeding. *Front. Plant Sci.* **2016**, *7*, 1131. [CrossRef] [PubMed]
60. Yang, C.; Everitt, J.H.; Du, Q.; Luo, B.; Chanussot, J. Using High-Resolution Airborne and Satellite Imagery to Assess Crop Growth and Yield Variability for Precision Agriculture. *Proc. IEEE* **2013**, *101*, 582–592. [CrossRef]
61. Hall, A.; Wilson, M.A. Object-based analysis of grapevine canopy relationships with winegrape composition and yield in two contrasting vineyards using multitemporal high spatial resolution optical remote sensing. *Int. J. Remote Sens.* **2013**, *34*, 1772–1797. [CrossRef]
62. Junges, A.H.; Fontana, D.C.; Anzanello, R.; Bremm, C. Normalized difference vegetation index obtained by ground-based remote sensing to characterize vine cycle in Rio Grande do Sul, Brazil. *Ciênc. E Agrotecnol.* **2017**, *41*, 543–553. [CrossRef]
63. Er-Raki, S.; Rodriguez, J.C.; Garatuza-Payan, J.; Watts, C.J.; Chehbouni, A. Determination of crop evapotranspiration of table grapes in a semi-arid region of Northwest Mexico using multi-spectral vegetation index. *Agric. Water Manag.* **2013**, *122*, 12–19. [CrossRef]

2

High-Resolution Multi-Sensor Remote Sensing to Support Date Palm Farm Management

Maggie Mulley [1,*,†], Lammert Kooistra [1] and Laurens Bierens [2]

1 Laboratory of Geo-Information Science and Remote Sensing, Wageningen University and Research, P.O. Box 47, 6700 AA Wageningen, The Netherlands; lammert.kooistra@wur.nl

2 TEC-IB B.V., Oude Veiling 29, 2635 GK Den Hoorn (ZH), The Netherlands; l.bierens@tec-ib.com

* Correspondence: maggie.m.mulley@gmail.com

† Current employment: Aerovision BV, Stadsring 47, 3811 HN Amersfoort, The Netherlands.

Abstract: Date palms are a valuable crop in areas with limited water availability such as the Middle East and sub-Saharan Africa, due to their hardiness in tough conditions. Increasing soil salinity and the spread of pests including the red palm weevil (RPW) are two examples of growing threats to date palm plantations. Separate studies have shown that thermal, multispectral, and hyperspectral remote sensing imagery can provide insight into the health of date palm plantations, but the added value of combining these datasets has not been investigated. The current study used available thermal, hyperspectral, Light Detection and Ranging (LiDAR) and visual Red-Green-Blue (RGB) images to investigate the possibilities of assessing date palm health at two "levels"; block level and individual tree level. Test blocks were defined into assumed healthy and unhealthy classes, and thermal and height data were extracted and compared. Due to distortions in the hyperspectral imagery, this data was only used for individual tree analysis; methods for identifying individual tree points using Normalized Difference Vegetation Index (NDVI) maps proved accurate. A total of 100 random test trees in one block were selected, and comparisons between hyperspectral, thermal and height data were made. For the vegetation index red-edge position (REP), the R-squared value in correlation with temperature was 0.313 and with height was 0.253. The vegetation index—the Vogelmann Red Edge Index (VOGI)—also has a relatively strong correlation value with both temperature ($R^2 = 0.227$) and height ($R^2 = 0.213$). Despite limited field data, the results of this study suggest that remote sensing data has added value in analyzing date palm plantations and could provide insight for precision agriculture techniques.

Keywords: remote sensing; date palms; precision agriculture; plantation management; thermal; hyperspectral

---

1. Introduction

Across the globe, people are suffering from a lack of food availability and food security as populations grow and land becomes unsuitable for farming [1–3]. There is a need for crops with high nutritional value that can withstand the arid and semiarid conditions in countries with limited water resources. The date palm is one such plant [2]. In the Middle East and North Africa alone, 100 million palms are cultivated on one million hectares of cropland [4].

There are several biotic and abiotic factors that affect the management of a healthy date palm crop. Global climate change may pose a threat to plantations of date palm; climate models predict that regions suitable for date palm growth will shrink, especially in the Middle East [5]. Water is a limiting factor for growth, although excess water can also reduce yield. Studies investigating the effects of water on date palm growth have found that insufficient water application slows the growth of the plants [6,7].

Increasing salinity also appears to negatively impact the growth of date palms [8–10]. This is a growing problem in arid regions as groundwater reservoirs are depleted due to excess irrigation.

Plantations of date palm and other palm species are also under threat due to infestations by different pests, the most extensive of which is the red palm weevil, or *Rhynchophorus ferrugineus* (Olivier, 1790). *R. ferrugineus* is spread via the transport of cuttings of palm tree leaves [4,11], and has been detected in 50% of date palm growing countries [12]. The effects of *R. ferrugineus* are severe; in the later stages of infestation the trunk can be damaged to the extent that the tree collapses and will most certainly die. Other pests that affect date palms include the Dubas bug, lesser date palm moth, and many different borer species [13]. The Dubas bug feeds on leaf sap which results in yellowing and wilting of leaves [3]. Lesser date palm moths infest and damage the fruit of the date palm [13]. Borers have similar effects to *R. ferrugineus*, although they may infest different parts of the tree, such as the middle rib of the frond and the upper section of the tree near the bunches of fruit [13].

At present, a heavy focus is placed on monitoring plantations for *R. ferrugineus* infestation. Unfortunately, current methods of detection do not allow for effective, reliable identification of infested trees [14]. In fact, the detection of palms in the early stages of infestation has been stated as the main issue in controlling weevil infestation [15]. It is very difficult to recognise whether a tree is infested based on visual inspection as the symptoms vary depending on conditions such as the tree species and initial site of infestation [14,16]. All of the current detection methods, including acoustic detection, x-ray, and sniffer dogs are often ineffective on a large scale because each tree has to be inspected individually [14,16].

Remote sensing has been suggested as a potential detection method, as it allows for large scale monitoring and effective data analysis techniques [16]. Thermal imaging studies have shown it might be possible to detect plants infested with *R. ferrugineus* using aerial thermal photographs [17]. Multi- and hyperspectral images have successfully identified infested trees. WorldView-3 satellite imagery has been used for *R. ferrugineus* detection [18]. Hyperspectral imagery was conducted on oil palms to classify *R. ferrugineus*-infested palms based on vegetation indices [19]. Light Detection and Ranging (LiDAR) has not been applied to the palm assessment case, but it does have potential; for example, LiDAR has been used to classify eucalyptus tree health [20].

Remote sensing has also been applied to assess the effect of other stressors on palm plantations. Cohen et al. (2012) used thermal images to investigate how changes in irrigation affect the health of date palm plantations [6]. The effect of soil salinity on date palm health has been assessed by using Landsat Thematic Mapper (TM) and Enhanced Thematic Mapper Plus (ETM+) images to calculate the Soil Adjusted Vegetation Index (SAVI) for two different years [21].

The studies listed above have all explored how individual stressors affect date palm plantations. However, it is highly likely that a date palm plantation will be subject to multiple stressors at once and identifying the influence of these effects will allow for more comprehensive and efficient plantation management. Remote sensing approaches can provide the tools to conduct this form of management; analysis of the entire plantation is possible using automated techniques in order to differentiate between a combination of stressors.

Therefore, the goal of this study is to explore which remote sensing sensors and derived indicators can detect stressors on palm health at different spatial levels, in order to support a holistic solution for effective date palm management. Plantations are often organised in large rectangular sub-sections or blocks—this will be the first spatial level to be evaluated. The second level of analysis will be conducted on an individual tree level. For this study, several different remote sensing imagery sources have been used, in order to determine whether different techniques are most suitable for highlighting differences in palm status. The available imagery sources are thermal, hyperspectral, and LiDAR, with data collected using airborne platforms. High resolution imagery is achievable with airborne platforms, allowing for analysis of individual tree health.

2. Materials and Methods

2.1. Study Area and Data Description

The farm shown in Figure 1 is in the Al-Kharj region in Saudi Arabia, southeast of the capital Riyadh. Saudi Arabia was one of the earliest countries growing date palms to be affected by *R. ferrugineus* infestation outside of South East Asia, with first infestations occurring since the 1980s [22]. The country has the highest per capita consumption of dates in the world. Flood irrigation is making way for more modern irrigation methods such as drip irrigation in many farms [23].

Of the farm, an area of 168.8 hectares (outlined in blue in Figure 1) has been chosen as the key study area of this analysis. This farm is relatively well managed, with continuous maintenance and targeted irrigation practices in place. As is exemplified in Figure 1b, date palm plantations are commonly divided into regular-sized rectangles of approximately 10 hectares; these sub-sections are hereafter referred to as blocks. Figure 1c illustrates the uniform growth patterns of the date palms; the crowns are distinct and organised in a grid-like structure.

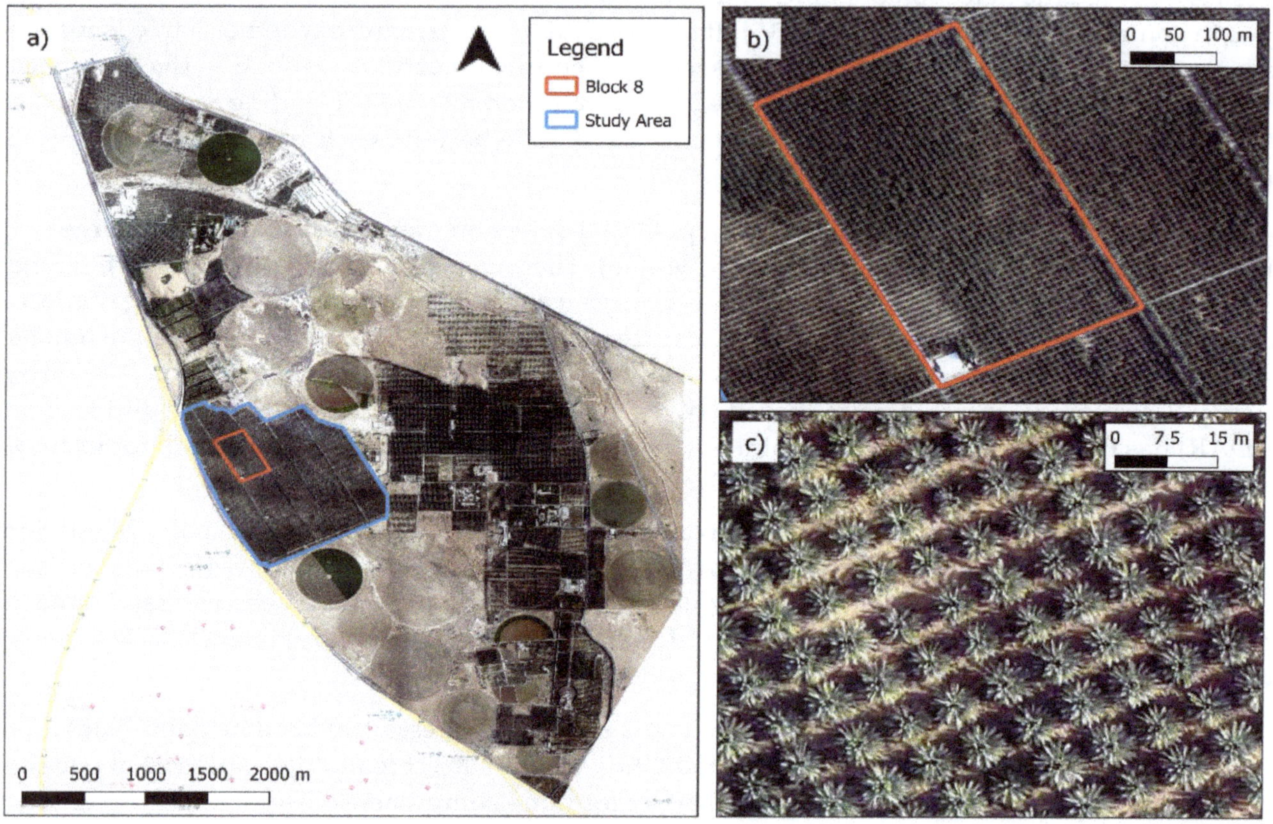

Figure 1. Visual red-green-blue (RGB) images with 15-cm resolution, taken using an aerial platform on 26 May 2016, of (**a**) the date palm plantation in the study area (outlined in blue), located south-east of Riyadh, the capital of Saudi Arabia; (**b**) an example of a block within the plantation (outlined in red) and (**c**) an example of individual palm trees.

For this study, four remote sensing data types were available for analysis: hyperspectral, thermal, LiDAR, and high-resolution visual red-green-blue (RGB) images. Each dataset contains different information that will be useful when assessing date palm health. The data has been collected using a Diamond DA42 MPP GEOSTAR (Diamond Aircraft Industries GmbH, Wiener Neustadt, Austria) aircraft, which is designed to carry many different remote sensing instruments. The flight was conducted on 26 May 2016 between approximately 6:00 and 9:00AM.

Table 1 shows information about the four sensors used for data collection. The spatial resolution refers to the pixel size of the data used in the analysis after preprocessing; the original LiDAR data was provided as a point cloud but has been processed into raster data.

Table 1. Characteristics of the four sensor systems used for data collection.

Data Type	RGB	LiDAR	Thermal	Hyperspectral	
Name of sensor [1]	Phase One iXA-R Camera [2]	RIEGL LMS-Q1560 [3]	VarioCAM®HD head [4]	HySpex VNIR-1800 [5]	Hy-Spex SWIR-384 [5]
Type of sensor	Medium-format camera system	Rotating polygon mirror	Uncooled microbolometer focal-plane array	Pushbroom camera actively cooled and stabilized scientific CMOS detector	Pushbroom camera Mercury cadmium telluride sensor
Spectral range	Visible	Near-infrared	7.5–14 μm	0.4–1 nm, 182 bands	0.93–2.5 nm, 288 bands
Spatial resolution	0.15 m, 0.6 m, 1.8 m	1 m, 2 m	1.8 m	1 m	1 m

[1] Manufacturer information; [2] Phase One Industrial, Frederiksberg, Demark; [3] RIEGL Laser Measurement Systems GmbH, Horn, Austria; [4] InfraTec, Dresden, Germany; [5] Norsk Elektro Optikk AS (NEO), Skedsmokorset, Norway.

2.2. *Methodology*

2.2.1. Preprocessing

Different preprocessing steps were necessary for each of the multiple remote sensing data sources available. These steps will be described briefly in the next few paragraphs. Steps were carried out in various software: R (version 3.3.1, R Foundation for Statistical Computing, Vienna, Austria), QGIS (version 2.1.4 'Essen', Open Source Geospatial Foundation, Chicago, IL, USA), ArcGIS (version 10.4, ESRI, Redlands, CA, USA), LAS tools (version 2016, rapidlasso GmbH, Gilching, Germany), and Agisoft PhotoScan (version 1.2, Agisoft LLC, St. Petersburg, Russia).

The RGB data was provided in JPEG format and was not geographically referenced. The files were given a geographic coordinate system and then projected to WGS 1984 UTM Zone 38N, and also converted to GeoTIFF. This projection was also used for all the other datasets.

The LiDAR dataset was originally received as a point cloud. To generate a tree height raster, Digital Surface Models (DSM) at different resolutions was generated; additionally, the mean, minimum, maximum and standard deviation values of the points within these pixels were used.

During acquisition, the VarioCAM thermal camera was controlled using the IRBIS®Thermography Software (InfraTec, Dresden, Germany). Within this software, the temperature range for acquisition was set to values between 0 and 50 degrees Celsius [24]. This range was broader than the actual temperature range which could be expected but was set by the flight operators to avoid saturation at low or high values. The raw thermal data was provided as individual snapshot images that were then stitched together using the structure-from-motion method [25] as implemented through Agisoft PhotoScan. In some areas where there are a lot of overlapping images, tie-points do not match perfectly, causing some blurring in the image; where this occurs, individual trees are not identifiable. This effect cannot be reduced without compromising the resolution of the entire image. The image was received as a (8 bit) greyscale image with 256 levels. To convert the values into a relative approximation of degrees Celsius, the acquired greyscale image was converted using the multiplication factor of 0.1953125 to fit the temperature range settings during acquisition (0 and 50 degrees Celsius).

The hyperspectral images were received without the inertial measurement unit (IMU) data due to an error in the acquisition procedure. This means that it was not possible to correct for image distortion due to plane movement. The images were also provided in an unknown multiplication factor, and thus the values are unusual. However, an evaluation of this data source is still considered worthwhile as it was acquired at the same time as the other remote sensing types. This meant that extensive steps were required to attain usable imagery.

By using the recorded times and positioning of the RGB imagery, the order in which hyperspectral images were acquired was determined. This meant that a representative image could be constructed and aligned with the study area. The image was then transformed using digitally assigned geometric control points as identified from the RGB images. The QGIS *Georeferencer* tool was used, with transformation type set as Thin Plate Spline which is useful for heavily distorted data, and Nearest Neighbour as a resampling method to maintain the original data values. The pixel data was multiplied by a scaling factor of 0.002 as this ensured they were in the relative range of standard reflectance values. As this is not the standard way to determine actual reflectance values, it is only possible to carry out relative comparisons within the dataset—this dataset cannot be numerically compared with any hyperspectral data collected in later studies.

2.2.2. Identification of Blocks and Trees

As analysis is carried out on two different spatial levels, namely blocks and trees, it is necessary to identify these regions. Block shapefiles were provided for the farm, and these were used for analysis of the thermal and LiDAR-derived height data. As the hyperspectral blocks were geometrically distorted, these were drawn manually.

Several methods were used in order to identify the trees for tree-level analysis. For the non-distorted data, the canopy height map as determined from LiDAR was used to locate individual tree points. This was done using the watershed method, which involves inverting the height rasters, calculating the sinks and converting them to points, and filtering these points at an appropriate buffer width. The software used was LAS tools and ArcMap.

It was also decided that deriving canopy polygons could be interesting as the area of the tree canopy could prove a useful indicator of tree health. The LiDAR data was analysed using an R package called Individual Tree Crown (ITC) Segment (version 0.6, Michele Dalponte, San Michele all'Adige, Italy) [26]. This tool selects local maxima within a height raster and then spreads outwards to select the surrounding pixels that are within a specific height range representing the canopy. A polygon shapefile is generated, for which the height and area are calculated. A 3-m width limit was selected to constrain the growth of the polygon outside of the general tree size. To investigate sensitivity to spatial resolution, the model was run for pixel sizes of both 1 m and 2 m. In this case, the raster's containing the maximum value of the point cloud within each pixel were used for analysis, as these data were available for both 1-m and 2-m resolution.

For the hyperspectral data, a different method for tree identification was used. This is necessary because the distortions in the hyperspectral dataset mean that it is not possible to use the same tree locations for each dataset. It was decided to use vegetation indices, specifically normalised difference vegetation index (NDVI) (see Equation (1)), to identify the trees. Once the NDVI was calculated, based on the values of Equation [1], the *TreeTopFinderTool* in the R package "Forest Tools" (version 0.1.5, Andrew Plowright, BC, Canada) was used to identify local maxima. It is hypothesised that the points where the NDVI is highest represent the top of the tree canopy. A width threshold of four was set, based on the upper limit of the average canopy size, which is 3 to 4 m. Additionally, a minimum NDVI value of 0.3 was set to ensure that maxima in treeless regions were excluded—this value was based on the bimodal distribution of the NDVI values that distinguishes between soil and tree points.

$$NDVI = \frac{NIR_{750} - R_{660}}{NIR_{750} + R_{660}} \quad (1)$$

2.2.3. Block-Level Analysis

The six blocks shown in Figure 2 have been selected for further analysis. These blocks were chosen based on the attributes provided in Table 2.

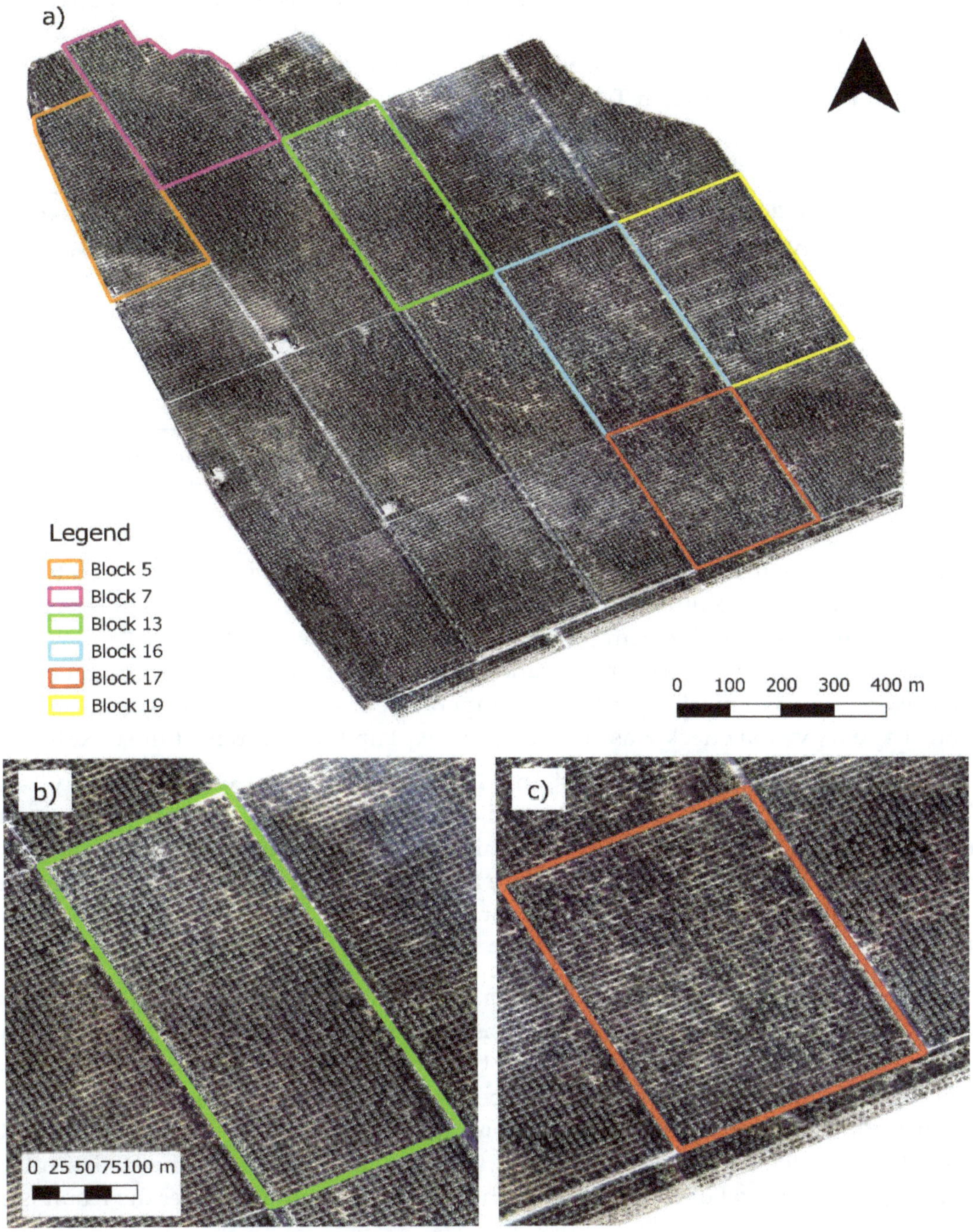

Figure 2. (**a**) Shows the selected test blocks; Blocks 5, 7, and 13 have been classified as healthy and Blocks 16, 17 and 19 have been classified as unhealthy. (**b**) and (**c**) shows an example of blocks classified as healthy (green) and unhealthy (red) respectively.

Two different groups of blocks have been determined based on their characteristics as described in Table 2. The data that was available from the farm manager mainly related to instances of *R. ferrugineus*, so this was a factor that was considered in block division. A visual assessment of the farm was made based on the homogeneity of canopy area and the sparseness of the block. The first group of trees was chosen with the requirements of having no instances of *R. ferrugineus* and of having high canopy homogeneity and low sparseness. The second group of blocks had relatively high instances of *R. ferrugineus*, and the appearance of the blocks revealed many gaps in trees and a wide variation of tree canopy widths. These two indicators were used to differentiate between potentially healthy and unhealthy groups. An additional factor that was considered was the amount of blurring in the thermal image, as discussed in Section 2.2.1; if the blurring in a block was visually estimated as being more than 70%, it was not considered.

The assumed healthy test blocks are 5, 7, and 13, and the unhealthy blocks are 16, 17, and 19. Their respective values are shown in Table 2.

Table 2. Criteria used to determine the selection for the six test blocks.

	May 2016	June 2016	RGB Check [1]	Thermal Image Distortion [2]
Block No.	No. RPW infested	No. RPW infested	1–5 rating	Percent distorted
5	0	0	1	20
7	0	0	1	30
13	0	0	2	30
16	8	4	5	30
17	10	4	5	30
19	1	9	5	20

[1] Indicator of block health status based on visual check for sparseness and heterogeneity, where 1 = homogenous and wide canopies, 5 = many patches and smaller canopies; [2] Visual check to determine how much of the image is affected by blurring due to structure-from-motion processing technique.

To summarise the statistical attributes of these blocks, boxplots were generated for temperature and height of the trees. These illustrate the spread of temperature and height across the test blocks, also in relation to all blocks in the farm. These boxplots illustrate the 25th, 50th and 75th percentiles, extreme values, and outlier values for the canopy pixels. For the temperature values, the derived canopy areas were used as a mask layer for the temperature raster. The height values are also based on the polygon shape. A visual check was conducted to explain the findings of the boxplot analysis.

For each block, a total of 10 'healthy' and 10 'unhealthy' trees were randomly selected after filtering based on their ratio between height and canopy area—based on discussions with the farm manager, it was hypothesised that trees with a high height area ratio (HAR), meaning they have a small canopy in relation to height, would be more likely to be unhealthy. Trees were selected based on whether they were above or below a threshold for this ratio and were also filtered based on height to attempt to select trees of similar height.

The mean temperatures of each group of trees (healthy and unhealthy) have been compared using two-factor analysis-of-variance (ANOVA) analysis. This statistical test is used to determine whether the differences in means of two or more groups at two or more levels are statistically significant; in this case, between the 'healthy' and 'unhealthy' trees, between the blocks, and the interaction effect between them. The post-hoc Tukey honest significant difference (HSD) test is applied following ANOVA—it is a comparison test that determines which pairs of means are responsible for the statistically significant test result. A significance level of $\alpha = 0.05$ has been used to determine whether the result is statistically significant or not.

2.2.4. Tree-Level Analysis

Different blocks were used for the individual tree analysis, as several of the blocks were not available in the hyperspectral dataset. Block 11 was selected for analysis for two reasons; firstly, it has the least visible blurring in the thermal dataset, and also it has a wide variety of trees with different canopy sizes.

As described in Section 2.2.2, the tree canopies have been identified based on NDVI. The next step was to select trees that can be used to extract more extensive information. In this case, trees were selected randomly from the hyperspectral dataset, and were matched manually to the trees in the thermal dataset. This was done by starting at the edges of the trees in each block and then counting vertically and the horizontally to identify the correct tree. A shapefile was created, and points were placed on the identified trees. A total of 100 trees was used for this block. As the trees are quite noticeable in each dataset, it is expected that this method was quite accurate at detection, although in some cases the tree may be misidentified.

Fourteen vegetation indices (VI) have been selected for further analysis (See Appendix A Table A1). These VIs can be split into general groups that focus on different regions of in the visible light spectrum, such as green, red, and the red-edge position. For each test tree, the correlation between each vegetation index and temperature and height has been derived, and R squared values have been calculated. A visual check has also been carried out.

3. Results

3.1. Block and Tree Identification Results

The results of the method which identifies the areas of entire tree canopies can be seen in Figure 3. The ITC segment tool was run on two different height rasters (spatial resolutions of 1 m and 2 m) derived from the LiDAR dataset. Table 3 shows the number of tree points identified using the two different height raster maps; these values have been compared with tree numbers provided by the farm manager. For the 1 m resolution data these values exceeded 90% in every block. Fewer trees are detected using the 2 m resolution data: 73–90% were detected when comparing the number of tree polygons to the number of tree points in each block.

Table 3. Number (#) of trees within the test blocks based on the farm managers records compared to results of the Individual Tree Crown (ITC) segment analysis for 1 m and 2 m height rasters.

Block No.	Recorded # of Trees	# of Trees (1 m Resolution)	# of Trees (2 m Resolution)
5	1188	1137	1043
7	1213	1197	1084
13	1417	1277	1131
16	1749	1572	1242
17	1422	1302	1013
19	1691	1617	1440

When comparing Figure 3a,b, it can be seen that in the denser parts of the plantation, using the 1m resolution LiDAR data has a higher detection rate of tree canopies.

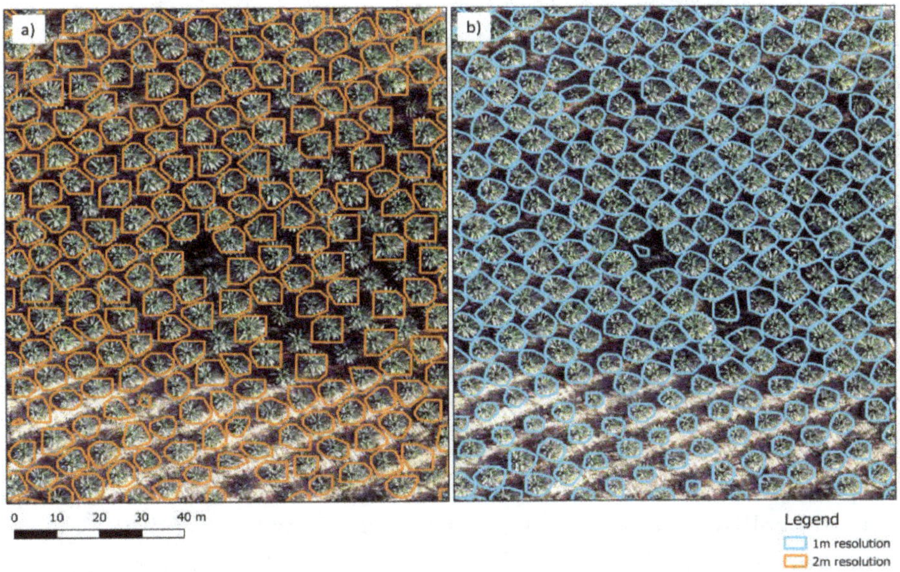

Figure 3. Section of an area of Block 5 in the farm. Canopy polygons derived using the ITC segment tool plotted in relation to the RGB data, for (**a**) the 2-m spatial resolution height raster (orange), and (**b**) 1-m spatial resolution (blue).

The second method used the NDVI image that was extracted from the hyperspectral dataset. Two different red bands were used to calculate the NDVI, and the comparisons can be seen in Figure 4. The outcome for both NDVI calculations is similar, although trees are identified more accurately in denser canopies when band 750 is used; for Block 16, for example, the number of trees recorded by the farm manager was 1749, compared to 1605 and 1601 for NDVI 750 and NDVI 800 respectively. For Block 19, the method using NDVI 750 identified 1661 trees, compared to 1656 (NDVI 800), and 1691 (trees recorded by farm manager). This indicates that in this case NDVI 750 is slightly more effective at identifying trees. Misidentification tends to be on the edges where the tree canopies are denser and therefore less distinct.

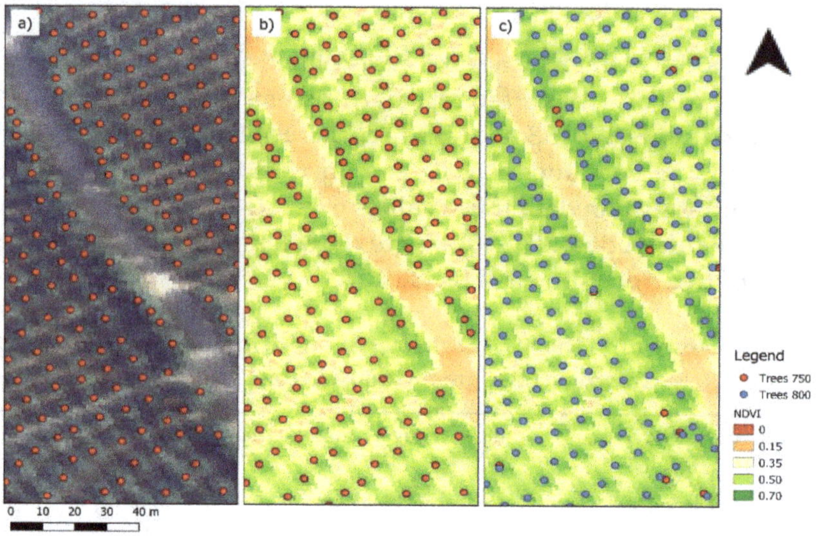

Figure 4. Tree points (red) derived using NDVI with band 750, overlaying (**a**) the hyperspectral image, and (**b**) the NDVI-derived using band 750, and (**c**) tree points derived using band 800 (blue) in comparison NDVI 750 points, overlying NDVI-derived using band 800.

For further analysis, the datasets with the highest detection rates were used. For analysis of the thermal and LiDAR datasets, the canopy polygons derived from the 1 m height rasters were used. The tree points resulting from the NDVI using spectral band 750 were chosen for the hyperspectral data analysis.

3.2. Block-Level Analysis

For block-level analysis, thermal data and height rasters were analysed. Six test blocks were chosen based on their attributes which have been discussed in the methods section. As the same six test blocks are not available for the hyperspectral dataset, this dataset has not been included in block-level analysis.

Boxplots of the thermal and height values for the six selected test blocks are presented in Figures 5 and 6. This illustrates the spread of heights across the test blocks, also in relation to all blocks in the farm. These boxplots illustrate the 25th, 50th and 75th percentiles, extreme values, and outlier values for the canopy pixels.

Figure 5 shows boxplots of the canopy temperature pixels of the test blocks. When treating block 5 as an outlier, it appears that two of the assumed healthy blocks have a mean temperature that is 0.5 °C lower than the unhealthy blocks. This is a positive result as the hypothesis was that unhealthy blocks would appear warmer than the healthy blocks. However, the mean temperature of block 5 is 0.5 °C higher than the mean temperature of the unhealthy test blocks. This could be because block 5 has been inaccurately classified as healthy. Feedback from the farm manager has indicated that the soil quality in this region is very poor so the trees there do not grow as well.

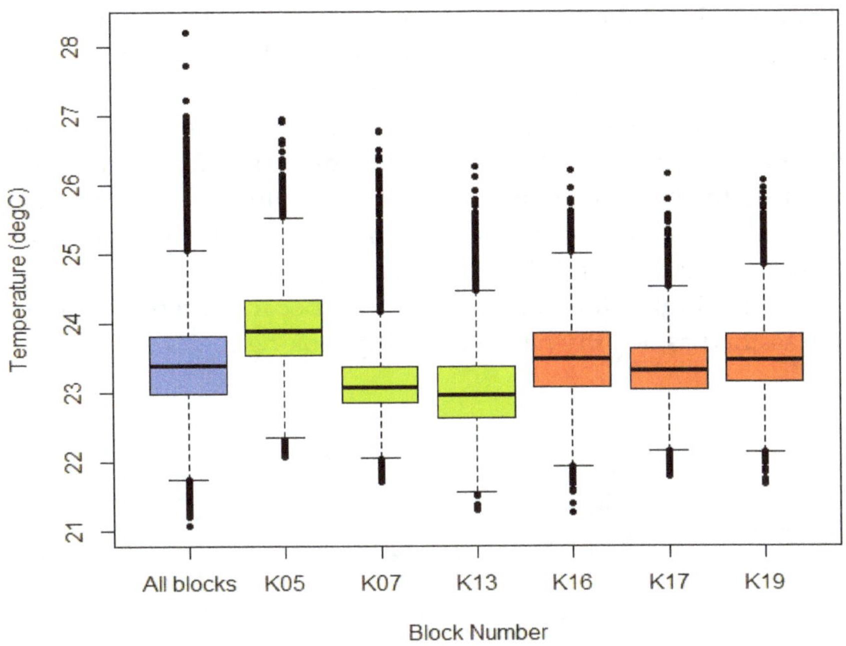

Figure 5. Boxplots of temperature for pixels defined as canopy using a buffer of 3 m and a height threshold of 4 m. The boxplot that is coloured blue shows the temperature for all blocks. The three test blocks in green (5, 7 and 13) are assumed healthy, and the three boxplots coloured red (16, 17 and 19) are unhealthy.

Figure 6 shows the outcome of the height analysis. The blocks classified as healthy appear to have similar heights to all blocks. Interestingly, the unhealthy blocks appear to deviate from the mean heights across all blocks. This is especially true for Block 19, which contains shorter trees than the other blocks. Block 16 appears to contain taller trees than the other blocks, and Block 17 has a wider spread within the upper and lower quartiles.

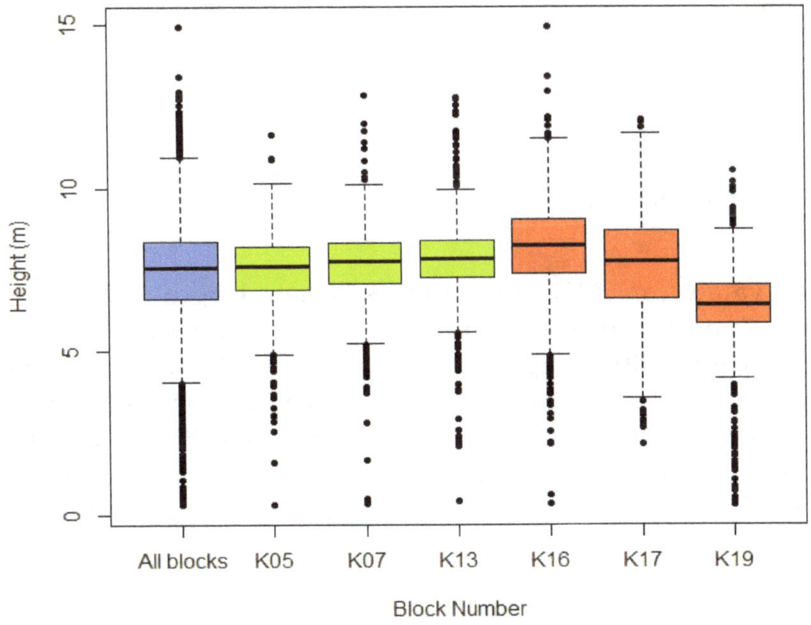

Figure 6. Boxplots for height of the blocks. The blue block shows the values for all blocks in the model farm. The healthy blocks are shown in green and the unhealthy blocks are shown in red.

The results of the boxplot analysis show some interesting trends, that are even more revealing when compared with the spatial plots of the blocks.

When looking at the spatial variation in the canopy pixels of the blocks (Figure 7), we can identify the areas that contribute to the varying values seen in the boxplots (Figure 5). Block 5 is evidently hotter than the other areas, especially in the south of the block, but also across the entire block as well. This is interesting, as the sparseness of the trees in the south of the block is similar to that in the sparse areas of the other blocks. It is difficult to determine the reason for this temperature difference. This area also corresponds to the high height area ratio (HAR) values seen in Figure 8. The unhealthy blocks clearly have more areas that are hotter. When comparing to the RGB data, these hotter areas tend to lie where the trees are sparser and thus more bare ground is exposed. There are also areas in the healthy blocks that have higher temperatures—this is because there is a high variability in all blocks and even those classified as healthy have unhealthy looking regions.

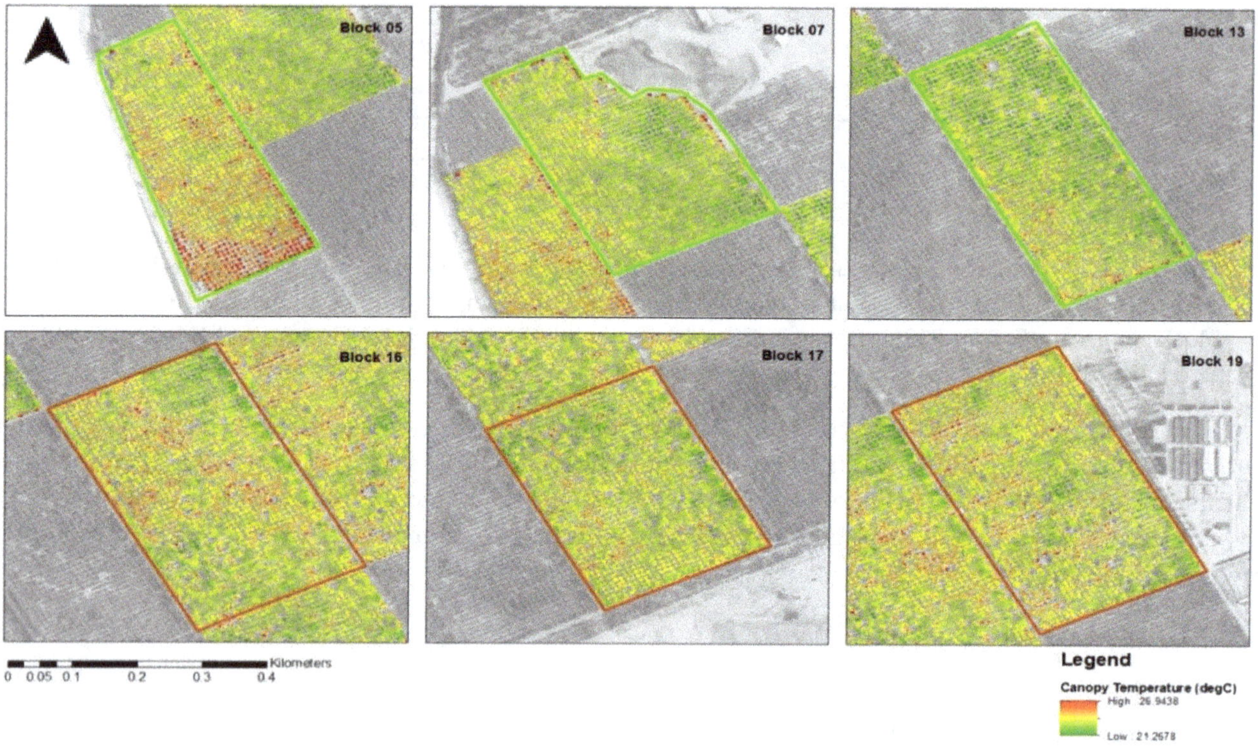

Figure 7. Temperature of the extracted canopy pixels representing individual trees for the test blocks.

Figure 8 shows the values for HAR in each block, where height is divided by canopy area. The healthy test blocks (5, 7 and 13) appear to have mostly values between 0.15 and 0.2 and Blocks 16 and 17 have more tree polygons with a larger height area ratio value, although one area of Block 5 contains trees with larger HAR values. This is to be expected as it indicates that there are more trees with abnormally small canopies, which one would expect in an unhealthier block. However, Block 19 has low HAR values, and in fact looks healthier than some of the "healthy" blocks. The trees in this block are shorter than in the other blocks, and the canopy area is similar, and thus it has the lowest mean height area ratio of all the blocks.

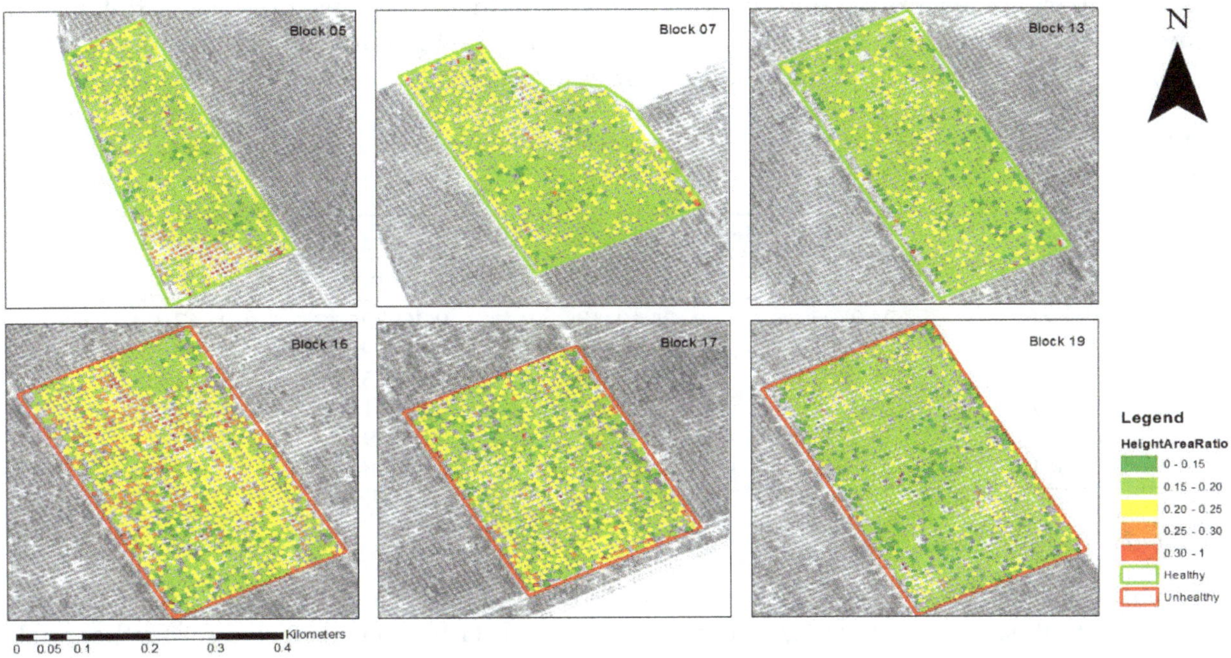

Figure 8. The height area ratios of the derived tree polygons representing individual trees for the test blocks.

Based on Figures 7 and 8, it is evident that within the blocks there is a lot of variation. This is especially true for Block 5, where there is a clear distinction in temperature and height between the north and south of the block.

ANOVA analysis was conducted on 10 randomly selected (within a height threshold; see Section 2.2.3) 'healthy' trees and 10 'unhealthy' trees per block, in order to evaluate whether the two factors of assumed health and the block containing the tree have an effect on tree temperature, and also whether there is interaction between the two (Table 4). There is a main effect of both health (probability (p) value = 0.000611) and block number (p value = 1.06E-14), and there is no interaction effect (p value = 0.834). The statistically significant outcome for block number indicates that the differences in the mean temperature of all the trees in a block are large enough to be considered as different. When comparing the mean temperature of all healthy trees and all unhealthy trees there is a statistically significant difference.

Table 4. Results of ANOVA analysis for temperature in relation to block number and the assumed block health.

	Df	Sum Sq	Mean Sq	F Value	Pr (>F)	Significance [1]
Block No.	5	20.515	4.103	21.161	1.06E-14	***
Health	1	2.417	2.41	12.465	0.000611	***
Block No.: Health	5	0.407	0.081	0.419	0.834311	
Residuals	108	20.94	0.194			

[1] Significance levels: 0 '***'; 0.001 '**'; 0.01 '*'; 0.05 '.'; 0.1 '·'; 1.

The results of post-hoc Tukey's HSD test reveal which mean differences contribute most to the statistical significance of the ANOVA results. A subset of results in Table 5 show six comparisons between healthy and unhealthy trees for each test block. When looking at the p value, we see that they are all greater than a significance level $\alpha = 0.05$, indicating no statistical significance. This suggests that when comparing the means of healthy and unhealthy trees within each block there are no statistically significant differences between these means. The differences in means of healthy and unhealthy trees range from 0.1 °C to 0.45 °C. This leads to the conclusion that the significant results seen in Table 4 are a result of comparing trees from different blocks. However, in all cases, we do see that the unhealthy trees have a mean temperature higher than the healthy trees, which does match the hypothesis that unhealthy trees have higher temperatures due to stressors [6,17].

Table 5. Selection of Tukey's HSD comparing mean temperatures of selected healthy (n = 10) and unhealthy trees (n = 10) per block.

	Mean Temperature (°C)			p Value
	Unhealthy	Healthy	Difference	
Block 5	24.37052	24.10113	0.26939	0.967108
Block 7	23.30163	23.04954	0.25209	0.979962
Block 13	23.02733	22.77498	0.25235	0.979803
Block 16	23.80391	23.4071	0.39681	0.682276
Block 17	23.40484	23.31995	0.08489	0.999999
Block 19	23.68493	23.23745	0.44748	0.50232

3.3. Individual Tree-Level Analysis

Individual tree analysis has been carried out in order to compare the temperature and height values per tree to the results of the vegetation analysis. Block 11 has been selected for further analysis as there are a wider variety of trees, such as those with wider and smaller canopies. The 100 randomly selected test trees are shown in Figure 9, overlaid on the RGB image, the Chlorophyll Index Green (CIG-Table A1) vegetation index and thermal imagery. The hyperspectral image has geometrical distortion across the whole image, and it is especially notable in the northern region, and along the edges. However, when comparing with the visual (RGB) image, definite patterns are visible when comparing the vegetation index values and the tree canopy density that is visible in Figure 9a. In the south of the block, for example, there is more dense vegetation, and on the right side on the bottom there are patches of sparser vegetation. The regions of sparse vegetation in the top right and dense vegetation in the central region are also visible in the thermal imagery. The randomly selected test points cover a good range of these areas, with points in both the dense and sparse regions of canopy.

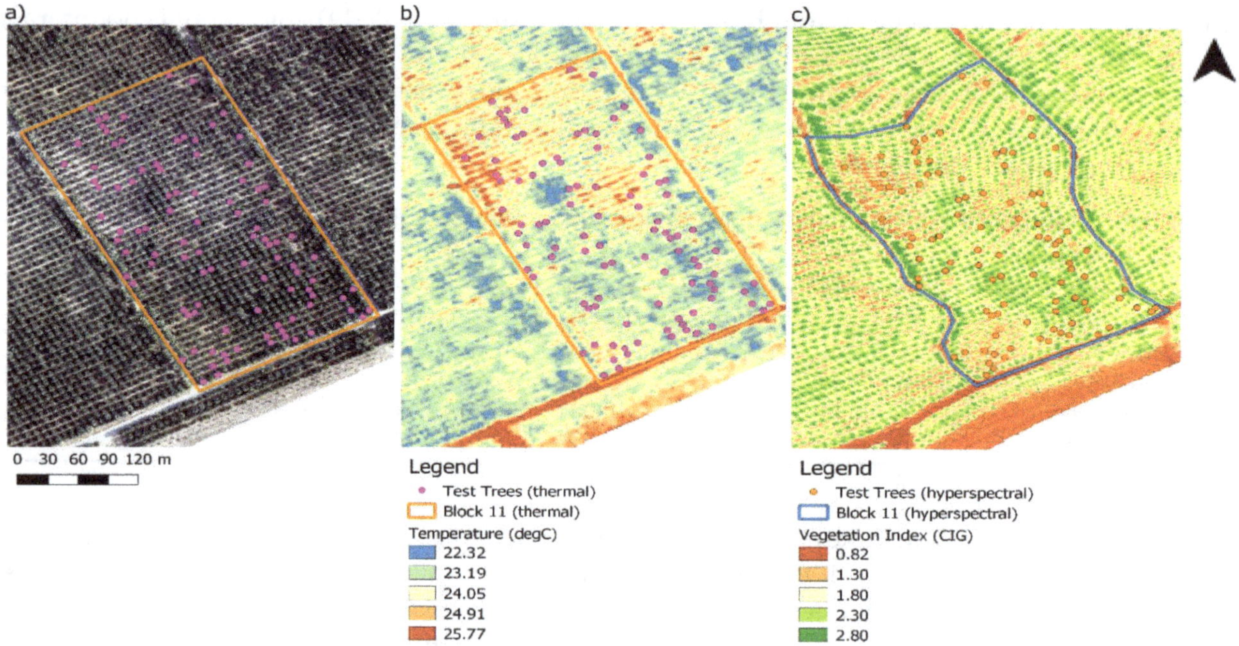

Figure 9. Illustrations of the selected test trees in Block 11 for (**a**) the visual RGB imagery, (**b**) the thermal imagery, and (**c**) the hyperspectral imagery represented as calculated Chlorophyll Index Green (CIG) (Table A1).

The correlation analysis between the vegetation indices and the height and temperature data show some interesting results. The R-squared values reach a maximum of 0.313 between the vegetation index red-edge position (REP) and temperature, and 0.253 between REP and height (Figure 10). The vegetation index Vogelmann red edge index (VOGI) also has a relatively strong correlation value with both temperature (R^2 = 0.227) and height (R^2 = 0.213) (Figure 10) and these indices both use bands in the red edge positions. Green NDVI (gNDVI) and Chlorophyll Index Green (CIG) also show relatively high correlation values (Table A2). These R-squared values are quite low in general, but as Figure 10 shows there is a trend between the two most highly correlated VIs (REP and VOGI) plotted against temperature and height for the 100 test trees. It is evident that temperature has a negative correlation with the VIs, indicating that at higher temperatures, these VI values tend to be lower. Height has the opposite correlation—taller trees have higher VI values.

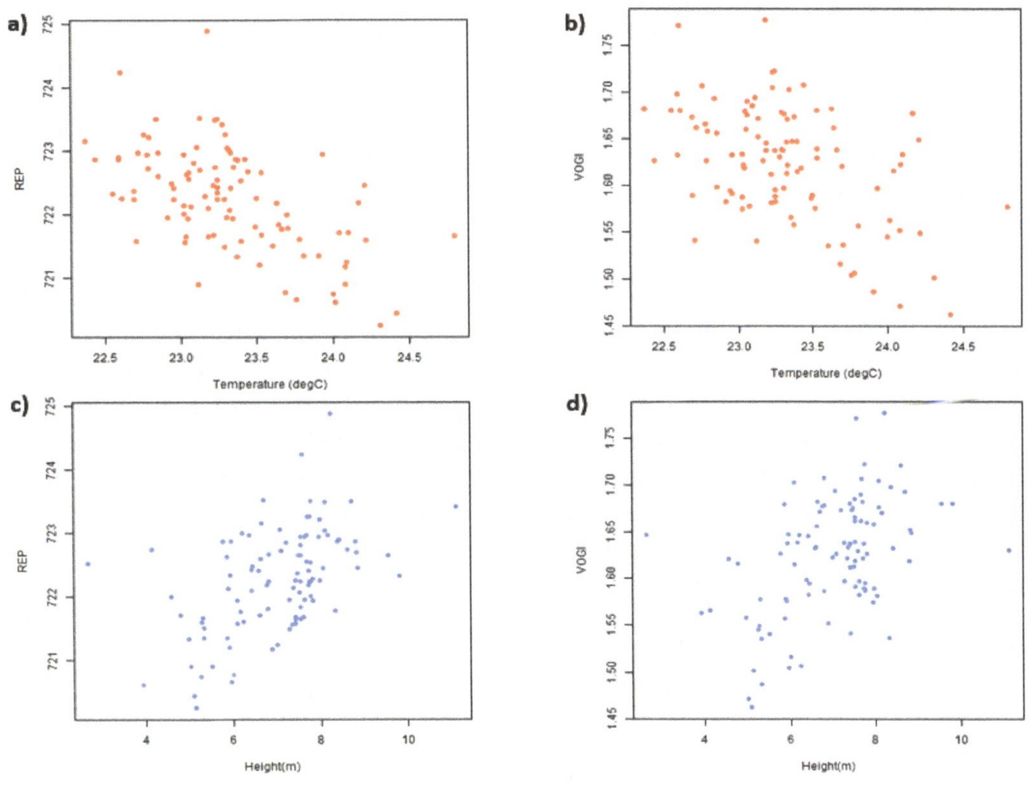

Figure 10. Correlation analysis for 100 test trees in Block 11: (**a**) and (**b**) show the correlation between temperature and two selected VIs, red-edge position (REP) and Vogelmann red edge index (VOGI), and (**c**) and (**d**) show the relationship between these indices and height.

In Figure 11 the temperature and height of the test trees are plotted overlaying the VOGI vegetation index image. It is evident that the regions with lower VI values tend to have higher temperatures and shorter trees; the opposite is generally true in regions with high VI values.

To get a clearer impression of the relation between the temperature and height values and VIs, the values have been visually displayed in relation to the VOGI vegetation index. It can be noted that the regions with lower VI values tend to have higher temperatures and shorter trees; the opposite is generally true in regions with high VI values.

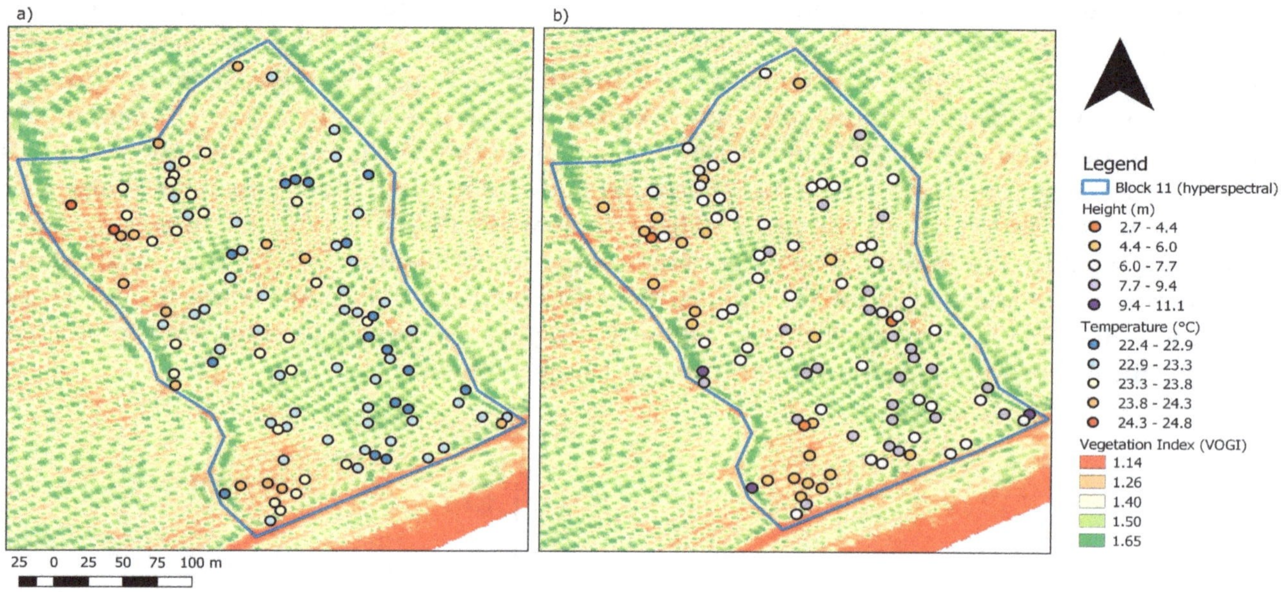

Figure 11. Illustration of the spatial patterns between tree-based point observations of (**a**) temperature and (**b**) height, with the vegetation index VOGI visualized as background map for comparison.

4. Discussion

This study explored various options for assessing vegetation health of date palm plantations using a combination of remote sensing data sources, specifically LiDAR, thermal, hyperspectral and visual RGB imagery. In order to differentiate between more general stressors and those that only affect individual palms, the plantations have been assessed on both a block and individual tree level. Although no direct field data was available for validation at tree level, exchange with the plantation manager provided support for the identified spatial patterns. The main contribution that this research can make is as a suggestion for methods to be used in further research. That being said, the preliminary results do show an ability to successfully differentiate between groups of trees and can be compared to earlier studies [6,16,17,19].

4.1. Exploring the Main Findings of the Study

On a block level, it is evident that the remote sensing images do highlight differences between the blocks; the results of the thermal box plots, for example, show up to a maximum of 1 °C difference between the blocks (Figure 5). These box plots show variation across the data—the spatial maps in Figures 7 and 8 illustrate how this variation is spatially distributed. Block 5 is a clear example, as it shows two very distinct areas in the north and the south of the block, where the trees are both much warmer and shorter. These inter-block variations can be seen to a lesser extent within each block. Therefore, it could be useful to add a further level to the current analysis, namely the development of management zones within the blocks that are based on tree characteristics.

ANOVA analysis was carried out to compare the temperature of individual healthy and unhealthy trees across each block, a method that has been applied in previous studies [6,17]. In the current study, the ANOVA results were statistically significant (Table 4) when comparing block number (p value = 1.06E-14) and tree health (p value = 0.000611) independently of one another. Tukey's HSD test (Table 5) shows the most significant differences come from comparing the trees in different blocks. There is no statistically significant result when comparing means of healthy trees in one block with means of unhealthy trees in that same block. The differences in means when comparing the unhealthy and healthy trees of one block range between 0.1 and 0.45 °C, which may not be practically significant to the extent that significant differences between trees can be identified based solely on temperature. When comparing the individual tree differences found in this study with those found by Cohen et al. (2012), who noted that trees under an 80% water deficit were warmer than the control treatment by approximately 1 °C following four months of the assigned water treatment, the differences in temperature seen in the current study seem notable [6]. Cohen et al. (2012) used comparable thermal images with a resolution of 1.8 m [6]. Golomb et al. (2015), using thermal images with a resolution of 0.5 m, also found statistically significant results when comparing mean Crop Water Stress Indices (CWSI) values of control trees and infested trees using ANOVA, with a statistical significance of $p = 0.001$ [17]. When looking at the absolute temperature difference, a preliminary study by Soroker et al. (2013) noted a mean temperature difference of 1.5 °C between trees infested with *R. ferrugineus* and control trees [16].

Spatial resolution for the thermal data was 1.8 m, which is comparable to the study of Cohen et al. (2012), but lower than for Golomb et al. (2015), where the pixel size was 0.5 by 0.5 m [6,17]. This spatial resolution could have several implications for the results. Based on the RGB images, a large palm tree canopy is approximately 6–8 m in diameter and is covered by approximately 9 to 16 pixels for 1.8 m resolution. However, due to the leaf structure of the date palm canopy there are areas where the background soil will be visible in the pixel, and temperature of the background soil will also be included in the pixel value—it is probable that only one or two pixels include only pure canopy. In studies looking for a large-scale pattern such as the effect of irrigation on an entire plantation, these changes could be visible at a resolution of 1.8 m because every tree in a block is affected by the lowered irrigation [6]. However, for analysis on an individual tree level this resolution may be too coarse to detect whether a tree has a markedly different temperature in comparison to the surrounding trees. Therefore, a maximum resolution of 0.5 m such as in the thermal images acquired by Golomb et al. (2015) might be required to be able to effectively assess individual tree stressors [17].

Considering the results of the hyperspectral data assessment several conclusions can be drawn for individual tree analysis based on the vegetation index analysis. In comparison with other studies, VOGI was one of the most indicative VIs of palm stress in the study by Shafri et al. (2012) on oil palms [19]. gNDVI was also found to identify *R. ferrugineus* infestation of date palms [18]. These were also two of the most indicative VIs when looking at correlations on an individual tree level in this study (Figure 10), indicating that VIs which focus on red-edge and green bands may be most suitable for date palm health analysis.

4.2. *Added Value of Combined Datasets and New Indicators*

Evaluating the combination of remote sensing data sources recorded at the same timestep makes this study unique in the sphere of date palm management. Several studies have shown that the combination of remote sensing sources can provide multivariate indicators with improved characterisation of the vegetation health status [20,26].

From a methodological perspective, the ability to combine datasets has proven to be very relevant when identifying trees. Using the tree objects derived from the LiDAR data to select the canopy pixels for temperature analysis provided an alternative to the method proposed in Cohen et al. (2012), which utilised the watershed algorithm to extract canopy pixels [6]. The LiDAR data again proved useful for temperature data analysis as the pixels below a certain height could be removed to reduce the chance of soil pixels being included in the analysis. A drawback is that an extra sensor is required, which will add costs to the data acquisition process. The high-resolution RGB data was compared with the outcome of the canopy extraction method to check whether the trees were delineated well (Figure 3).

Correlation analysis between thermal and hyperspectral datasets showed promising relationships (Figure 10). The expected trends are visible, with negative correlation between temperature and several of the VIs. A positive correlation was observed for height and the VIs. Whilst the R-squared values between the variables are quite low, there is a clear spatial pattern between the indicators and VIs as seen in Figure 10. This suggests that these data, especially thermal and hyperspectral imagery, could be combined to identify stressed trees. The temperatures of the trees range from 22.5 °C to 24.5 °C (Figure 10), a difference of 2 °C. Cohen et al. (2012) were able to illustrate differences in temperature of stressed and healthy palms of 1 °C [6]. This provides an indication that using thermal data could prove useful for diagnosis of individual tree health. Based on the correlation results, the hyperspectral data should be useful for reinforcing this diagnosis.

Based on the results of the tree point extraction using NDVI, it can be concluded that this method has high potential for identifying tree objects. Over 90% of trees were identified based on a comparison of field measured trees and NDVI, assuming accurate tree detection (Section 3.1). Misidentifications occur along the edges of the block where the trees are often close together and individual tree canopies are not distinct. As these results are based on distorted data one can assume that, when using a geometrically corrected dataset, identification accuracy will be higher. Using this identification method could prove to be an alternative to expensive LiDAR techniques. It should be noted that one of the reasons this method works well is because the canopies are distinct in a data palm plantation; however, in denser canopies such as the edges of date palm blocks, identification may be less accurate.

In this study, the height area ratio (HAR) index was used in order to assess the health of the blocks. When conducting analysis on the height of the trees, it became apparent that there were large differences in height of trees in certain blocks. For example, Block 19 had lower trees than the other blocks (Figure 6). From this information, one could conclude that either Block 19 is less healthy than the other blocks, or that the trees are younger in that block. Therefore, adding canopy area as a variable when considering tree health adds another consideration to the analysis. By determining the ratio between height and canopy area, one can determine whether a tree has an abnormally small canopy in relation to height—this could be an indication that the tree is experiencing growth limitations. In a study by Shendryk et al. (2016), a width height ratio was included amongst hyperspectral and LiDAR indicators to determine the most "useful" indicator for assessing tree health using Principal Component Analysis (PCA) [20]. Of the indicators derived from the recorded tree height, the width height ratio was the most important in their study. The conclusion that taller trees have smaller canopies in comparison to shorter trees indicates that using HAR alone does not give an accurate representation of tree health, as the growth of the tree affects this value for reasons other than health. Therefore, using an index that normalises height area ratio based on the height of a tree could prove more meaningful when making conclusions about tree health.

4.3. Data Limitations

As has been discussed in the Section 2.2.1 on preprocessing, there are some issues with data quality, especially for the hyperspectral dataset used in this study. Both the geometric distortions and the unknown scaling factor mean that the values used in the analysis might not be fully representative and certainly cannot be directly compared to other hyperspectral datasets that could be used in the future. For example, one step involved using bilinear interpolation for geo-correction, which resulted in multiple raster points with the same value because the hyperspectral image was effectively stretched to fit the corrected dataset.

Due to the distortions of the hyperspectral analysis, it was not possible to compare the trees on an individual level without manually matching the points. This meant that a relatively small sample size of 100 trees has been used as representative of over 10,000 trees. It is also possible that errors have been made when manually identifying the corresponding trees, as in some cases it was difficult to count rows and columns accurately within the distorted data.

Additionally, there are issues within the thermal dataset; specifically, for some limited areas blurring was noticed. This is due to the ortho-mosaicking technique used to merge the snapshot images. Trees are not identifiable in some regions, meaning that extracting the temperature per tree will result in values that are not an accurate representation of tree temperature. This was considered when selecting a block for individual tree analysis carried out in Section 2.2.4, as it was important that the blurring effect was minimal.

Finally, without ground truth data, it is not possible to determine a measure of actual tree health. This means that all assumptions made about tree health are based on indirect field observations and the relative health indicated by the RGB imagery (Table 2). The remote sensing datasets have been compared in various ways to determine whether the expected relations for healthy and unhealthy trees can be seen, and in some cases the initial assumptions were correct. However, as was seen in Figure 5, Block 5 had a much higher temperature than expected despite being considered as a healthy block. This suggests that there were factors not considered in classifying the blocks, and further field information was required. Following the analysis, further field information was provided, and the farm manager indicated that Block 5 had relatively poor soil quality in the southern region. Validation with field data is an essential step for future studies to explore the methods discussed in this paper further. As a result of the limited field data, especially on an individual tree level, the conclusions that can be made are limited to identifying that there are differences in date palms or date palm groups, and these differences cannot be attributed to a specific stressor.

4.4. Future Developments

Overall, this study reiterates the findings of earlier research indicating that different remote sensing datasets could prove useful for management of date palm plantations [6,16,17,19]. Thermal and hyperspectral data are the most promising, especially since identifying trees using NDVI is comparable to the results of LiDAR analysis. For future studies the opportunities of high-resolution RGB imagery to identify variation in palm canopy geometry could be investigated by using a combination of RGB based vegetation indices [27] and object-based image analysis.

Remote sensing data will prove most useful when provided as part of an integrated plantation management platform, where field data is being collected as the flight is taken, and ideally field data will be available over an extended period of time. The imagery could provide useful insights at all steps of an integrated system, including; field registration, field surveys, data acquisition, and monitoring. For example, the step of tree identification using NDVI or LiDAR would be useful for field registration. A topic that could be interesting to explore would be identifying within-block parcels with differing characteristics to be classified as management zones, which would allow for the adoption of precision agriculture techniques. Regular monitoring using a combination of satellite and aerial, or even drone imagery [28], potentially using the methods as explored in this article, could add value to the current monitoring systems that are in place. Based on a time-series analysis approach, changes in vegetation reflectance properties can be indicative of palm health, production and disease infestation processes over the growing season. For example, Jimenez-Brenes et al. [29] showed how multi-temporal UAV-based images can be adopted to support pruning management in olive tree orchards. The ground monitoring system should also be well designed and could include ground sensors [30] and standardised methods of recording tree characteristics. Based on these spatial-temporal data, spatial analytics and geostatistical functions can be adopted to evaluate infestation processes of pests and diseases both at farm and regional scale-level [31]. A design based on the DateGIS platform approach could prove useful, as it aims to integrate imagery data with a ground data recording system where farm employees can update information about specific trees in the field [32].

5. Conclusions

This research explored the options of using multiple types of high-resolution remote sensing imagery in order to assess its potential for detecting various health aspects of date palm plantations. The factors which affect health were considered on two levels; those that impact large areas of the plantation such as a block, and those that affect the individual trees. Based on the spatial variation within blocks, it is suggested to also consider a third level in future studies; namely, areas within blocks with similar characteristics. Despite issues with the hyperspectral and thermal data preprocessing and quality, the methods developed in this study provide new options for future analysis and indicate that remote sensing data could aid plantation management and supplement precision agriculture techniques. Specifically, using a combination of high resolution thermal and hyperspectral imagery can give an indication of individual tree health, and by using these indicators together it could be possible to define a status for each tree. There are many new developments to be made in this area of research—combining remote sensing with detailed field data could provide an early indication of *R. ferrugineus* infestation; and comparing flights at different time steps could provide insight into the rates of change of date palm health. Overall, this study adds to the currently small body of research regarding using remote sensing imagery for date palm plantation management by using a combination of data sources and by providing suggestions for future research in this topic.

Author Contributions: Conceptualization: M.M., L.K., L.B.; Methodology: M.M. and L.K.; Software: M.M.; Formal analysis: M.M..; Writing—Original draft preparation: M.M..; Writing—Review and editing: L.K and L.B.; Visualisation: M.M.; Supervision: L.K. and L.B.; Project administration: L.B.; Funding acquisition: L.B.

Acknowledgments: The authors want to acknowledge Mohamed Ali Bob, farm manager of the Al Mohamadia date farm (www.mohamadia.com.sa/en/) for his valuable support on date farm management and ground truthing feedback, and TAQNIA ETSC for provisioning and operating of the Diamond DA42 MPP GEOSTAR aircraft. Further, the authors want to acknowledge Harm Bartholomeus (Wageningen University and Research) for his support on LiDAR data preprocessing, Juha Suomalainen (Finnish Geospatial Research Institute) for support with the preprocessing to generate a mosaic for the thermal dataset, and Daniel Iordache (VITO) for support in preprocessing of the hyperspectral dataset.

Appendix A

Table A1. Formulae for all hyperspectral vegetation indices used for the hyperspectral analysis. R_n refers to the band number used.

Vegetation Index	Formula	Reference
SRI	$\frac{R_{800}}{R_{670}}$	[19,33]
NDVI 800	$\frac{R_{800}-R_{660}}{R_{800}+R_{660}}$	[19,34]
MSR	$\frac{R_{800}/R_{670}-1}{(R_{800}/R_{670})^{-1}+1}$	[19,35]
NDVI 750	$\frac{R_{750}-R_{660}}{R_{750}+R_{660}}$	[36]
VOGI	$\frac{R_{740}}{R_{720}}$	[19,37]
CIR	$\frac{R_{780}}{R_{710}}-1$	[38,39]
gNDVI	$\frac{R_{801}-R_{550}}{R_{801}+R_{550}}$	[18,40,41]
CIG	$\frac{R_{780}}{R_{550}}-1$	[38]
PRI	$\frac{R_{531}-R_{570}}{R_{531}+R_{570}}$	[42,43]
Optimum NDVI	$\frac{R_{738}-R_{610}}{R_{738}+R_{610}}$	[19]
SIPI	$\frac{R_{800}-R_{445}}{R_{800}+R_{680}}$	[18,44]
MCARI/OSAVI [705,750]	$\frac{[R_{750}-R_{705}-0.2(R_{750}-R_{550})](R_{750}/R_{705})}{(1+0.16)(R_{750}-R_{705})/(R_{750}+R_{705}+0.16)}$	[45–47]
TCARI/OSAVI [705,750]	$\frac{3[R_{750}-R_{705}-0.2(R_{750}-R_{550})(R_{750}/R_{705})]}{(1+0.16)(R_{750}-R_{705})/(R_{750}+R_{705}+0.16)}$	[46,47]
REP	$700+40\frac{R_{670}+R_{780}/2-R_{700}}{R_{740}-R_{700}}$	[48]

Appendix B

Table A2. Results of R-squared analysis of the vegetation indices contained in Table A1 and their relations to temperature and height.

Vegetation Index	Temperature (R-Squared)	Height (R-Squared)
REP	0.313	0.253
VOGI	0.227	0.213
gNDVI	0.206	0.189
CIG	0.196	0.204
CIR	0.196	0.202
TCOS750	0.191	0.103
SIPI	0.137	0.161
SRI	0.119	0.147
NDVI800	9.106	0.146
NDVI750	0.088	0.136
MSR	0.08	0.14
NDVI730	0.071	0.112
MCOS750	0.054	0.134
PRI	0.001	0.014

References

1. Stringer, L.C. Reviewing the links between desertification and food insecurity: From parallel challenges to synergistic solutions. *Food Secur.* **2009**, *1*, 113–126. [CrossRef]
2. Godfray, H.C.J.; Beddington, J.R.; Crute, I.R.; Haddad, L.; Lawrence, D.; Muir, J.F.; Pretty, J.; Robinson, S.; Thomas, S.M.; Toulmin, C. Food security: The challenge of feeding 9 billion people. *Science* **2010**, *327*, 812–818. [CrossRef] [PubMed]
3. Arias, E.; Hodder, A.J.; Oihabi, A. FAO support to date palm development around the world: 70 years of activity. *Emir. J. Food Agric.* **2016**, *28*, 1–11.
4. Al-Dosary, N.M.N.; Al-Dobai, S.; Faleiro, J.R. Review of the management of red palm weevil *Rhynchophorus ferrugineus* olivier in date palm *Phoenix dactylifera* L. *Emir. J. Food Agric.* **2016**, *28*, 34–44. [CrossRef]
5. Shabani, F.; Kumar, L.; Nojoumian, A.H.; Esmaeili, A.; Togheyani, M. Projected future distribution of date palm and its potential use in alleviating micronutrient deficiency. *J. Sci. Food Agric.* **2015**, *96*, 1132–1140. [CrossRef] [PubMed]
6. Cohen, Y.; Alchanatis, V.; Prigojin, A.; Levi, A.; Soroker, V.; Cohen, Y. Use of aerial thermal imaging to estimate water status of palm trees. *Precis. Agric.* **2012**, *13*, 123–140. [CrossRef]
7. Elshibli, S.; Elshibli, E.M.; Korpelainen, H. Growth and photosynthetic CO_2 responses of date palm plants to water availability. *Emir. J. Food Agric.* **2016**, *28*, 58–65. [CrossRef]
8. Tripler, E.; Ben-Gal, A.; Shani, U. Consequence of salinity and excess boron on growth, evapotranspiration and ion uptake in date palm (*Phoenix dactylifera* L., cv. Medjool). *Plant Soil* **2007**, *297*, 147–155. [CrossRef]
9. Askri, B.; Ahmed, A.T.; Abichou, T.; Bouhhlila, R. Effects of shallow water table, salinity and frequency of irrigation water on the date palm water use. *J. Hydrol.* **2014**, *513*, 81–90. [CrossRef]
10. Yaish, M.W.; Kumar, P.P. Salt tolerance research in date palm tree (*Phoenix dactylifera* L.), past, present, and future perspectives. *Front. Plant Sci.* **2015**, *6*, 348. [CrossRef]
11. Abraham, V.A.; Al Shuaibi, M.A.; Faleiro, J.R.; Abozuhairah, R.A.; Viyasagar, P.S.P.V. An integrated management approach to for red palm weevil *Rhynchophorus ferrugineus* Oliv. A key pest of date palm in the Middle East. *Agric. Sci.* **1998**, *3*, 77–84. [CrossRef]
12. Faleiro, J.R. A review of the issues and management of the Red Palm Weevil *Rhynchophorus ferrugineus* (Coleoptera: Rhynchophoridae) in coconut and date palm during the last one hundred years. *Int. J. Trop. Insect Sci.* **2006**, *26*, 135–154.
13. Ali, A.-S.A.; Hama, N.N. Integrated management for major date palm pests in Iraq. *Emir. J. Food Agric.* **2016**, *28*, 24–33. [CrossRef]

14. Gutiérrez, A.; Ruiz, V.; Moltó, E.; Tapia, G.; del Mar Téllez, M. Development of a bioacoustic sensor for the early detection of Red Palm Weevil (*Rhynchophorus ferrugineus* Olivier). *Crop Prot.* **2010**, *29*, 671–676. [CrossRef]
15. Dembilio, O.; Jacques, J.A. *Sustainable Pest Management in Date Palm: Current Status and Emerging Challenges*, 1st ed.; Springer International Publishing: Basel, Switzerland, 2015; pp. 13–36.
16. Soroker, V.; Suma, P.; la Pergola, A.; Cohen, Y.; Cohen, Y.; Alchanatis, V.; Golomb, O.; Goldshtein, E.; Hetzroni, A.; Galazan, L.; et al. Early detection and monitoring of Red Palm Weevil: Approaches and challenges. In Proceedings of the Palm Pest Mediterranean Conference, Nice, France, 19–18 July 2013.
17. Golomb, O.; Alchanatis, V.; Cohen, Y.; Levin, N.; Cohen, Y.; Soroker, V. *Precision Agriculture '15*, 1st ed.; Wageningen Academic Publishers: Wageningen, The Netherlands, 2015; pp. 643–650.
18. Bannari, A.; Khurshid, K.S.; Staenz, K.; Schwarz, J. Potential of Hyperion EO-1 hyperspectral data for wheat crop chlorophyll content estimation. *Can. J. Remote Sens.* **2008**, *34*, 139–157. [CrossRef]
19. Shafri, H.Z.M.; Hamdan, N.; Anuar, M.I. Detection of stressed oil palms from an airborne sensor using optimized spectral indices. *Int. J. Remote Sens.* **2012**, *33*, 4293–4311. [CrossRef]
20. Shendryk, I.; Broich, M.; Tulbure, M.G.; McGrath, A.; Keith, D.; Alexandrov, S.V. Mapping individual tree health using full-waveform airborne laser scans and imaging spectroscopy: A case study for a floodplain eucalypt forest. *Remote Sens. Environ.* **2016**, *187*, 202–217. [CrossRef]
21. Alhammadi, M.S.; Glenn, E.P. Detecting date palm trees health and vegetation greenness change of the eastern coast of the United Arab Emirates using SAVI. *Int. J. Remote Sens.* **2008**, *29*, 1745–1765. [CrossRef]
22. Vidyasagar, P.S.P.V.; Al Saihati, A.; Al Mohanna, O.E.; Subbei, A.I.; Mohsin, A.M.A. Management of Red Palm Weevil *Rhynchophorus ferrugineus* oliv., a serious pest of date palm in Al Qatif, Kindgdom of Saudi Arabia. *J. Plant. Crop.* **2000**, *28*, 35–43.
23. Aleid, S.M.; Al-Kharyi, J.M.; Al-Bahrany, A.M. *Date Palm Genetic Resources and Utilization: Volume 2: Asia and Europe*, 1st ed.; Springer: Dordrecht, The Netherlands, 2015; pp. 49–95.
24. InfraTec. *VarioCam HD Head: User Manual*; InfraTec: Dresden, Germany, 2016; Available online: https://www.infratec.at/downloads/en/thermography/manuals/infratec-manual-variocam-hd-head.pdf (accessed on 11 January 2019).
25. Westoby, M.J.; Brasington, J.; Glasser, N.F.; Hambrey, M.J.; Reynolds, J.M. 'Stucture-from-Motion' photogrammetry: A low-cost, effective tool for geoscience applications. *Geomorphology* **2012**, *179*, 300–314. [CrossRef]
26. Dalponte, M.; Reyes, F.; Kandare, K.; Gianelle, D. Delineation of Individual Tree Crowns from ALS and Hyperspectral data: A comparison among four methods. *Eur. J. Remote Sens.* **2015**, *48*, 365–382. [CrossRef]
27. Hamuda, E.; Glavin, M.; Jones, E. A survey of image processing techniques for plant extraction and segmentation in the field. *Computelectron. Agric.* **2016**, *125*, 184–199. [CrossRef]
28. Gevaert, C.M.; Suomalainen, J.; Tang, J.; Kooistra, L. Generation of spectral-temporal response surfaces by combining multispectral satellite and hyperspectral UAV imagery for precision agriculture applications. *IEEE J.-Stars* **2015**, *8*, 3140–3146. [CrossRef]
29. Jiménez-Brenes, F.M.; López-Granados, F.; de Castro, A.I.; Torres-Sánchez, J.; Serrano, N.; Peña, J.M. Quantifying pruning impacts on olive tree architecture and annual canopy growth by using UAV-based 3D modelling. *Plant Methods* **2017**, *13*, 55. [CrossRef] [PubMed]
30. Massimo, P.; Alberto, R.A.; Roberto, M.; Khalid, A.; Ali, A. Devices to detect red palm weevil infestation on palm species. *Precis. Agric.* **2018**, *19*, 1049. [CrossRef]
31. Al-Kindi, K.M.; Kwan, P.; Andrew, N.R.; Welch, M. Modelling spatiotemporal patterns of dubas bug infestations on date palms in northern Oman: A geographical information system case study. *Crop Prot.* **2017**, *93*, 113–121. [CrossRef]
32. DateGIS Demo—An Integrated Geo-Information System (GIS) for Precision Agriculture of Date Palm Farming. Available online: https://business.esa.int/projects/dategis-demo (accessed on 9 December 2018).
33. Jordan, C.F. Derivation of Leaf Area Index from quality of light on the forest floor. *Ecology* **1996**, *90*, 663–666. [CrossRef]
34. Baret, F.; Guyot, G. Potentials and limits of vegetation indices for LAI and APAR assessment. *Remote Sens. Environ.* **1991**, *35*, 161–173. [CrossRef]
35. Chen, J.M. Evaluation of vegetation indices and a modified simple ratio for boreal applications. *Can. J. Remote Sens.* **1996**, *22*, 229–242. [CrossRef]

36. Apan, A.; Held, A.; Phinn, S.; Markley, J. Detecting sugar cane 'orange rust' disease using EO-1 Hyperion imagery. *Int. J. Remote Sens.* **2004**, *25*, 489–498. [CrossRef]
37. Vogelmann, J.E.; Rock, B.N.; Moss, D.M. Red edge spectral measurements from sugar maple leaves. *Int. J. Remote Sens.* **1993**, *14*, 1563–1575. [CrossRef]
38. Gitelson, A.A.; Gritz, Y.; Merzlyak, M.N. Relationships between leaf chlorophyll content and spectral reflectance and algorithms for non-destructive chlorophyll assessment in higher plant leaves. *J. Plant Physiol.* **2003**, *160*, 271–282. [CrossRef] [PubMed]
39. Clevers, J.G.P.W.; Kooistra, L. Using Hyperspectral Remote Sensing Data for Retrieving Canopy Chlorophyll and Nitrogen Content. *IEEE J.-Stars* **2012**, *5*, 574–583. [CrossRef]
40. Gitelson, A.A.; Kaufman, Y.J.; Merzlyak, M.N. Use of a green channel in remote sensing of global vegetation from EOS-MODIS. *Remote Sens. Environ.* **1996**, *58*, 289–298. [CrossRef]
41. Blackburn, G.A. Relationships between spectral reflectance and pigment concentrations in stacks of deciduous broadleaves. *Remote Sens. Environ.* **1999**, *70*, 224–237. [CrossRef]
42. Gamon, J.A.; Peñuelas, J.; Field, C.B. A narrow-waveband spectral index that tracks diurnal changes in photosynthetic efficiency. *Remote Sens. Environ.* **1992**, *14*, 35–44. [CrossRef]
43. Barton, C.V.M. Advances in remote sensing of plant stress. *Plant Soil* **2012**, *354*, 41–44. [CrossRef]
44. Peñuelas, J.; Baret, J.; Filella, I. Semi-empirical indices to assess carotenoids/chlorophyll a ratio from leaf spectral reflectance. *Photosynthetica* **1995**, *31*, 221–230.
45. Daughtry, C.S.T.; Walthall, C.L.; Kim, M.S.; Brown de Colstoun, E.; McMurtrey, J.E., III. Estimating corn leaf chlorophyll concentration from leaf and canopy reflectance. *Remote Sens. Environ.* **2000**, *74*, 229–239. [CrossRef]
46. Haboudane, D.; Miller, J.R.; Tremblay, N.; Zarco-Tejada, P.J.; Dextraze, L. Integrated narrowband vegetation indices for prediction of crop chlorophyll content for application to precision agriculture. *Remote Sens. Environ.* **2002**, *81*, 416–426. [CrossRef]
47. Wu, C.; Niu, Z.; Tang, Q.; Huang, W. Estimating chlorophyll content from hyperspectral vegetation indices: Modelling and validation. *Agric. For. Meteorol.* **2008**, *148*, 1230–1241. [CrossRef]
48. Clevers, J.G.P.W.; de Jong, S.M.; Epema, G.; van der Meer, F.; Bakker, W.H.; Skidmore, A.K.; Addink, E.A. MERIS and the red-edge position. *Int. J. Appl. Earth Obs. Geoinf.* **2001**, *3*, 313–320. [CrossRef]

3

Combined use of Low-Cost Remote Sensing Techniques and $\delta^{13}C$ to Assess Bread Wheat Grain Yield under Different Water and Nitrogen Conditions

Salima Yousfi [1,2], Adrian Gracia-Romero [1,2], Nassim Kellas [3], Mohamed Kaddour [3], Ahmed Chadouli [3], Mohamed Karrou [4], José Luis Araus [1,2] and Maria Dolores Serret [1,2,*]

[1] Section of Plant Physiology, University of Barcelona, 08028 Barcelona, Spain; yousfisalima@hotmail.com (S.Y.); adriangraciaromero@hotmail.com (A.G.-R.); jaraus@ub.edu (J.L.A.)

[2] AGROTECNIO Center, University of Lleida, 25198 Lleida, Spain

[3] Institut Technique des Grandes Cultures, Alger 16016, Algeria; nassim.gfield@yahoo.fr (N.K.); kaddour.mohamed47@yahoo.fr (M.K.); chadouli_ahmed@yahoo.fr (A.C.)

[4] International Center for Agricultural Research in the Dry Areas (ICARDA), Avenue Hafiane Cherkaoui, Rabat 10112, Morocco; m_karrou@yahoo.fr

* Correspondence: dserret@ub.edu

Abstract: Vegetation indices and canopy temperature are the most usual remote sensing approaches to assess cereal performance. Understanding the relationships of these parameters and yield may help design more efficient strategies to monitor crop performance. We present an evaluation of vegetation indices (derived from RGB images and multispectral data) and water status traits (through the canopy temperature, stomatal conductance and carbon isotopic composition) measured during the reproductive stage for genotype phenotyping in a study of four wheat genotypes growing under different water and nitrogen regimes in north Algeria. Differences among the cultivars were reported through the vegetation indices, but not with the water status traits. Both approximations correlated significantly with grain yield (GY), reporting stronger correlations under support irrigation and N-fertilization than the rainfed or the no N-fertilization conditions. For N-fertilized trials (irrigated or rainfed) water status parameters were the main factors predicting relative GY performance, while in the absence of N-fertilization, the green canopy area (assessed through GGA) was the main factor negatively correlated with GY. Regression models for GY estimation were generated using data from three consecutive growing seasons. The results highlighted the usefulness of vegetation indices derived from RGB images predicting GY.

Keywords: wheat; canopy temperature depression; NDVI; RGB images; grain yield; $\delta^{13}C$

1. Introduction

Bread wheat is one of the most cultivated herbaceous crops in the Mediterranean region [1], with water stress and low nitrogen fertility being the main constraints limiting productivity [2]. These limitations are likely to increase in the future because climatic change is expected to decrease precipitation and increase evapotranspiration in the Mediterranean region [3]. Increasing productivity in these semi-arid environments depends on the efficiency of crop management [2] and breeding [4], where efficient and affordable methodologies to monitor crop performance, or to assess phenotypic variability for breeding, are needed. Remote sensing techniques at the canopy level have become valuable tools for precision agriculture and high throughput phenotyping [5–7]. Thus, both spectral and thermal approaches have been proposed as potential solutions to identify crop N status and water stress across large areas [8,9]. In this way, these techniques can help farmers to practice more sustainable agriculture, minimizing risks of losing the harvest by providing (whenever possible) the resources (e.g., water and fertilizer) needed to secure yield. However, the adoption of new technologies often requires much up-front investment and is therefore restricted to large-scale production and/or farmers with substantial economic resources. This limitation is particularly evident for smallholder farmers from developing countries. Nevertheless, satellite-derived indices can be used in local management to support farmers' decision making, including the rate of irrigation and fertilizer application, and eventually yield prediction in wheat [10,11] and other crops [12,13]. While satellite images are often freely available, as in the case of Sentinel-2 [14], the resolution (not higher than 100 square meters per pixel), together with the periodicity of image acquisition and weather constraints (e.g., clouds) and the need for computing support and trained staff makes this form of remote sensing is unattainable for smallholder farmers. Different approaches to small-scale-tailored crop management have been proposed. For example, site-specific nitrogen management using leaf color charts has been proposed in irrigated wheat [15,16]. However, the interaction of the water regime with nitrogen status may affect leaf color, making this method impractical for rainfed or deficit-irrigation crops. A more flexible alternative uses optical sensors such as portable spectroradiometers (like, for example, the GreenSeeker) [17,18]. However, the cost of the equipment may limit the uptake of this approach. In this sense, the use of low-cost remote sensing methods to schedule irrigation and fertilization and predict yield, such as digital conventional imagery and/or infrared thermometry [19], may contribute to more sustainable agriculture in arid and semi-arid regions of the Mediterranean where irrigation and fertilization are not optimized in terms of timing and amount. While remote sensing has been regarded as a potentially useful approach in predicting grain yield, an inherent limitation of remote sensing methods is that the relationships between yield and vegetation indices may be site and season specific, changing between sites and years. Thus, for example, in the case of sensor calibration for N management, site-year characteristics have a critical impact [20]. While new methods for sensor-based site-specific N management are probably needed, it is likely that the best approaches will arise from the use of multiple sensors [20,21], therefore increasing the cost of deployment. Even when low cost remote sensing approaches using single sensors have shown great potential in experimental trials with wheat [22], their practical application needs to be proven.

The normalized difference vegetation index (NDVI) is one of the most well-known multispectral vegetation indices. The NDVI has been used extensively to estimate plant biomass [23–25], nitrogen status [26] and yield in wheat and other cereals [27–29]. The leaf chlorophyll content measured with a portable chlorophyll meter, which uses the same principle as the NDVI, but on the basis of the light transmitted through the leaf, has also been used extensively [30]. As an alternative, information derived from conventional digital Red-Green-Blue (RGB) images to formulate canopy vegetation indices is a low-cost and an easy proximal sensing approach to assess grain yield in cereals [31–33], even when limitations related to shadows and changes in ambient light conditions need to be taken into consideration [34]. Information derived from RGB images allows estimation of a wide range of crop traits in durum and bread wheat, such as early vigor, leaf area index, leaf senescence, aerial biomass and grain yield [31,33]. The green area and the greener area are two indices derived from conventional digital images [31]. The first parameter describes the amount of green biomass in the picture, while the second one excludes the more yellowish-green pixels. In fact, greener area is aimed at capturing active photosynthetic area and plant senescence [31]. Such indices are formulated using open access software [19,35].

It has been long recognized that plant temperature may represent a valuable index to detect differences in plant water regimes [36–38]. Reynolds et al. (2007) [39] have reported that wheat canopy temperature is a relative measure of plant transpiration associated with water uptake from the soil. Under water limited conditions, transpiration and its associated evaporative cooling are reduced, resulting in higher leaf temperatures. Given that a major role of transpiration is leaf cooling, canopy temperature and its depression relative to ambient air temperature is an indicator of the degree to which transpiration cools leaves under a demanding environmental load [40]. In that sense, infrared thermometry has been proposed as a low-cost approach in crop management to enable scheduling of support irrigation [41], to assess spatial soil heterogeneity [42], or to evaluate genotypic performance to drought [43]. However, potential interaction effects between N fertilization and water regime should be considered. In particular, how does N fertilization affect the water by temperature relationships and even how does water affect the interaction of N fertilization with vegetation indices. This is not trivial because haying-off, which is the negative effect of nitrogen fertilization on productivity caused by an imbalance between transpired biomass and the available water, is regarded as a potential problem for wheat cultivation in Mediterranean regions [44]. As a consequence, unexpected relationships between remote sensing readings and grain yield may occur.

Similarly, other physiological characteristics related to plant water status, such as stable carbon isotope composition ($\delta^{13}C$; frequently measured as discrimination from the surrounding air, $\Delta^{13}C$) are also often used for evaluating genotypic performance under water stress [40,44] or even to monitor spatial variability and water status [45]. The natural ^{13}C abundance in plant matter provides time-integrated information of the effects of water stress on the photosynthetic carbon assimilation of C3 species, including wheat [46–48]. Conditions inducing stomatal closure (e.g., water deficit or salinity) restrict the CO_2 supply to carboxylation sites, which then decreases the $\Delta^{13}C$ (or increases the $\delta^{13}C$) of plant matter [47,49]. Under Mediterranean conditions the $\delta^{13}C$ of mature kernels is better correlated with grain yield than the $\delta^{13}C$ of other plant parts [50]. The costs of these analyses have decreased throughout the years, making their analysis increasingly feasible.

The objective of this study was to assess the grain yield performance of wheat under a range of water and fertilization conditions in the Mediterranean, using different low-cost remote sensing approaches to assess canopy green biomass (NDVI and vegetation indices derived from conventional RGB images), and characteristics associated with plant water status, (canopy temperature depression), together with additional traits informing on the water status ($\delta^{13}C$ of mature grains and the stomatal conductance of the flag leaf). The novelty of the study centers on (i) testing how different low-cost, user-friendly remote sensing techniques may contribute to site-specific wheat management and eventually to the prediction of yield across seasons; and (ii) how interactions between growing conditions (water regime and N fertilization) may affect the predictive strength of these techniques. Moreover, to better explore the potential usefulness of our study for wheat phenotyping we developed a conceptual model of how the combination of these different traits explains genotypic variability in grain yield under different combinations of water regimes and nitrogen fertilization.

2. Materials and Methods

2.1. Plant Material and Growing Conditions

Field trials were conducted during the 2014–2015 crop season at Bir Ould Khelifa in the area of Khemis Miliana commune, approximately 230 km to the south west of Algiers (Algeria) and with geographical coordinates 36°11′50.23″ N and 2°13′17.69″ E. This commune receives an average rainfall of between 400 and 450 mm, and it is characterized by clay-silty fertile soils, with high organic matter content and high levels of total and mineral nitrogen (Table S1). Monthly total accumulated rainfall and temperatures for the study region for the 2014–2015 crop season are presented in Table S2. Four bread wheat (*Triticum aestivum* L.) genotypes were planted on 14 December 2014. The wheat genotypes were "Ain-Abid" and "Arz" (modern varieties) and "Wifak" and "Maaouna" (local varieties). The experimental design was a split-split-plot, with the main-plot factor being water regime, the subplot factor was the N amount and the sub-subplot factor was genotype (Figure S1). A total of 108 plots (four genotypes, three replicates per genotype, three water regimes, and three nitrogen fertilization treatments) each with a size of 10 m × 1.2 m and six rows, 20 cm apart, were studied. The three water regimes consisted of one rainfed and two support irrigation treatments of 30 mm (SI-1, a single amount of supplemental irrigation) and 60 mm (SI-2, a double amount of supplemental irrigation) aimed at providing water in the typical range for the agronomic practices in this area. Supplementary irrigation was applied with sprinklers around the beginning of stem elongation. For SI-1, one irrigation (30 mm) was applied at the beginning of stem elongation (31 Zadoks stage), whereas for SI-2, a second irrigation was also delivered at heading (58 Zadoks stage). Nitrogen fertilization was applied using urea fertilizer, and treatments consisted of no fertilizer (N0) and 60 kg ha^{-1} and 120 kg ha^{-1} of nitrogen fertilizer (N60, N120 respectively). Application of N fertilization was achieved at two growing stages, tillering and jointing (26 and 31 Zadoks stages). Plots were harvested with a sickle after physiological maturity and grain yield was estimated. Thousand kernel weight and the number of kernels per square meter (Kernel m^{-2}) were evaluated.

2.2. *Vegetation Indices*

Remote sensing measurements were performed on a sunny day around anthesis (26 March 2015) between 10 h and 15 h solar time. The NDVI was determined with a portable spectroradiometer (GreenSeeker handheld crop sensor, Trimble, Sunnyvale, CA, USA). The NDVI is formulated using the following equation: (NIR − R)/(NIR + R), where R is the reflectance in the red band and NIR is the reflectance in the near-infrared band. The distance between the sensor and the plots was kept constant at around 50–60 cm above and perpendicular to the canopy. Additionally, one conventional digital picture was taken per plot, holding the camera about 80 cm above the plant canopy in a zenithal plane and focusing near the center of each plot. The images were acquired with an Olympus E-M10 camera (Olympus Corporation, Tokyo, Japan), using a 14 mm lens, triggered at a speed of 1/125 seconds with the aperture programmed in automatic mode. The size of the images was 4608 × 3456 pixels stored in JPG format using RGB color standard [51]. Pictures were analyzed with the free-access BreedPix 0.2 software, now integrated within the CerealScanner plugin (https://integrativecropecophysiology.com/software-development/cerealscanner/), from the Mediterranean Crop Ecophysiology Group, University of Barcelona [22], which was developed for digital image processing. This software quickly provides digital values from different color properties. The vegetation indices measured were the green area (GA) and the greener area (GGA). GA is the portion (as a %) of pixels with $60 < Hue < 120$ from the total amount of pixels, whereas greener area is formulated as the % of pixels with $80 < Hue < 120$ [31,52]. GGA is designed to capture the active photosynthetic area excluding senescent leaves. In addition, the leaf chlorophyll content of five flag leaf blades per plot was measured using a Minolta SPAD-502 portable meter (Spectrum Technologies Inc., Plainfield, IL, USA).

2.3. *Canopy Temperature Measurements*

Canopy temperature (CT) was measured at noon (12 h–14 h), on the same day around anthesis as the vegetation indices, using an infrared thermometer (PhotoTempTM MXSTMTD, Raytek®, Santa Cruz, CA, USA). Measurements were taken above the plants, pointing towards the canopy at a distance of about one meter and having the sun towards the rear. The air temperature was measured simultaneously for each plot with a temperature humidity meter (Testo 177-H1 Logger, Testo, Lenzkirch, Germany) and employed for the calculation of the canopy temperature depression (CTD) as the difference between the ambient and the canopy temperature.

2.4. *Stomatal Conductance*

Stomatal conductance was measured on the flag leaves on the same days as the remote sensing traits. Two measurements per plot were taken around noon (12 h–14 h), using a Decagon SC-1 Leaf Porometer (Decagon Devices Inc., Pullman, WA, USA).

2.5. *Stable Carbon Isotope Composition*

Carbon isotope composition was analyzed in mature grains using an Elemental Analyzer (Flash 1112 EA; ThermoFinnigan, Bremen, Germany) coupled with an Isotope Ratio Mass Spectrometer (Delta C IRMS, ThermoFinnigan, Bremen, Germany) operating in continuous flow mode to determine the stable carbon ($^{13}C/^{12}C$) isotope ratios. Samples of about 0.7 mg of dry matter and reference materials were weighed into tin capsules, sealed, and then loaded into an automatic sampler (ThermoFinnigan, Bremen, Germany) prior to EAIRMS analysis. Measurements were carried out at the Scientific Facilities of the University of Barcelona. The $^{13}C/^{12}C$ ratios were expressed in δ notation determined by: $\delta^{13}C = (^{13}C/^{12}C)_{sample}/(^{13}C/^{12}C)_{standard} - 1$ [53], where sample refers to plant material and standard to Pee Dee Belemnite (PDB) calcium carbonate. International isotope secondary standards of known $^{13}C/^{12}C$ ratios (IAEA CH_7 polyethylene foil, IAEA CH_6 sucrose, and USGS 40 L-glutamic acid) were used for calibration to a precision of 0.1‰.

2.6. Statistical Analysis

Data were subjected to factorial ANOVA to test the effects of the growing conditions (water regime and nitrogen fertilization), genotype, and their interaction. Mean comparisons were performed using Tukey's honestly significant difference (HSD) test. Pearson correlation coefficients between grain yield and all different traits were calculated. Multiple linear regression analysis (stepwise) was used to analyze grain yield under different growing conditions. Data were analyzed using IBM SPSS Statistics 24 (SPSS Inc., Chicago, IL, USA). Figures were created using Sigma-Plot 11.0 for Windows (Systat Software Inc., Point Richmond, CA, USA).

The performance of the different remote sensing traits in predicting yield performance across the seasons was evaluated using data from the present study, together with data already published by our team from the two previous seasons (2012–2013 and 2013–2014) related to grain yield and different remote sensing traits measured in a set of genotypes grown under different water regimes [54]. The equations of the linear relationships between these traits and grain yield (GY) determined in the first crop season (2012–2013) were further tested during the two following crop seasons 2013–2014 and 2014–2015 (the latter of these conducted during the present study). Predicted and measured grain yields were expressed as relative values; grain yields were normalized with regard to the highest yield combination (genotype and growing condition), then the means of the three replicates per genotype were calculated.

Finally, we performed path analyses to quantify the relative contributions of direct and indirect effects of water status ($\delta^{13}C$, g_s and CTD) and vegetation indices (GA, GGA) on grain yield. This methodology offers the possibility of building associations between variables on the basis of prior knowledge. Mechanisms that play potential roles in grain yield variation and involving traits that exhibited genotypic differences have been proposed, as detailed in the conceptual model displayed in Figure S2. This model was aimed at understanding grain yield responses to genotypic differences under different levels of nitrogen fertilization (N0, N60, N120) and under different water regimes (supplementary irrigation and rainfed). A model with a comparative fit index (CFI) [42] with values > 0.9 was taken as indicative of a good fit. Data were analyzed using IBM, SPSS, Amos 21 (SPSS Inc., Chicago, IL, USA).

3. Results

3.1. Irrigation and Fertilization Effects on Grain Yield

Both irrigation and nitrogen fertilization significantly affected the grain yield (GY) and the agronomic yield components (Table 1). The doubled amount of supplemental irrigation (SI-2 with all nitrogen fertilization combined) and the highest nitrogen fertilization N120 (120 kg N ha^{-1} of fertilizer with all irrigations combined) were the growing conditions that exhibited the highest grain yield. Significant interactions only existed for grain yield between genotype and water regime ($p = 0.049$).

Table 1. Mean values of genotypes, water regimes, nitrogen fertilization levels for grain yield (GY), thousand kernel weight (TKW) and kernels m^{-2} and the corresponding ANOVA.

	GY (T ha^{-1})	TKW (g)	Kernels m^{-2}
Genotypes			
Ain Abid	2.20 ^{a}	21.73 ^{a}	275.9 ^{a}
Arz	2.35 ^{ab}	23.70 ^{b}	298.3 ^{ab}
Maaouna	2.35 ^{ab}	24.35 ^{b}	307.9 ^{b}
Wifak	2.50 ^{b}	24.83 ^{b}	308.6 ^{b}
Water regime			
RF	2.07 ^{a}	22.42 ^{a}	270.6 ^{a}
SI-1	2.29 ^{b}	22.72 ^{a}	308.8 ^{b}
SI-2	2.68 ^{c}	25.83 ^{b}	313.7 ^{b}
Nitrogen fertilization			
N0	2.05 ^{a}	22.60 ^{a}	291.3 ^{a}
N60	2.40 ^{b}	23.43 ^{a}	307.6 ^{a}
N120	2.60 ^{c}	24.93 ^{b}	294.1 ^{a}
Level of significance			
Genotype (G)	0.026	0.000	0.029
Water regime (WR)	0.000	0.000	0.000
Nitrogen fertilization (N)	0.000	0.001	0.256
G × WR	0.049	0.126	0.388
G × N	0.065	0.141	0.688
N × WR	0.070	0.110	0.840
G × WR × N	0.801	0.050	0.863

Means followed by different letters are significantly different ($p < 0.05$) according to Tukey's honestly significant difference (HSD) test. For more details, including the acronyms for the treatments, see Materials and Methods.

3.2. *Vegetation Indices*

The water regime significantly affected the NDVI ($p < 0.001$), as well as the green area (GA) ($p < 0.001$) and the greener area (GGA) ($p < 0.001$) indices (Table 2). The values of these three vegetation indices were highest under SI-2 compared to the single amount of supplemental irrigation (SI-1) and rainfed conditions. However, no difference was observed in vegetation indices across fertilization treatments except for GGA ($p = 0.005$). Moreover, leaf chlorophyll content (LC) slightly increased (Table 2) under SI-1 and rainfed conditions compared to SI-2 and under N60 and N120 compared to N0. Interactions were not significant, whatever the combination (genotypes, water regimes and fertilization levels) or variables considered.

Table 2. Mean values of genotypes, water regimes, and nitrogen fertilization levels for the vegetation indices NDVI (Normalized Difference Vegetation Index), GA (Green Area), GGA (Greener Area) and LC (Leaf chlorophyll content) and the corresponding ANOVA. Parameters were measured around anthesis. Means followed by different letters are significantly different ($p < 0.05$) according to Tukey's honestly significant difference (HSD) test.

	NDVI	GA	GGA	LC
Genotypes				
Ain Abid	0.70 ^{c}	0.88 ^{c}	0.74 ^{d}	46.88 ^{a}
Arz	0.64 ^{b}	0.82 ^{b}	0.58 ^{b}	52.32 ^{c}
Maaouna	0.57 ^{a}	0.74 ^{a}	0.50 ^{a}	50.04 ^{b}
Wifak	0.58 ^{a}	0.81 ^{b}	0.64 ^{c}	50.57 ^{bc}
Water regime				
RF	0.56 ^{a}	0.73 ^{a}	0.52 ^{a}	50.52 ^{b}
SI-1	0.62 ^{b}	0.78 ^{b}	0.60 ^{b}	50.35 ^{ab}
SI-2	0.70 ^{c}	0.93 ^{c}	0.73 ^{c}	49.00 ^{a}

Table 2. *Cont.*

	NDVI	GA	GGA	LC
Nitrogen fertilization				
N0	0.62 [a]	0.80 [a]	0.58 [a]	49.08 [a]
N60	0.63 [a]	0.82 [a]	0.64 [b]	50.75 [b]
N120	0.63 [a]	0.82 [a]	0.63 [b]	50.03 [ab]
Level of significance				
Genotype (G)	0.000	0.000	0.000	0.000
Water regime (WR)	0.000	0.000	0.000	0.037
Nitrogen fertilization (N)	0.699	0.304	0.005	0.038
G × WR	0.071	0.073	0.381	0.144
G × N	0.729	0.423	0.562	0.438
N × WR	0.765	0.721	0.598	0.495
G × WR × N	0.975	0.518	0.492	0.257

3.3. Canopy Temperature Depression, Stable Carbon Isotope Composition and Stomatal Conductance

Water regime affected significantly the canopy temperature depression (CTD) ($p < 0.001$), the stomatal conductance (g_s) ($p < 0.001$) and the stable carbon isotope composition ($\delta^{13}C$) ($p < 0.001$) of mature grains. Rainfed conditions decreased g_s and CTD, whereas $\delta^{13}C$ increased compared to support irrigation conditions (Table 3). Nevertheless, fertilization treatments did not affect any of these three parameters (Table 3). No significant interactions were observed except for the $\delta^{13}C$ between the water regime and nitrogen fertilization ($p = 0.022$).

Table 3. Mean values of genotypes, water regimes, and nitrogen fertilization levels for canopy temperature depression (CTD), stomatal conductance (g_s) of the flag leaves and the stable carbon isotope composition ($\delta^{13}C$) of the mature grains. Means followed by different letters are significantly different ($p < 0.05$) according to Tukey's honestly significant difference (HSD) test.

	CTD (°C)	g_s (μmol CO_2 m^{-2} s^{-1})	$\delta^{13}C$ (‰)
Genotypes			
Ain Abid	1.92 [a]	143.71 [a]	−24.07 [b]
Arz	1.48 [a]	119.07 [a]	−24.09 [b]
Maaouna	1.20 [a]	119.51 [a]	−24.43 [a]
Wifak	1.07 [a]	149.90 [a]	−24.64 [a]
Water regime			
RF	−0.49 [a]	71.18 [a]	−23.50 [c]
SI-1	1.68 [b]	95.31 [a]	−24.19 [b]
SI-2	3.07 [c]	232.65 [b]	−25.23 [a]
Nitrogen fertilization			
N0	1.67 [a]	132.37 [a]	−24.28 [a]
N60	1.40 [a]	146.86 [a]	−24.34 [a]
N120	1.19 [a]	119.93 [a]	−24.31 [a]
Level of significance			
Genotype (G)	0.066	0.102	0.000
Water regime (WR)	0.000	0.000	0.000
Nitrogen fertilization (N)	0.269	0.142	0.841
G × WR	0.743	0.689	0.371
G × N	0.307	0.963	0.502
N × WR	0.336	0.950	0.022
G × WR × N	0.658	0.106	0.932

3.4. *Genotypic Effect on Grain Yield, Vegetation Indices and Water Status Traits*

The genotypic effect was significant ($p < 0.05$) for grain yield (GY) and ($p < 0.001$) for vegetation indices (Tables 1 and 2) under the growing conditions analyzed together, whereas only the $\delta^{13}C$ (in the case of water status traits) was significantly ($p < 0.001$) different between genotypes (Table 3). The genotypic difference was also examined within each of the nine growing conditions, resulting from the combination of the three water regimes and the three nitrogen fertilization levels (Table 4). GY only showed a genotypic effect under SI-2 combined with nitrogen fertilization (either N120 or N60). However, under the SI-1 and rainfed conditions, GY did not show genotypic differences, regardless of the N fertilization regime. The vegetation indices, LC content, $\delta^{13}C$ and g_s also showed genotypic effects in some specific growing conditions while CTD did not (Table 4).

Table 4. Genotype effect on the Normalized Difference Vegetation Index (NDVI) and the Green Area and the Greener Area (GA and GGA) vegetation indices, the leaf chlorophyll content (LC), the stomatal conductance (g_s) of the flag leaf, the canopy temperature depression (CTD), the stable carbon isotope composition ($\delta^{13}C$) of the mature grains and the grain yield (GY).

Growing Conditions	NDVI	GA	GGA	LC	g_s	CTD	$\delta^{13}C$	GY
SI-2 with N120	0.002 **	0.089 ns	0.010 **	0.037 *	0.061 ns	0.493 ns	0.308 ns	0.048 *
SI-2 with N60	0.000 ***	0.009 **	0.002 **	0.043 *	0.970 ns	0.564 ns	0.216 ns	0.024 *
SI-2 without N	0.003 **	0.000 ***	0.001 ***	0.008 **	0.796 ns	0.111 ns	0.112 ns	0.144 ns
SI-1 with N120	0.127 ns	0.650 ns	0.299 ns	0.118 ns	0.624 ns	0.703 ns	0.292 ns	0.279 ns
SI-1 with N60	0.050 *	0.109 ns	0.006 **	0.071 ns	0.000 ***	0.076 ns	0.205 ns	0.708 ns
SI-1 without N	0.129 ns	0.519 ns	0.455 ns	0.883 ns	0.973 ns	0.610 ns	0.027 *	0.070 ns
Rainfed with N120	0.064 ns	0.150 ns	0.056 ns	0.035 *	0.212 ns	0.738 ns	0.594 ns	0.732 ns
Rainfed with N60	0.034 *	0.005 **	0.034 *	0.002 **	0.416 ns	0.682 ns	0.019 *	0.751 ns
Rainfed without N	0.000 ***	0.002 **	0.001 ***	0.103 ns	0.242 ns	0.519 ns	0.086 ns	0.149 ns

Significance levels—ns, not significant; * $p < 0.05$; ** $p < 0.01$ and *** $p < 0.001$.

3.5. *Relationships of the Grain Yield with the Vegetation Indices*

Combining both irrigation (SI-2 and SI-1) conditions, the two RGB vegetation indices (GA and GGA) were positively correlated with GY at N120 and only GA at N60, while the NDVI was not correlated with GY (Figure 1). Moreover, under rainfed conditions GA, GGA and the NDVI were correlated with GY at N60 (Figure 1). In the absence of nitrogen fertilization, no correlation was found between GY and the different vegetation indices (Table S3).

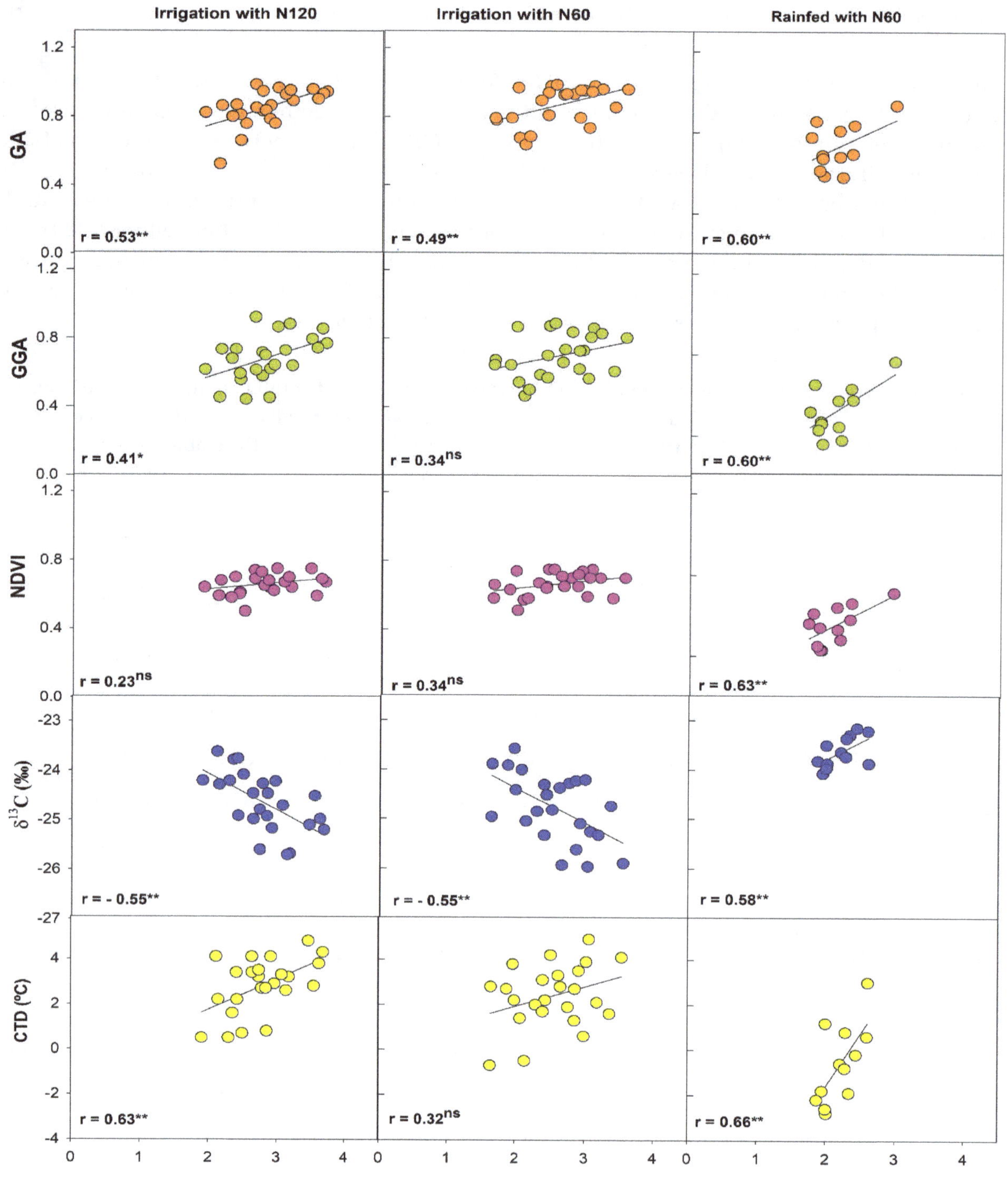

Figure 1. Relationships of grain yield (GY) with NDVI, GA, and GGA, $\delta^{13}C$ of mature grains and the canopy temperature depression (CTD) under (left column) irrigation with N120, (central column) irrigation with N60, and (central column) rainfed conditions with N60. The irrigation data combines the two support-irrigation regimes (SI-1, SI-2). Significance levels—ns, not significant; * $p < 0.05$ and ** $p < 0.01$. Abbreviations for variables and growing conditions are as defined in Tables 1 and 2.

3.6. Relationships of the Grain Yield with the Canopy Temperature Depression and the $\delta^{13}C$

Grain yield was negatively correlated with the $\delta^{13}C$ of mature grains under irrigation at both N120 and N60, and positively correlated with the $\delta^{13}C$ of rainfed conditions at N120 (Figure 1). Moreover, GY was positively correlated with CTD under either irrigation (SI-1 and SI-2 combined) or rainfed conditions at N120, while under N60 no association was found (Figure 1 and Table S3). In the absence of N fertilizer, no correlations were found (Table S3).

3.7. Relationships of Vegetation and Water Status Indices

The NDVI correlated highly and positively with both the GA and GGA vegetation indices when all genotypes, growing conditions, and replicates analyzed were combined (Table 5). Moreover, the CTD was also significantly ($p < 0.01$) correlated with the other water status parameters; negatively with $\delta^{13}C$ and positively with g_s. Likewise, both vegetation and water status indices were significantly ($p < 0.01$) correlated; the NDVI, GA and GGA were positively associated with CTD and g_s and negatively correlated with $\delta^{13}C$ (Table 5).

Table 5. Correlation coefficients of the relationships between different indices used to measure crop biomass and water status. Parameters were measured at anthesis. Treatments and genotypes were analyzed together. Significance levels—** $p < 0.010$; *** $p < 0.000$. Abbreviations of variables as in Tables 1 and 3.

Vegetation Indices	Correlation Coefficients
NDVI vs. GA	0.77 ***
NDVI vs. GGA	0.73 ***
Water status indices	
CTD vs. $\delta^{13}C$	−0.57 **
CTD vs. g_s	0.63 **
Vegetation and water status indices	
NDVI vs. CTD	0.63 **
GA vs. CTD	0.55 **
GGA vs. CTD	0.54 **
NDVI vs. g_s	0.51 **
GA vs. g_s	0.58 **
GGA vs. g_s	0.55 **
NDVI vs. $\delta^{13}C$	−0.43 **
GA vs. $\delta^{13}C$	−0.49 **
GGA vs. $\delta^{13}C$	−0.44 **

3.8. Grain Yield Estimation Using Vegetation Indices and Canopy Temperature

A general model using linear regression of the NDVI, GA and CTD with GY in the 2012–2013 crop season was developed in this study to estimate GY in the two successive crop seasons (2013–2014 and 2014–2015). Only the NDVI, GA and CTD were involved in this model because they were the three variables included in the stepwise models to explain the difference in GY in the present study (see Section 3.9 below) and were significantly ($p < 0.01$) correlated with GY in the first crop season [54]. With all growing conditions and genotypes analyzed together, the predicted and measured grain yields were positively and highly significantly ($p < 0.01$) correlated in 2013-2014 and 2014–2015 (Figure 2) using any of the three parameters studied alone (NDVI, GA, CTD). However, the range of normalized values predicted was smaller, in general, using the NDVI (at 2014–2015 crop season) than either of the other two indices. Predicted and measured GY for the 2013–2014 crop season was also positively and highly correlated using the NDVI, GA and CTD (Figure 3) under irrigation (SI-2 and SI-1 combined) with N fertilization (N120 and N60 combined) and without N fertilization (N0). For the 2014–2015 crop season, predicted and measured grain yields were also highly significantly ($p < 0.01$) correlated using either the NDVI, GA or CTD (Figure 3), but only under irrigation (SI-2 and SI-1 combined) with N fertilizer (N120 and N60 combined) and with no correlation under N0. In the absence of irrigation (rainfed conditions) and regardless of the N fertilization conditions, we did not find any association between the predicted and the measured grain yields (Figure 3). Grain yield estimation was also examined within each of the nine growing conditions resulting from the combination of the three water regimes and the three nitrogen fertilization levels (Figure S3). Under any of the two irrigation conditions with and without N fertilizer and the rainfed conditions with N fertilizer, the predicted GY was positively correlated with the measured GY in both 2013–2014 and 2014–2015 using at least one of the NDVI, GA or CTD parameters (Figure S3). Under rainfed conditions and N0 the predicted and measured GYs were not correlated in either crop season (Figure S3).

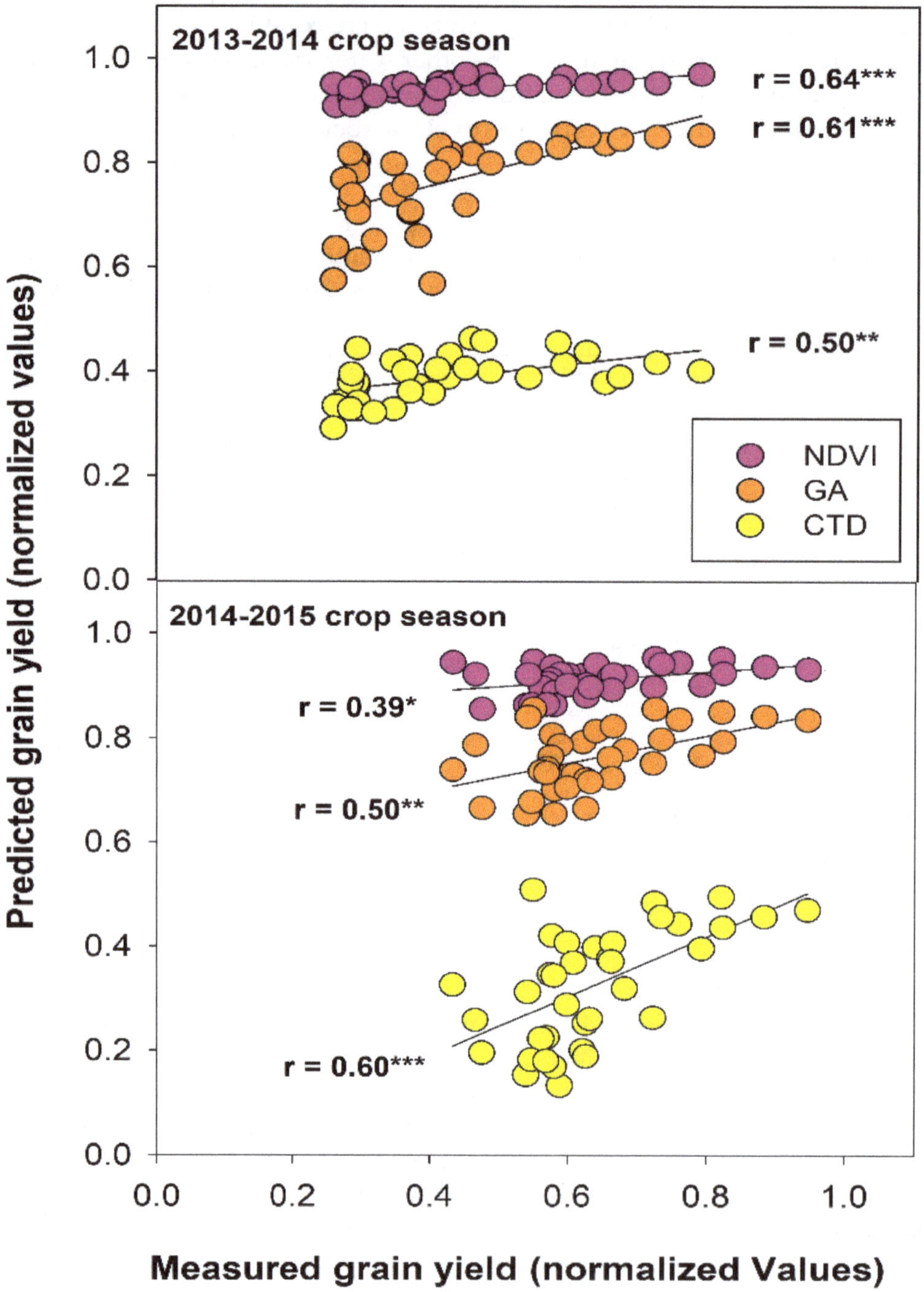

Figure 2. Relationships of the measured versus the predicted grain yields of wheat achieved during two successive crop seasons (2013–2014 and 2014–2015). Predicted grain yield values were calculated using the linear relationships of the grain yield with two vegetation indices, the Normalized Difference Vegetation Index (NDVI), the relative Green Area index (GA), as well as the canopy temperature depression (CTD). All variables were evaluated during the 2012–2013 crop season. Measurements were performed in the same region as the present study. For each crop season, different combinations of wheat genotypes under different water regimes and nitrogen fertilization levels are plotted together. Significance levels—ns, not significant; * $p < 0.05$; *** $p < 0.01$ and *** $p < 0.001$.

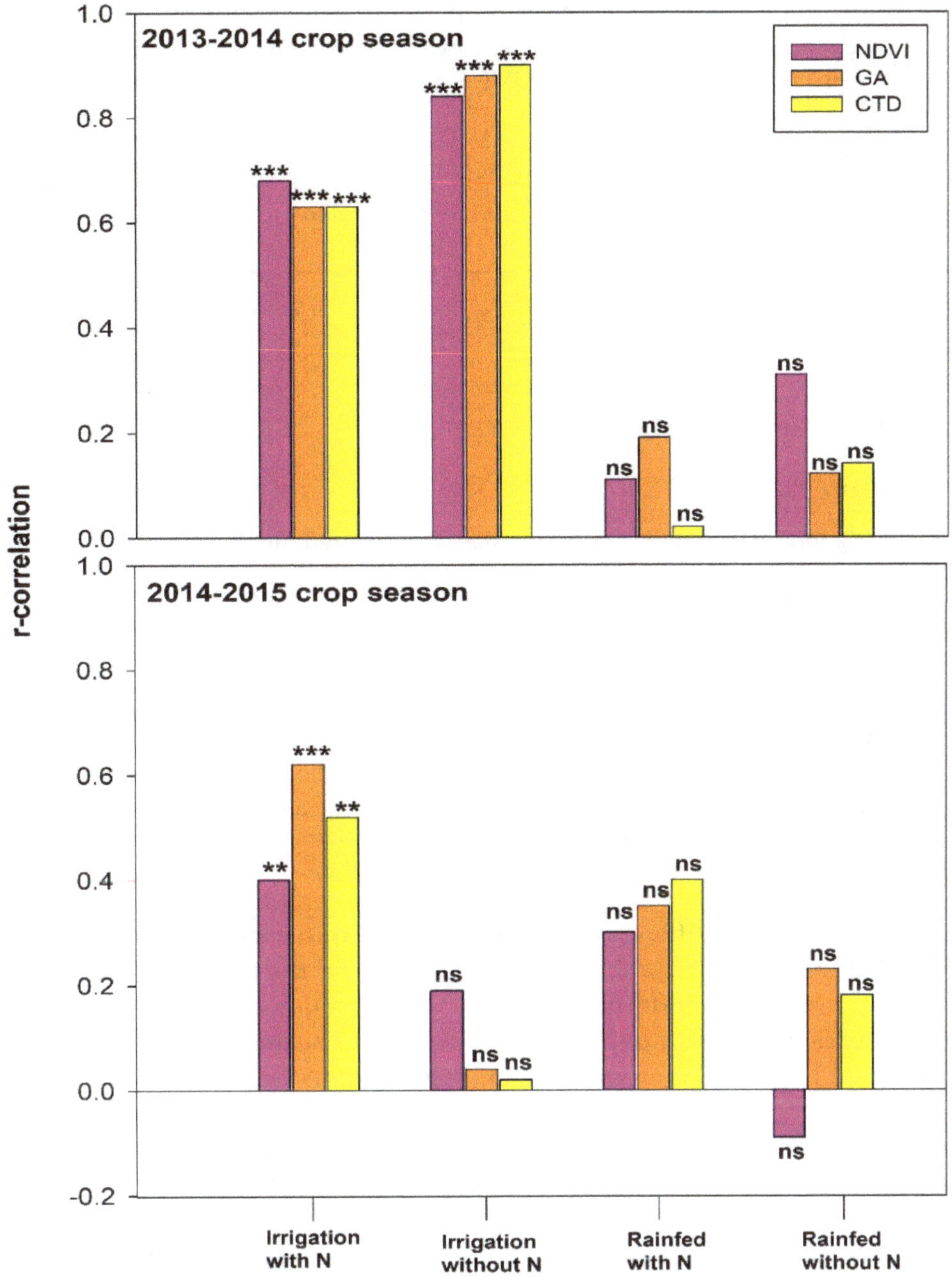

Figure 3. Correlation coefficients of the relationships of the measured versus the predicted grain yields achieved during two successive crop seasons (2013–2014 and 2014–2015). Predicted grain yield values were calculated using the linear relationships of the grain yield with the NDVI, (GA) and CTD evaluated during the 2012–2013 crop season in the same region as the present study. Predicted and measured grain yield were correlated under irrigation and rainfed conditions both with and without N. The irrigation with N values combine both irrigation regimes and the N120 and N60 treatments, irrigation without N combines the two irrigation regimes without N fertilization, and the rainfed conditions with N combine the two levels of N fertilization. Significance levels—ns, not significant; ** $p < 0.01$ and *** $p < 0.001$. Abbreviations of variables are as in Tables 1–3.

3.9. The Combined Effect of Remote Sensing Indices and Physiological Traits in Explaining Grain Yield

Stepwise regressions were performed for the irrigation conditions combined (SI-1 and SI-2) and the rainfed conditions, and considered both groups under each of the three N fertilization levels using grain yield (GY) as the dependent variable and the vegetation indices (NDVI, GA, GGA and LC) and water status traits (CTD, g_s, and $\delta^{13}C$) as independent variables (Table 6). Except for the rainfed conditions with N60, the first trait selected by the model to explain GY was related to the plant water status, while under rainfed conditions the NDVI was the first and only trait chosen by the model (Table 6).

Table 6. Multiple linear regressions (stepwise) explaining grain yield (GY) variation as a dependent variable and the NDVI, GA, GGA, LC, CTD, g_s and $\delta^{13}C$ as independent variables.

Dependent Variable	Growing Conditions	Variable Chosen	Correlation Coefficients	Final Stepwise Model
GY	Irrigation with N120	CTD	0.63 ***	0.24 CTD + 2.14
GY	Irrigation with N60	$\delta^{13}C$	0.55 **	$-0.42\ \delta^{13}C - 7.91$
GY	Irrigation without N	No variables were entered into the equation for this treatment		
GY	Rainfed with N120	CTD	0.66 ***	0.09 CTD + 2.27
GY	Rainfed with N60	NDVI	0.63 **	0.80 NDVI + 2.33
GY	Rainfed without N	No variables were entered into the equation for this treatment		

Data were analyzed under each level of N fertilizer for the irrigation (combined SI-1 and SI-2) and the rainfed treatments. Significance levels—** $p < 0.01$ and *** $p < 0.001$. Abbreviations for variables and growing conditions as defined in Tables 1–3.

Furthermore, the genotypic differences in grain yield within each of the nine growing conditions (resulting from the combination of the three water regimes and the three nitrogen fertilization levels) were assessed through a stepwise model having GY as the dependent variable and any of the vegetation indices and water status traits as independent variables. The model identified at least one trait positively correlated with GY in only four of the nine growing conditions (Table S4).

Additionally, a conceptual model based on a path analysis was proposed (Figure S2) that separated direct acclimation responses in grain yield related to water status traits and vegetation indices (through GA and GGA). The three water status traits were included in the model because they represent different scales: Temporal ($\delta^{13}C$), individual organ (g_s) and canopy (CTD). Concerning the vegetation indices, the NDVI was discarded because GA (whole photosynthetic biomass) and GGA (non-senescent biomass) already tracked the same parameter. The final objective of the model was to dissect how these physiological traits may have directly or indirectly assessed GY performance within different growing conditions.

The four path models proposed (Figure 4) provided an acceptable fit to the data (CFI > 0.9 in all cases). Under irrigation and with and without N fertilizer, g_s had a strong and negative association with $\delta^{13}C$ and strong and positive associations with CTD and GA. Significant paths corresponding to a direct (negative) association of $\delta_{13}C$ with GY were also observed in irrigation conditions without N fertilizer and without, while CTD had a positive association with GY only under irrigation with N fertilizer. A direct positive and strong association of GA with GGA was observed under irrigation. GGA was in turn strongly and negatively associated with GY only under irrigation without N fertilizer. Under rainfed conditions, GY was not associated with $\delta^{13}C$, but was positively associated with CTD and GGA with N fertilizer, and negatively associated with GGA without fertilizer.

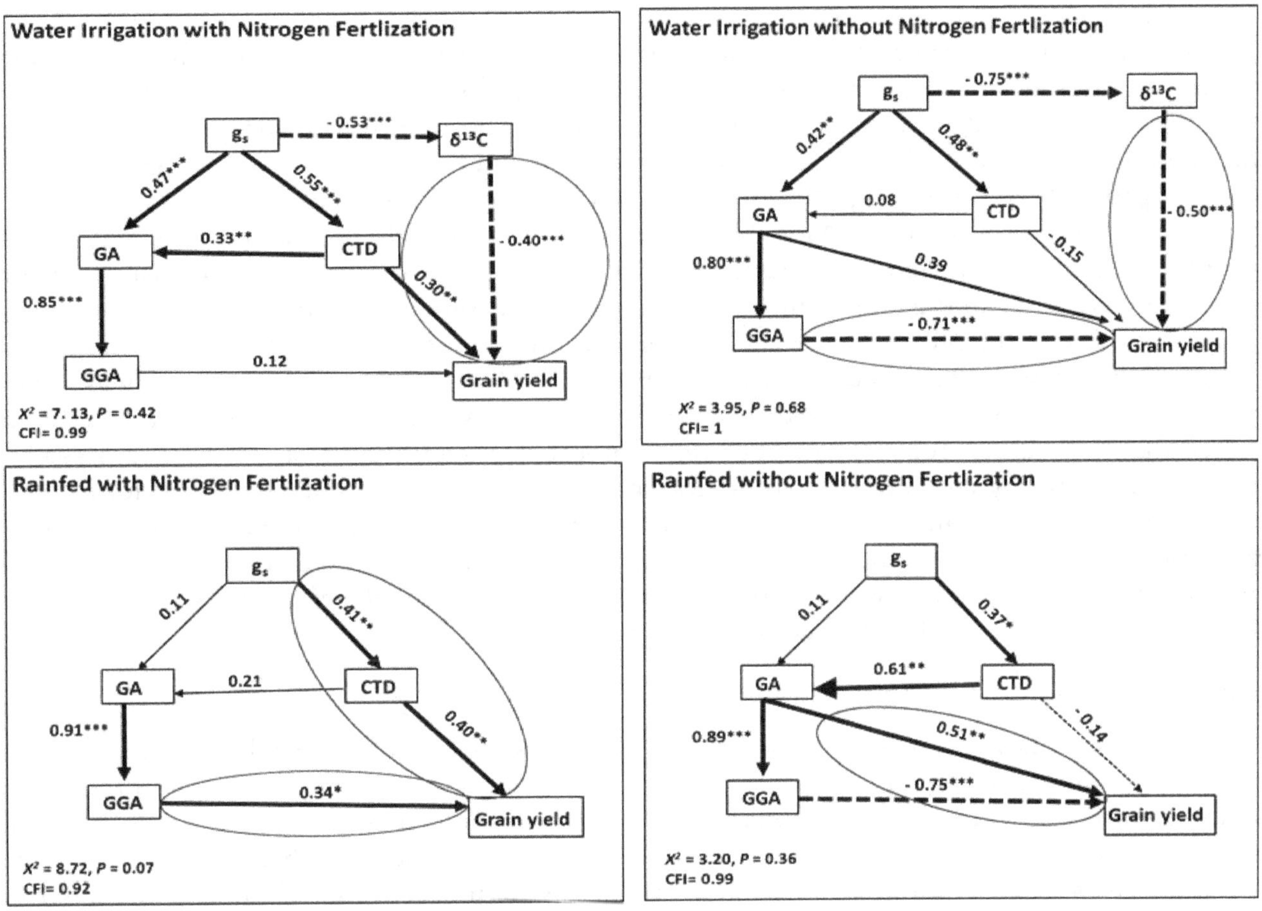

Figure 4. Path analyses of four wheat genotypes grown under different combinations of nitrogen fertilization and water regimes. The irrigation with N values combine both the SI-1 and SI-2 irrigation regimes and the N120 and N60 nitrogen treatments, irrigation without N combines the two irrigation regimes (SI-1 and SI-2) without N fertilization, and the rainfed conditions with N combine the two levels of N fertilization (N60 and N120). Physiological parameters included in the model are: The stomatal conductance (g_s), the stable carbon isotope composition ($\delta^{13}C$) of mature grains, the Relative Green Area (GA) and the Relative Greener Area (GGA), and the canopy temperature depression (CTD). The width of the arrows is proportional to the path coefficient values. Dashed lines indicate negative relationships. CFIs with values > 0.9 are taken as indicative of a good fit. Significance levels—* $p < 0.05$; ** $p < 0.01$ and *** $p < 0.001$.

Path analysis was also examined within each of the nine growing conditions (resulting from the combination of the three water regimes and the three nitrogen fertilization levels (Figure S4). Water irrigation conditions with and without N fertilization indicated the dependence of grain yield on the water status traits $\delta^{13}C$ and CTD (and the latter also under rainfed conditions with N fertilizer). In the absence of N fertilization under both irrigation and rainfed conditions, GY seemed strongly and negatively dependent on the vegetation index GGA.

4. Discussion

The grain yields achieved, even under the best growing conditions (SI-2 with N120) are below 3 tonnes per hectare, and this clearly corresponds with moderate–low yielding conditions in the Mediterranean [44,54]. The vegetation indices tested in this study, generated either spectroradiometrically (NDVI) or derived from RGB images, performed well when assessing differences in water conditions. The efficacy of these indices in capturing differences in growth and senescence in response to water regime has been reported for wheat already [32,33,54]. The choice of anthesis as the phenological stage for the remote sensing measurements was decided based on the results of Yousfi et al. (2016) [54] under similar agro-ecological conditions as in the present study, together with an additional study under Mediterranean conditions where NDVI and RGB indices were measured periodically during the crop cycle [22], with robust correlations between those indices measured at anthesis and grain yield being reported. While the argument may be valid that any action on N management at anthesis is probably too late to significantly affect yield, particularly for fully irrigated wheat, it may still positively affect grain quality. More importantly, scheduling irrigation at anthesis may be fully relevant for cereals under Mediterranean conditions where drought increases progressively during the reproductive stage of the crop. Thus, following previous reports in wheat [54–56], a significant association between the NDVI, GA and GGA vegetation indices and the water status parameters ($\delta^{13}C$, g_s and CTD) was also found in this study. The NDVI, GA and GGA were positively associated with CTD and g_s and negatively associated with $\delta^{13}C$. In this context, Lopes and Reynolds (2012) [57] reported that the relationship observed between chlorophyll retention or 'stay-green' (assessed via the NDVI) and canopy temperature would confirm the functionality of stay-green in terms of gas exchange and would explain a better capacity to use water by the stay-green genotypes under stressful environments related to low fertilization and the lack of water. Our data confirm the close association between vegetation indices and water status parameters and identified the canopy greenness as a good indicator of crop water status and irrigation management.

4.1. Vegetation Indices and Nitrogen Fertilization

Digital images have been used to evaluate the nitrogen status of crops [58,59]. Our results showed that GGA was the only vegetation index exhibiting a significant difference between N treatments, with lower values found under N0 compared with N60 and N120. The absence of N fertilizer limits plant growth, whereas it may accelerate plant senescence during the reproductive stage of the crop, therefore decreasing GGA compared to plants fertilized with nitrogen. In this context, [31] described the greener area index (GGA) derived from RGB images is a good parameter for capturing active photosynthetic area and plant senescence because it is formulated with green pixels alone. Furthermore, the NDVI failed to assess differences under different N treatments. Digital pictures provide information that is not currently acquired through spectral reflectance measurements, such as the portion of yellow leaves in wheat growing under field conditions [31,32,52]. In the case of GA, this index, which takes into account yellow/green pixels, is less stringent in terms of excluding non-senescent parts of the plant. This may explain why GA measured during anthesis was not affected by N fertilization.

4.2. Canopy Temperature and Water Status in Wheat

Many studies have recognized canopy temperature depression (CTD) as an indicator of overall plant water status [36,60] and a potential tool for irrigation management [9,61]. In our study, CTD measured with an infrared thermometer was lower (and even negative) under rainfed conditions compared to support irrigation. A priori, a higher CTD indicates a greater capacity for transpiration, for taking up water from the soil, and therefore for maintaining a better plant water status [36]. In the case of the rainfed trials, the fact that the leaf temperature was higher than the air temperature indicates that rainfed plants were subjected to severe water stress that closed the stomata. In fact, the stomatal conductance measured in the rainfed plants was very low and one third of that measured under the best support-irrigation regime.

In addition, Gutierrez et al. (2010) [62] reported that the association between canopy temperature and the normalized difference water index confirmed that canopy temperature is a good indicator of hydration status. According to this, our results showed highly significant associations of CTD with $\delta^{13}C$ (negative) and g_s (positive). Furthermore, CTD seems to be a better indicator of the water status at the crop level than other traits related to water status, such as leaf g_s [63]. In our study, CTD was strongly associated with $\delta^{13}C$ (r = 0.84 ***) under SI-1 at N60, with the NDVI (r = 0.67 **) under rainfed conditions without N, and with grain yield (r = 0.66 **) under rainfed conditions at N120, while g_s

was not correlated with any of these parameters. These results confirm the close association between canopy temperature and other water status parameters and identify the canopy temperature as a good indicator of crop water status.

4.3. Relationship of Vegetation Indices and Water Status Traits with GY

As found in previous studies in wheat, the RGB canopy indices measured at flowering were strongly correlated with GY [31,32,64]. For wheat under Mediterranean conditions, the reproductive stage is usually the best period for crop monitoring, since the crop is exposed to increasing stress (drought) conditions during the last part of the crop cycle. Following on from this, the present study revealed a positive relationship between GA and GGA with grain yield under irrigation. Additionally, stepwise analysis reinforced the evidence for the usefulness of the RGB vegetation indices to assess GY. The GA vegetation index was chosen by the model as the first independent variable, explaining 66% of GY variability under SI-2 without N fertilizer. Moreover, various studies have reported that RGB-based indices may perform far better than the NDVI for GY prediction in wheat [32,56,65]. In our study, the NDVI failed to assess GY under irrigation. In contrast, the NDVI was correlated (positively) with GY under rainfed conditions and was also the only variable chosen by the stepwise model, explaining 63% of GY variation under rainfed conditions. In this context, Casadesus et al. 2007 [18] reported that the NDVI measured at anthesis in durum wheat correlated positively with GY under severe water stress conditions, but failed to correlate under well-watered conditions. Verhulst and Govaerts (2010) [66] have also reported that the NDVI has been correlated with long-term water stress. The reason for the low correlation of NDVI against GY under well-watered conditions is because plant canopies during anthesis are very dense and the measured NDVI values become saturated. NDVI is an index based on the strong contrast between the near infrared and the red band reflectance of a vegetation canopy, and this difference becomes wider as the canopy cover increases. Thus, NDVI works better with stress conditions where canopies are sparse and/or early senescence is present [27,31]. In any case, anthesis proved to be the correct phenological stage for remote sensing evaluations when crops under different levels of stress were compared, which may be the case for crops exposed to a range of different combinations of water and nitrogen fertilization conditions.

Furthermore, the water status of plants can also be associated with grain yield. In our study, CTD (positively) and $\delta^{13}C$ (negatively) correlated with GY and both parameters were chosen by the stepwise model as the first variables to explain GY variation under different irrigated and rainfed growing conditions. In this context, previous studies have shown that a higher CTD is associated with increased wheat yield under irrigated, hot environments [38,67], but also under dryland environments [68].

However, in our study, and regardless of the water regime (rainfed or irrigation), neither the vegetation indices (NDVI, GA and GGA) nor the water status indices (CTD and $\delta^{13}C$) were associated with GY in the absence of N fertilizer. A lack of variability in green biomass and grain yield associated with the lack of nitrogen fertilization might explain this outcome.

4.4. Phenotyping Parameters under Different Water and N Supplies

The results of our study have shown the usefulness of vegetation indices with low implementation costs as a means to identify genetic variability under different growing conditions in the field. From nine of the growing conditions studied (resulting from the combination of the three water regimes and the three nitrogen fertilization levels), GY was significantly different between genotypes under only two of the growing conditions (SI-2 at N120 and N60). In contrast, two of the vegetation indices (NDVI, GGA) measured at anthesis were able to distinguish between the genotypes growing under six of the nine growing conditions (including rainfed), regardless of the N fertilization conditions. In this context, multispectral ground-based portable spectroradiometric devices have been used in wheat phenotyping [24,27]. The conventional RGB images have also been proposed as a selection tool for cereal breeding [18–20].

The genotypic differences observed using vegetation indices possibly reflect differences in canopy stay green during the reproductive stage. Lopes and Reynolds (2012) [57] reported that stay-green is regarded as a key indicator of stress adaptation. Thus, our study revealed the usefulness of the vegetation indices to select the most tolerant genotypes in terms of retaining a greener biomass during the last part of the crop cycle. The three vegetation indices assayed were able to identify genotypic differences, even under the most severe growing conditions, such as rainfed with and without N fertilizer and where GY failed to detect differences among genotypes. Phenotyping wheat genotypes for water and N fertilization deficit at anthesis using these vegetation indices should permit the formulation of the best crosses between genotypes. However, by comparison, canopy temperature performed much worse as a phenotyping parameter in our study. It has been reported that CTD is a poor indicator of plant performance when the yield is highly dependent on limited amounts of soil-stored water [69,70]. Moreover, the canopy in these trials, particularly during the reproductive stages, frequently leaves areas of bare soil exposed that may affect the canopy temperature readings. Leaf chlorophyll content measured by a portable device was perfect for distinguishing among genotypes, regardless of the water status (irrigation or rainfed), but only when trials were provided with nitrogen fertilizer. Therefore, for the agronomic conditions of our study, the vegetation indices assessed at the canopy level performed better as phenotyping tools than canopy temperature and chlorophyll content measures.

4.5. Grain Yield Prediction across Crop Seasons Using Low-Cost Remote Sensing Techniques

The results of data combining the growing conditions, genotypes and replicates support the use of different affordable remote sensing techniques to estimate grain yield across crop seasons. However, in agreement with Clevers (1997) [71], estimates of crop growth and yield using crop growth models often lost accuracy as the growing conditions became more stressed. The loss of accuracy may be the consequence of a very narrow range of variability in grain yield associated with stressed growing conditions. Moreover, vegetation indices derived from RGB images performed comparatively better than the NDVI, probably because GA was less saturated than the NDVI. The application timing could have played a critical role here—i.e., saturated NDVI at anthesis is indeed not expected to perform well, while the saturation pattern of RGB indices is less evident. In fact, the acquisition of high-resolution RGB images is fast and its dependence on atmospheric conditions (e.g., sunny versus cloudy days) is minimal [22,32]. Therefore, the availability, cost and practicality of digital cameras make them an ideal tool for the management of crop water and fertilization status [19,22]. However, in agreement with previous reports, the relationships between yield and vegetation were site and season specific [20,21]. In our study models were not able to predict absolute yields, but rather relative differences in yield, which makes the approach unfeasible for yield forecasting, but it is still useful in terms of crop management and even phenotyping. The strong relationships of these vegetation and water status indices with grain yield expressed in relative units support the effectiveness of these low-cost indices in crop management. Nevertheless, whereas the evaluations in the three successive seasons were performed in the same region, crop management conditions (water and fertilization regimes) affected the performance of the models. Hence, to make yield predictions more holistic and effective across different environments, it is necessary to use more robust calibration; for example, incorporating site-year covariates or a multi sensor approach [21].

4.6. An Integrated Model to Predict Grain Yield That Combines Remote-Sensing Traits, Canopy Reflectance Measurements and Grain $\delta^{13}C$

In this study, we performed a path analysis to dissect how the vegetation indices (GA and GGA) and the water status (CTD, $\delta^{13}C$ and g_s) indices directly or indirectly assessed GY performance within each of the growing conditions assayed. The following parameters may provide suitable coverage of the factors affecting GY performance under a given water regime and nitrogen fertilization supply: Vegetation indices as indicators of photosynthetic capacity (GA) and the effect of early senescence (GGA) on the canopy; CTD which informs about the current water status of the canopy; the $\delta^{13}C$

in mature grains as a time-integrated indicator of photosynthetic and transpirative gas exchange of the crop; and the water status at the single organ level (assessed as the g_s of the flag leaf). Under irrigation and nitrogen fertilization, both $\delta^{13}C$ and CTD (indicators of photosynthetic and water status) had a direct association with GY. Better grain yield performance is associated with higher CTD and lower $\delta^{13}C$ under supplementary irrigation. The association of $\delta^{13}C$ (negative) and CTD (positive) with GY is probably due the higher stomatal conductance and transpiration, (which increases CTD), therefore increasing the photosynthetic capacity even at the expense of a lower water use efficiency (and thus $\delta^{13}C$) and the consequent increases in GY [47,72]. Under rainfed conditions with N fertilizer, the transpiration/water status (assessed through CTD) and photosynthetic potential (evaluated through vegetation indices) affected GY. Under rainfed conditions with N120, CTD had a positive effect on GY. We suggest that N fertilization only had a positive effect on grain yield providing that there was water available to maintain transpiration (higher CTD), stomatal conductance and thus photosynthesis in the available canopy (higher vegetation indices).

In the absence of nitrogen fertilization, and despite the water conditions (irrigated or rainfed), the total canopy area (evaluated through GA) has a positive effect on GY. However, an excess of young (not senescing) leaf area (evaluated through GAA) had a strong negative association with GY. Stay-green character may have a negative effect on GY in the absence of nitrogen fertilization because it limits the retranslocation of N to the inflorescences and ultimately affects grain filling. In the case of the rainfed crop fertilized with a limited amount of nitrogen (N60), the active canopy area (evaluated through GGA) had a positive effect on grain yield, which may indicate that the limitation is imposed by the amount of photosynthetic area rather than by the availability of N to reproductive tissues. Under conditions of high nitrogen fertilization (N120) and irrespective of the water regime (irrigated or rainfed), GY is not affected by the size of the canopy or even by its greenness (assessed through GA or GGA), but by the water status of the crop (evaluated through CTD and $\delta^{13}C$).

5. Conclusions

This study demonstrated the potential of low-cost RGB vegetation indices and the canopy temperature for the management of growing conditions (essentially the water and nitrogen regimes) under Mediterranean conditions. Although the models did not predict absolute yields, they are still useful in terms of crop management and even phenotyping. Nevertheless, grain yield estimation performs better under irrigation than under the low-yielding conditions of rainfed cultivation in the absence of nitrogen fertilization and this illustrates one of the potential limitations associated with the remote sensing-based yield predictions; they are affected by specific environmental conditions. Even though a multispectral vegetation index, such as the NDVI is a widely accepted approach to monitor changes in growth under different conditions, in this study we have shown that vegetation indices derived from conventional images like the GA and GAA indices provided a similar if not better prediction of grain yield and at a comparatively lower cost than the NDVI. The use of vegetation indices derived from RGB images to assess GY could be popularized in the near future via apps installed on mobile phones.

The vegetation indices have also proven their suitability for differentiating among genotypes. Furthermore, these traits may contribute through path analysis to develop physiological models for assessing wheat ideotypes best suited to different water and nitrogen regimes. The models showed that for nitrogen fertilized trials, and regardless of the water regime imposed, the water status parameters were the main factors determining GY performance. Moreover, a larger green area at anthesis may also contribute to a larger yield. In the absence of nitrogen fertilization, a large greener canopy area (assessed through GGA) at anthesis is a factor that negatively affects grain yield.

Supplementary Materials
Figure S1: Scheme detailing the different plots of the experimental design, Figure S2: Conceptual model of the path analyses quantifying the relative strengths of the direct and indirect relationships of the different physiological traits and grain yield. Physiological parameters included in the model are: The stomatal conductance (g_s), the stable carbon isotope composition ($\delta^{13}C$) of mature grains, the Relative Green Area (GA) and the Relative Greener Area (GGA) indices calculated from digital pictures, and the canopy temperature depression (CTD) measured with an infrared thermometer, Figure S3: Relationships of the measured versus the predicted grain yields of wheat achieved during two successive crop seasons (2013–2014 and 2014–2015). Predicted grain yield values were calculated using the linear relationships of the grain yield with the two vegetation indices, the Normalized Difference Vegetation Index (NDVI) measured with a portable spectroradiometer, and the relative Green Area index (GA) calculated from digital images, and the canopy temperature depression (CTD) measured with an infrared thermometer. All variables were evaluated during the 2012–2013 crop season. Measurements were performed in the same region as the present study. Grain yield and vegetation index data of the two first crop seasons have been reported in Yousfi et al. (2016). For each crop season, the nine different growing conditions (resulting from the combination of the three water regimes and the three nitrogen fertilization levels) were analysed. The different combinations of nitrogen fertilization and water regimes for the 2013–2014 crop season were as follows: Supplementary irrigation (SI-2) with nitrogen fertilization N120 and N60 and without N fertilization; Supplementary irrigation (SI-1) with nitrogen fertilization N120 and N60 and without N fertilization; Rainfed with nitrogen fertilization N120 and N60 and without N fertilization. In addition, for the 2014–2015 crop season the combinations were as follows: Supplementary irrigation (SI-2) with nitrogen fertilization N120 and N60 and without N fertilization; Supplementary irrigation (SI-1) with nitrogen fertilization N120 and N60 and without N fertilization; Rainfed with nitrogen fertilization N120 and N60 and without N fertilization. Significance levels—ns, not significant; * $p < 0.05$; ** $p < 0.01$ and *** $p < 0.001$, Figure S4: Path analyses of four wheat genotypes grown under different combinations of nitrogen fertilization and water regimes. The different combinations of nitrogen fertilization and water regimes are as follows: (A) SI-2 with high fertilization (N120); (B) SI-2 with medium nitrogen fertilization (N60); (C) SI-2 without nitrogen fertilization; (D) SI-I with high nitrogen fertilization (N120); (E) SI-I without nitrogen fertilization (N60); (F) Rainfed with high fertilization (N120); (G) Rainfed with medium nitrogen fertilization (N60); (H) Rainfed without nitrogen fertilization. Physiological parameters included in the model are: The stomatal conductance (g_s), the stable carbon isotope composition ($\delta^{13}C$) of mature grains, the Relative Green Area (GA) and the Relative Greener Area (GGA) indices calculated from digital pictures, and the canopy temperature depression (CTD) measured with an infrared thermometer. The width of the arrows is proportional to the path coefficient values. Dashed lines indicate negative relationships. Overall fit statistics for each path model (chi-squared, the probability and comparative fit index, CFI) are shown at the bottom left of each panel. CFIs with values > 0.9 were taken as indicative of a good fit. Significance levels—* $p < 0.05$; ** $p < 0.01$ and *** $p < 0.001$; Table S1: Soil chemical characteristics at different depths, Table S2: Monthly total accumulated rainfall (PP), minimum air temperature (T min), maximum air temperature (T max) and average air temperature (T aver) for the 2014–2015 crop season. Values were collected at the meteorological station of Khemis Miliana (Algeria), Table S3: Correlation coefficients of the linear relationships of grain yield (GY) with NDVI, GA, GGA, CTD and $\delta^{13}C$ under irrigation without N fertilization and rainfed conditions with N120, N60 and without N, Table S4. Multiple linear regressions (stepwise) across genotypes and replicates explaining grain yield (GY) variation as a dependent variable and the NDVI, GA, GGA, LC, CTD, g_s and $\delta^{13}C$ as independent variables. Data were analyzed within each level of nitrogen fertilization and water regime. Significance levels—* $p < 0.05$; ** $p < 0.01$ and *** $p < 0.001$. Abbreviations for variables and growing conditions as defined in Tables 1 and 2.

Author Contributions: S.Y., J.L.A., M.D.S., M.K. (Mohamed Karrou), N.K. and A.C. conceived and designed the experiment. N.K., M.K. (Mohamed Kaddour) and A.C. contributed to the experimental work. N.K., M.K. (Mohamed Kaddour), A.C., M.K. (Mohamed Karrou), S.Y., J.L.A. and M.D.S. contributed to the experimental measures and field sampling. A.G.-R. and M.D.S. performed stable isotope analyses. S.Y., J.L.A. and M.D.S. contributed to the data analysis and interpreted the results. S.Y. wrote the paper under the supervision of J.L.A. and M.D.S. and all three revised the manuscript. All authors read and approved the final manuscript.

Acknowledgments: We acknowledge the support of the staff from the ITGC (Institut Technique des Grandes Cultures, Algeria). J. L Araus acknowledges the support of the ICREA Academia Award.

References

1. FAO. Statistical Year Book. 2012. Available online: http://www.fao.org/docrep/015/i2490e/i249e00.htm (accessed on 14 June 2015).
2. Oweis, T.; Zhang, H.; Pala, M. Water use efficiency of rainfed and irrigated bread wheat in a Mediterranean environment. *Agron. J.* **2000**, *92*, 231–238. [CrossRef]
3. Lobell, D.B.; Burke, M.B.; Tebaldi, C.; Mastrandrea, M.D.; Falcon, W.P.; Naylor, R.L. Prioritizing climate change adaptation needs for food security in 2030. *Science* **2008**, *319*, 607–610. [CrossRef] [PubMed]

4. Araus, J.L.; Slafer, G.A.; Reynolds, M.P.; Royo, C. Plant breeding and water relations in C3 cereals: What should we breed for? *Ann. Bot.* **2002**, *89*, 925–940. [CrossRef] [PubMed]
5. Seelan, S.K.; Laguette, S.; Casady, G.M.; Seielstad, G.A. Remote sensing applications for precision agriculture: A learning community approach. *Remote Sens. Environ.* **2003**, *88*, 157–169. [CrossRef]
6. Chapman, S.C. Use of crop models to understand genotype by environment interactions for drought in real-world and simulated plant breeding trials. *Euphytica* **2008**, *161*, 195–208. [CrossRef]
7. Araus, J.L.; Cairns, J.E. Field high-throughput phenotyping: the new crop breeding frontier. *Trends Plant Sci.* **2014**, *19*, 52–61. [CrossRef] [PubMed]
8. Glenn, E.P.; Huete, A.R.; Nagler, P.L.; Nelson, S.G. Relationship between remotely-sensed vegetation indices, canopy attributes and plant physiological processes: what vegetation indices can and cannot tell us about the landscape. *Sensors* **2008**, *8*, 2136–2160. [CrossRef] [PubMed]
9. Jones, H.G.; Serraj, R.; Loveys, B.R.; Xiong, L.Z.; Wheaton, A. Thermal infrared imaging of crop canopies for the remote diagnosis and quantification of plant responses to water stress in the field. *Funct. Plant Biol.* **2009**, *36*, 978–989. [CrossRef]
10. Sui, J.; Qin, Q.; Ren, H.; Sun, Y.; Zhang, T.; Wang, J.; Gong, S. Winter wheat production estimation based on environmental stress factors from satellite observations. *Remote Sens.* **2018**, *10*, 962. [CrossRef]
11. Zhang, C.; Pattey, E.; Liu, J.; Cai, H.; Shang, J.; Dong, T. Retrieving leaf and canopy water content of winter wheat using vegetation water indices. *IEEE J. Sel. Top. Appl. Earth Obs. Remote Sens.* **2018**, *11*, 112–126. [CrossRef]
12. Segovia-Cardozo, D.A.; Rodríguez-Sinobas, L.; Zubelzu, S. Water use efficiency of corn among the irrigation districts across the Duero river basin (Spain): Estimation of local crop coefficients by satellite images. *Agric. Water Manag.* **2019**, *212*, 241–251. [CrossRef]
13. Serrano, J.; Shahidian, S.; Marques da Silva, J. Evaluation of normalized difference water index as a tool for monitoring pasture seasonal and inter-annual variability in a Mediterranean agro-silvo-pastoral system. *Water* **2019**, *11*, 62. [CrossRef]
14. Zhang, T.; Su, J.; Liu, C.; Chen, W.H.; Liu, H.; Liu, G. Band selection in Sentinel-2 satellite for agriculture applications. In Proceedings of the 2017 23rd International Conference on Automation and Computing (ICAC), Huddersfield, UK, 7–8 September 2017; IEEE: Huddersfield, UK, 2017; pp. 1–6.
15. Varinderpal-Singh; Bijay-Singh; Yadvinder-Singh; Thind, H.S.; Gobinder-Singh; Satwinderjit-Kaur; Kumar, A.; Vashistha, M. Establishment of threshold leaf colour greenness for need-based fertilizer nitrogen management in irrigated wheat (*Triticum aestivum* L.) using leaf colour chart. *Field Crops Res.* **2012**, *130*, 109–119. [CrossRef]
16. Varinderpal-Singh; Bijay-Singh; Yadvinder-Singh; Thind, H.S.; Buttar, G.S.; Kaur, S.; Kaur, S.; Bhowmik, A. Site-specific fertilizer nitrogen management for timely sown irrigated wheat (*Triticum aestivum* L. and *Triticum turgidum* L. ssp. durum) genotypes. *Nutr. Cycl. Agroecosyst.* **2017**, *109*, 1–16. [CrossRef]
17. Bija-Singh; Varinderpal-Singh; Yadvinder-Singh; Thind, H.S.; Kumar, A.; Choudhary, O.P.; Gupta, R.K.; Vashistha, M. Site-specific fertilizer nitrogen management using optical sensor in irrigated wheat in the Northwestern India. *Agric. Res.* **2017**, *6*, 159–168. [CrossRef]
18. Van Loon, J.; Speratti, A.; Govaerts, B. Precision for smallholder farmers: a small-scale-tailored variable rate fertilizer application kit. *Agriculture* **2018**, *8*, 48. [CrossRef]
19. Araus, J.L.; Kefauver, S.C. Breeding to adapt agriculture to climate change: Affordable phenotyping solutions. *Curr. Opin. Plant Biol.* **2018**, *45*, 237–247. [CrossRef]
20. Colaço, A.F.; Bramley, R.G. Do crop sensors promote improved nitrogen management in grain crops? *Field Crops Res.* **2018**, *218*, 126–140. [CrossRef]
21. Bramley, R.G.V.; Ouzman, J. Farmer attitudes to the use of sensors and automation in fertilizer decision-making: nitrogen fertilization in the Australian grains sector. *Precis. Agric.* **2019**, *20*, 157–175. [CrossRef]
22. Fernandez-Gallego, J.A.; Kefauver, S.C.; Vatter, T.; Gutiérrez, N.A.; Nieto-Taladriz, M.T.; Araus, J.L. Low-cost assessment of grain yield in durum wheat using RGB images. *Eur. J. Agron.* **2019**, *105*, 146–156. [CrossRef]
23. Hansen, P.M.; Schjoerring, J.K. Reflectance measurement of canopy biomass and nitrogen status in wheat crops using normalized difference vegetation indices and partial least squares regression. *Remote Sens. Environ.* **2003**, *86*, 542–553. [CrossRef]
24. Babar, M.A.; Reynolds, M.P.; Van Ginkel, M.; Klatt, A.R.; Raun, W.R.; Stone, M.L. Spectral reflectance indices as a potential indirect selection criteria for wheat yield under irrigation. *Crop Sci.* **2006**, *46*, 578–588. [CrossRef]

25. Marti, J.; Bort, J.; Slafer, G.A.; Araus, J.L. Can wheat yield be assessed by early measurements of normalized difference vegetation index? *Ann. Appl. Biol.* **2007**, *150*, 253–257. [CrossRef]
26. Wright, D.L.; Rasmussen, V.P.; Ramsey, R.D. Comparing the use of remote sensing with traditional techniques to detect nitrogen stress in wheat. *Geocarto. Int.* **2005**, *20*, 63–68. [CrossRef]
27. Aparicio, N.; Villegas, N.; Casadesus, J.; Araus, J.L.; Royo, C. Spectral vegetation indices as non-destructive tools for determining durum wheat yield. *Agron. J.* **2000**, *92*, 83–91. [CrossRef]
28. Filella, I.; Serrano, L.; Serra, J.; Penuelas, J. Evaluating wheat nitrogen status with canopy reflectance indices and discriminant analysis. *Crop Sci.* **1995**, *35*, 1400–1405. [CrossRef]
29. Royo, C.; Aparicio, N.; Villegas, D.; Casadesus, J.; Monneveux, P.; Araus, J.L. Usefulness of spectral reflectance indices as durum wheat yield predictors under contrasting Mediterranean conditions. *Int. J. Remote Sens.* **2003**, *24*, 4403–4419. [CrossRef]
30. Araus, J.L.; Amaro, T.; Zuhair, Y.; Nachit, M.M. Effect of leaf structure and water status on carbon isotope discrimination in field-grown durum wheat. *Plant Cell Environ.* **1997**, *20*, 1484–1494. [CrossRef]
31. Casadesus, J.; Kaya, Y.; Bort, J.; Nachit, M.M.; Araus, J.L.; Amor, S.; Ferrazzano, G.; Maalouf, F.; Maccaferri, M.; Martos, V.; et al. Using vegetation indices derived from conventional digital cameras as selection criteria for wheat breeding in water-limited environments. *Ann. Appl. Bot.* **2007**, *150*, 227–236. [CrossRef]
32. Vergara-Diaz, O.; Kefauver, S.C.; Elazab, A.; Nieto-Taladriz, M.T.; Araus, J.L. Grain yield loss assessment for winter wheat associated with the fungus *Puccinia striiformis* f.sp. *tritici* using digital and conventional parameters under field conditions. *Crop J.* **2015**, *3*, 200–210. [CrossRef]
33. Mullan, D.J.; Reynolds, M.P. Quantifying genetic effects of ground cover on soil water evaporation using digital imaging. *Funct. Plant. Biol.* **2010**, *37*, 703–712. [CrossRef]
34. Deery, D.; Jimenez-Berni, J.; Jones, H.; Sirault, X.; Furbank, R. Proximal remote sensing buggies and potential applications for field-based phenotyping. *Agronomy* **2014**, *4*, 349–379. [CrossRef]
35. Araus, J.L.; Kefauver, S.C.; Zaman-Allah, M.; Olsen, M.S.; Cairns, J.E. Translating high throughput phenotyping into genetic gain. *Trends Plant Sci.* **2018**, *23*, 451–466. [CrossRef] [PubMed]
36. Blum, A.; Mayer, J.; Gozlan, G. Infrared thermal sensing of plant canopies as a screening technique for dehydration avoidance in wheat. *Field Crops Res.* **1982**, *5*, 137–146. [CrossRef]
37. Amani, I.; Fischer, R.A.; Reynolds, M.P. Canopy temperature depression association with yield of irrigated spring wheat cultivars in hot climate. *J. Agron. Crop Sci.* **1996**, *176*, 119–129. [CrossRef]
38. Reynolds, M.P.; Singh, R.P.; Ibrahim, A.; Ageeb, O.A.A.; Larque-Saavedra, A.; Quick, J.S. Evaluating physiological traits to complement empirical selection for wheat in warm environments. *Euphytica* **1998**, *100*, 84–95. [CrossRef]
39. Reynolds, M.; Dreccer, F.; Trethowan, R. Drought-adaptive traits derived from wheat wild relatives and landraces. *J. Exp. Bot.* **2007**, *58*, 177–186. [CrossRef]
40. Araus, J.L.; Slafer, G.A.; Royo, C.; Serret, M.D. Breeding for yield potential and stress adaptation in cereals. *Crit. Rev. Plant Sci.* **2008**, *27*, 377–412. [CrossRef]
41. Inoue, Y. Remote detection of physiological depression in crop plants with infrared thermal imagery. *Jpn. J. Crop Sci.* **1990**, *59*, 762–768. [CrossRef]
42. O'Shaughnessy, S.A.; Evett, S.R.; Colaizzi, P.D.; Howel, T.A. Using radiation thermography and thermometry to evaluate crop water stress in soybean and cotton. *Agric. Water Manag.* **2011**, *98*, 1523–1535. [CrossRef]
43. Idso, S.B.; Jackson, R.D.; Pinter, P.J.; Reginato, R.J.; Hatfield, J.L. Normalizing the stress-degree-day parameter for environmental variability. *Agric. Meteorol.* **1981**, *24*, 45–55. [CrossRef]
44. Araus, J.L.; Cabrera-Bosquet, L.; Serret, M.D.; Bort, J.; Nieto-Taladriz, M.T. Comparative performance of $\delta^{13}C$, $\delta^{18}O$ and $\delta^{15}N$ for phenotyping durum wheat adaptation to a dryland environment. *Funct. Plant Biol.* **2013**, *40*, 595–608. [CrossRef]
45. Van Leeuwen, C.; Tregoat, O.; Chone, X.; Bois, B.; Pernet, D.; Gaudillere, J.P. Vine water status is a key factor in grape ripening and vintage quality for red Bordeaux wine. How can it be assessed for vineyard management purposes? *J. Int. Sci. Vig.* **2009**, *43*, 121–134. [CrossRef]
46. Farquhar, G.D.; O'Leary, M.H.; Berry, J.A. On the relationship between carbon isotope discrimination and the intercellular carbon dioxide concentration in leaves. *Aust. J. Plant. Physiol.* **1982**, *9*, 121–137. [CrossRef]
47. Farquhar, G.D.; Richards, R.A. Isotopic composition of plant carbon correlates with water-use-efficiency of wheat genotypes. *Aust. J. Plant Physiol.* **1984**, *11*, 539–552. [CrossRef]

48. Condon, A.G.; Richards, R.A. Broad sense heritability and genotypes-environment interaction for carbon isotope discrimination in field-grown wheat. *Aust. J. Agric. Res.* **1992**, *43*, 921–934. [CrossRef]
49. Yousfi, S.; Serret, M.D.; Voltas, J.; Araus, J.L. Effect of salinity and water stress during the reproductive stage on growth, ion concentrations, Δ^{13}C, and δ^{15}N of durum wheat and related amphiploids. *J. Exp. Bot.* **2010**, *61*, 3529–3542. [CrossRef] [PubMed]
50. Araus, J.L.; Amaro, T.; Casadesus, J.; Asbati, A.; Nachit, M.M. Relationships between ash content, carbon isotope discrimination and yield in durum wheat. *Aust. J. Plant Physiol.* **1998**, *25*, 835–842. [CrossRef]
51. Susstrunk, S.; Robert-Buckley, R.; Swen, S. Standard RGB color spaces. In Proceedings of the Color and Imaging Conference Final Program and Proceedings, Scottsdale, AZ, USA, 16–19 November 1999; Society for Imaging Science and Technology: Springfield, MO, USA, 1999; pp. 127–134.
52. Gracia-Romero, A.; Vergara-Díaz, O.; Thierfelder, C.; Cairns, J.E.; Kefauver, S.C.; Araus, J.L. Phenotyping Conservation Agriculture Management Effects on Ground and Aerial Remote Sensing Assessments of Maize Hybrids Performance in Zimbabwe. *Remote Sens.* **2018**, *10*, 349. [CrossRef]
53. Farquhar, G.D.; Ehleringer, J.R.; Hubick, K.T. Carbon isotope discrimination and photosynthesis. *Annu. Rev. Plant Physiol.* **1989**, *40*, 503–537. [CrossRef]
54. Yousfi, S.; Kellas, N.; Saidi, L.; Benlakhel, Z.; Chaou, L.; Siad, D. Comparative performance of remote sensing methods in assessing wheat performance under Mediterranean conditions. *Agric. Water Manag.* **2016**, *164*, 137–147. [CrossRef]
55. Arbuckle, J.L. *Amos Users' Guide, Version 3.6*; Small Waters Corporation: Chicago, IL, USA, 1997.
56. Elazab, A.; Bort, J.; Zhou, B.; Serret, M.D.; Nieto-Taladriz, M.T.; Araus, J.L. The combined use of vegetation índices and stable isotopes to predict durum wheat grain yield under contrasting water conditions. *Agric. Water Manag.* **2015**, *158*, 196–208. [CrossRef]
57. Lopes, M.S.; Reynolds, M.P. Stay-green in spring wheat can be determined by spectral reflectance measurements (normalized difference vegetation index) independently from phenology. *J. Exp. Bot.* **2012**, *63*, 3789–3798. [CrossRef] [PubMed]
58. Li, Y.; Chen, D.; Walker, C.N.; Angus, J.F. Estimating the nitrogen status of crops using a digital camera. *Field Crop Res.* **2010**, *118*, 221–227. [CrossRef]
59. Vergara-Díaz, O.; Zaman-Allah, M.; Masuka, B.; Hornero, A.; Zarco-Tejada, P.; Prasanna, B.M.; Cairns, J.E.; Araus, J.L. A novel remote sensing approach for prediction of maize yield under different conditions of nitrogen fertilization. *Front. Plant Sci.* **2016**, *7*, 666. [CrossRef] [PubMed]
60. Jackson, R.D.; Idso, S.B.; Reginato, R.J.; Pinter, P.J., Jr. Canopy temperature as a crop water stress indicator. *Water Resour. Res.* **1981**, *17*, 1133–1138. [CrossRef]
61. Jones, H.G. Irrigation scheduling: advantages and pitfalls of plant-based methods. *J. Exp. Bot.* **2004**, *55*, 2427–2436. [CrossRef]
62. Gutierrez, M.; Reynolds, M.P.; Raun, W.R.; Stone, M.L.; Klatt, A.R. Spectral water indices for assessing yield in elite bread wheat genotypes grown under well irrigated, water deficit stress, and high temperature conditions. *Crop Sci.* **2010**, *50*, 197–214. [CrossRef]
63. Balota, M.; Amani, I.; Reynolds, M.P.; Acevedo, E. *Evaluation of Membrane Thermostability and Canopy Temperature Depression as Screening Traits for Heat Tolerance in Wheat*; Wheat Special Report No.20; CIMMYT: Mexico, D.F., Mexico, November 1993.
64. Morgounov, A.; Gummadov, N.; Belen, S.; Kaya, Y.; Keser, M.; Mursalova, J. Association of digital photo parameters and NDVI with winter wheat grain yield in variable environments. *Turk. J. Agric. For.* **2014**, *38*, 624–632. [CrossRef]
65. Zhou, B.; Elazab, A.; Bort, J.; Vergara, O.; Serret, M.D.; Araus, J.L. Low-cost assessment of wheat resistance to yellow rust through conventional RGB images. *Comput. Electron. Agric.* **2015**, *116*, 20–29. [CrossRef]
66. Verhulst, N.; Govaerts, B. *The Normalized Difference Vegetation Index (NDVI) GreenSeekerTM Handheld Sensor: Toward the Integrated Evaluation of Crop Management. Part A: Concepts and Case Studies*; CIMMYT: Mexico, D.F., Mexico, 2010.
67. Fischer, R.A.; Rees, D.; Sayre, K.D.; Lu, Z.M.; Condon, A.G.; Larque-Saavedra, A. Wheat yield progress associated with higher stomatal conductance and photosynthetic rate, and cooler canopies. *Crop Sci.* **1998**, *38*, 1467–1475. [CrossRef]
68. Balota, M.; Payne, W.A.; Evett, S.R.; Lazar, M.D. Canopy temperature depression sampling to assess grain yield variation and genotypic differentiation in winter wheat. *Crop Sci.* **2007**, *47*, 1518–1529. [CrossRef]

69. Winter, S.R.; Musick, J.T.; Porter, K.B. Evaluation of screening techniques for breeding drought-resistant Winter wheat. *Crop Sci.* **1998**, *28*, 512–516. [CrossRef]
70. Royo, C.; Villegas, D.; Garcia del Moral, L.F.; Elhani, S.; Aparicio, N.; Rharrabti, Y.; Araus, J.L. Comparative performance of carbon isotope discrimination and canopy temperature depression as predictors of genotypes differences in durum wheat yield in Spain. *Aust. J. Agric. Res.* **2002**, *53*, 561–569. [CrossRef]
71. Clevers, J.G.P.W. A simplified approach for yield prediction of sugar beet on optical remote sensing data. *Remote Sens. Environ.* **1997**, *61*, 221–228. [CrossRef]
72. Cabrera-Bosquet, L.; Molero, G.; Nogués, S.; Araus, J.L. Water and nitrogen conditions affect the relationships of $\Delta^{13}C$ and $\Delta^{18}O$ to gas exchange and growth in durum wheat. *J. Exp. Bot.* **2009**, *60*, 1633–1644. [CrossRef] [PubMed]

Development of Chlorophyll-Meter-Index-Based Dynamic Models for Evaluation of High-Yield Japonica Rice Production in Yangtze River Reaches

Ke Zhang [1,2], Xiaojun Liu [1,2], Syed Tahir Ata-Ul-Karim [1,2,3], Jingshan Lu [1,2], Brian Krienke [4], Songyang Li [1,2], Qiang Cao [1,2], Yan Zhu [1,2], Weixing Cao [1,2] and Yongchao Tian [1,2,*]

1 National Engineering and Technology Center for Information Agriculture, Key Laboratory for Crop System Analysis and Decision Making, Ministry of Agriculture, 1 Weigang Road, Nanjing 210095, China; 2017201080@njau.edu.cn (K.Z.); liuxj@njau.edu.cn (X.L.); ataulkarim@issas.ac.cn (S.T.A.-U.-K.); 2016201019@njau.edu.cn (J.L.); 2016101006@njau.edu.cn (S.L.); qiangcao@njau.edu.cn (Q.C.); yanzhu@njau.edu.cn (Y.Z.); caow@njau.edu.cn (W.C.)

2 Key Laboratory for Information Agriculture, Jiangsu, Collaborative Innovation Center for Modern Crop Production, Nanjing Agricultural University, 1 Weigang Road, Nanjing 210095, China

3 Key Laboratory of Soil Environment and Pollution Remediation, Institute of Soil Science, Chinese Academy of Science, Nanjing 210008, China

4 Department of Agronomy and Horticulture, University of Nebraska-Lincoln, Lincoln, NE 68583, USA; krienke.brian@unl.edu

* Correspondence: yctian@njau.edu.cn

Abstract: Accurate estimation of the nitrogen (N) spatial distribution of rice (*Oryza sativa* L.) is imperative when it is sought to maintain regional and global carbon balances. We systematically evaluated the normalized differences of the soil and plant analysis development (SPAD) index (the normalized difference SPAD indexes, NDSIs) between the upper (the first and second leaves from the top), and lower (the third and fourth leaves from the top) leaves of Japonica rice. Four multi-location, multi-N rate (0–390 kg ha^{-1}) field experiments were conducted using seven Japonica rice cultivars (9915, 27123, Wuxiangjing14, Wunyunjing19, Wunyunjing24, Liangyou9, and Yongyou8). Growth analyses were performed at different growth stages ranging from tillering (TI) to the ripening period (RP). We measured leaf N concentration (LNC), the N nutrition index (NNI), the NDSI, and rice grain yield at maturity. The relationships among the NDSI, LNC, and NNI at different growth stages showed that the NDSI values of the third and fourth fully expanded leaves more reliably reflected the N nutritional status than those of the first and second fully expanded leaves (LNC: $NDSI_{L3,4}$, $R^2 > 0.81$; $NDSI_{others}$, $0.77 > R^2 > 0.06$; NNI: $NDSI_{L3,4}$, $R^2 > 0.83$; $NDSI_{others}$, $0.76 > R^2 > 0.07$; all $p < 0.01$). Two new diagnostic models based on the $NDSI_{L3,4}$ (from the tillering to the ripening period) can be used for effective diagnosis of the LNC and NNI, which exhibited reasonable distributions of residuals (LNC: relative root mean square error (RRMSE) = 0.0683; NNI: RRMSE = 0.0688; $p < 0.01$). The relationship between grain yield, predicted yield, and $NDSI_{L3,4}$ were established during critical growth stages (from the stem elongation to the heading stages; $R^2 = 0.53$, $p < 0.01$, RRMSE = 0.106). An $NDSI_{L3,4}$ high-yield change curve was drawn to describe critical $NDSI_{L3,4}$ values for a high-yield target (10.28 t ha^{-1}). Furthermore, dynamic-critical curve models based on the $NDSI_{L3,4}$ allowed a precise description of rice N status, facilitating the timing of fertilization decisions to optimize yields in the intensive rice cropping systems of eastern China.

Keywords: SPAD; leaf nitrogen concentration; nitrogen nutrition index; grain yield; dynamic model

1. Introduction

Nitrogen (N) is one of the most important yield-limiting factors [1]. Appropriate N management is essential to achieve relatively high yields with low N input, particularly to ensure maximum rice yield and quality [2,3]. Yield-target-based N fertilization plays an important role when developing profitable and environmentally friendly rice production systems (which is good in environment protection during field production) [4,5]. Accurate and remote in-season estimation of crop N status and its site-specific applications in intensive rice cropping systems is challenging [6]. Rapid, non-destructive, and cost-effective N diagnostic tools are imperative for accurate and timely diagnosis of rice N status at critical growth stages—it is imperative to match N requirements to soil N supply [2,3].

Non-destructive diagnostic strategies use various devices to monitor crop growth and N status [7,8]. A chlorophyll meter (Soil Plant Analysis Development, SPAD-502, Minolta Camera Co., Osaka, Japan) has been widely used for simple, rapid, and non-destructive assessment of leaf chlorophyll concentrations [9], however, the readings are significantly influenced by growth stage, plant leaf position, leaf measurement location, leaf thickness, leaf weight, the cultivar, solar radiation, and environmental stress [1,10,11].

Previous studies sought to correlate specific leaf weight (SLW) with SPAD values and leaf N concentrations (LNC) [12,13], which range from 2–3.2% [4]; multiplying upper leaf SPAD values by the leaf area index (LAI) [14,15], or linking sensor data to the product of SPAD and height [16]. Many researchers have developed SPAD indices [17], including SPAD positional difference sufficiency index [18], a relative SPAD index [19], and a normalized SPAD index [20]. Chlorophyll meter readings have also been linked with digital still canopy color images to improve chlorophyll meter data [13]. However, linking SPAD values to tissue N concentrations remains challenging due to controversies in their reliability [4].

During the plant growth cycle, N and carbon (C) levels are in dynamic balance in crops; this is particularly significant when paddy rice leaves turn from green to yellow [21]. Either an N deficit or excess will retard crop growth. Cropping duration is controlled principally by genotype and N nutritional status. Various rice canopy leaves reflect N status, the color difference between 4LFT (the fourth fully expanded leaf from the top) and 3LFT (the third fully expanded leaf from the top) can be used to diagnose the critical N concentration at the booting stage; this is 27 g kg^{-1} dry matter weight (DW) for Japonica rice and 25 g kg^{-1} DW for Indica rice [22]. However, other indices, such as the relative SPAD index (RSI) [23], the difference SPAD index (DSI) [24], the relative difference SPAD index (RDSI) [22,25], and the normalized difference SPAD index [24], have been developed using the SPAD readings of 1LFT (the first fully expanded leaf from the top), 2LFT (the second fully expanded leaf from the top), and 3LFT to assess in-season crop N status. Recently, Yuan et al. reported that the 2LFT, 3LFT, and 4LFT SPAD values were related to various N indicators (e.g., the N nutrition index (NNI) and leaf N accumulation (LNA); the LNC index). The cited authors concluded that the normalized SPAD index of 4LFT (the NSI4) could be used to increase grain yield and nitrogen efficiency [10]. Therefore, SPAD indices of lower leaves are better than those of upper leaves regarding diagnosing rice N status.

Several attempts have been made to establish quantitative relationships between SPAD indicators and the NNIs of C3 and C4 crops [26], such as wheat [24], rice [27], corn [28], barley [29], etc. Given the differences among environmental conditions and genotypes, the relationships between NNI and SPAD values vary greatly. However, the relative SPAD values are not notably influenced by the cultivar, growing season, or growth stage. Calculating relative SPAD values requires the use of a non-N-limiting treatment as a control, reducing the utility of the method regarding on-farm N diagnosis [30]. SPAD value differences among varieties, and variations in production levels, can be eliminated by normalizing experimental data before modeling [31]. Therefore, some studies have used different normalized SPAD index (NDSI) values to reduce the effects of variation. LNCs of lower leaves were more sensitive to increases in N application rate, and SPAD readings of lower leaves were closely related to tissue N concentrations [28,32]. Thus, LNC of lower leaves may better reflect crop N status when N application rates vary. Although efforts have been made to link crop N status to

NDSI, no attempt has yet been made to associate NDSI values with the NNI or grain yield of rice, or to establish the relationships between the NNIs and NDSIs of the four topmost fully expanded leaves, on the one hand, and grain yield, on the other.

In recent years, many authors have sought to simulate the crop tiller number (LAI) and other dynamic indicators [31]. Dynamic models of crop growth indices should ideally be universal [33], but labor- and time-consuming due to destructive sampling required for obtaining the LAI, dry matter (DM), tissue N concentration, and other growth parameters. As the LNC correlates strongly with chlorophyll content, SPAD meters have been used as real-time, portable, non-destructive devices for estimating N levels by assessing transmittance [7]. A previous study found that SPAD readings of flag leaves correlated strongly with grain yields at different wheat growth stages, and multiple regression implied that maintenance of optimal leaf chlorophyll content over the interval of 50–75 days after sowing was essential to obtain high yields [34]. To date, few dynamic models based on spectral indices are available. In 2017, Liu et al. reported a double logistic NDVI dynamic model for high-yield production in rice, which can be used to accurately predict canopy NDVI dynamic changes during the entire growth period [35]. Further studies on SPAD index variation regarding the establishment of a SPAD index-based dynamic model are essential for monitoring and diagnosing crop nutritional status in-season.

Therefore, we defined the relationships between NDSI, LNC, and NNI during different growth periods: (1) to accurately access N nutrition status using SPAD-based index, and (2) to draw a dynamic-critical curve showing when yields were lower than required by the NDSI target to guide N fertilizer topdressing, thus aiding rice production in eastern China.

2. Materials and Methods

2.1. Sites and Experimental Design

Yangtze River Reaches is the main rice production region of China, which has a great influence on China's food security (Huang RH et al., 2002). Yangtze River Reaches is not only the major agricultural regions of China, but also the oldest niche of rice cultivation [36]. China contributes 29% of global rice production, and Yangtze River Reaches alone contributes more than 65% of the national rice production in China [37]. Thus, four field experiments using multi-N rates (0–390 kg N ha^{-1}; N0, N1, N2, N3, N4, N5, N6, and N7 were 0, 130, 150, 225, 260, 300, 375, and 390 kg N ha^{-1}, respectively) were conducted at Jiangning (E 118.98°, N 31.93°) and Wujiang (E 121.28°, N 31.33°) in eastern China from 2007 to 2009, and in 2013. Seven Japonica rice cultivars (9915, 27123, Wuxiangjing14, Wuyunjing19, Yongyou8, Wuyunjing24, Wuxiangjing19) used were the most widely cultivated cultivars in Jiangsu Province; with distinct subspecies and yield potentials. All experiments were conducted using a randomized complete block design with different N treatments and three replications. Details of the cultivars and N application rates are shown in Table 1.

Rice seedlings with three to four fully expanded leaves were transplanted on the 20 June 2007, 25 June 2008, and 26 June 2013, which were raised alone. The hill spacing was 0.30 m × 0.15 m (about 22 hills m^{-2}), with three seedlings per hill in all experiments. Each plot was of 3 m × 6 m in size. N treatment featured 30% N application at the pre-planting stage, and the remaining N was top dressed three times at the tillering (TI, 20%), booting (BT, 30%), and heading (HD, 20%) stages, as urea. In each experiment, 59 kg ha^{-1} phosphorus as P_2O_5 and 158 kg ha^{-1} potassium as K_2O were incorporated into each plot before transplantation following local standard rice production practices. Crop management practices at each site followed local recommendations to maximize grain yield (the only limiting factor was N fertilizer). Weeds, diseases, and insects were intensively controlled, as in conventional cultivation, throughout the growing period. Data from Exp. 1, Exp. 2, and Exp. 4 were used for model calibrating, while data of Exp. 3 were used to validate the models.

Table 1. Basic information about the four rice field experiments.

Experiment No.	Transplanting/Harvesting Date	Location	Cultivar	N Rate (kg N ha^{-1})	Soil Characteristic
EXP. 1 2007	20-Jun; 21-Oc	Jiangning, E 118.98°, N 31.93°	9915, 27123 (Japonica)	N0 (0) N1 (130) N4 (260) N7 (390)	Soil type = Fe-leachic -stagnic Anthrosols Soil pH = 6.5 OM = 13.5 g·kg^{-1} Total N = 1.13 g·kg^{-1} Available P = 45 mg·g^{-1} Available K = 82.6 mg·g^{-1}
EXP. 2 2008	25-Jun; 27-Oct	Jiangning, E 118.98°, N 31.93°	WXJ14, 27123 (Japonica)	N0 (0) N1 (130) N4 (260) N7 (390)	Soil type = Fe-leachic -stagnic Anthrosols Soil pH = 6.9 OM = 13.5 g·kg^{-1} Total N = 1.38 g·kg^{-1} Available P = 43 mg·g^{-1} Available K = 80 mg·g^{-1}
EXP. 3 2009	19-Jun; 20-Oct	Jiangning, E 118.98°, N 31.93°	WYJ19, YY8, WXJ14, WYJ24 (Japonica)	N0 (0) N1(130) N4 (260) N7 (390)	Soil type = Gley- stagnic Anthrosols Soil pH = 6.9 OM = 26.15 g·kg^{-1} Total N = 1.65 g·kg^{-1} Available P = 38 mg·g^{-1} Available K = 70 mg·g^{-1}
EXP. 4 2013	19-Jun; 20-Oct	Wujiang, E 121.28°, N 31°33′	WYJ19, WXJ19 (Japonica)	N0 (0) N2(150) N3 (225) N5 (300) N6 (375)	Soil type = Typic Endoaquepts Soil pH = 6.75 OM = 26.15 g·kg^{-1} Total N = 2.15 g·kg^{-1} Available P = 45.5 mg·g^{-1} Available K = 115.3 mg·g^{-1}

Note: "9915, 27123, WXJ14, WYJ19, WYJ24, YY8, and WXJ19" are rice cultivars; "WXJ14" is Wuxiangjing14; "WYJ19" is Wuyunjing19; "WYJ24" is Wuyunjing24; "YY8" is Yongyou8; and "WXJ19" is Wuxiangjing19. "N0 = 0 kg N ha^{-1}; N1 = 130 kg N ha^{-1}; N2 = 150 kg N ha^{-1}; N3 = 225 kg N ha^{-1}; N4 = 260 kg N ha^{-1}; N5 = 300 kg N ha^{-1}; N6 = 375 kg N ha^{-1}; N7 = 390 kg N ha^{-1}".

2.2. Plant Sampling and Measurement; Shoot Biomass, Nitrogen Concentration, and the NNI

Plants collected by randomly clipping 1 m^2 from the TI to the ripening period (RP) stages were separated into green leaf blades (leaves) and culm-plus-sheath (stems), oven-dried for 30 min at 105 °C to halt metabolic processes, and then dried at 80 °C in a forced-draft oven until constant weight was attained. Each component was then ground to powder, passed through a 1-mm-diameter sieve in a Wiley mill, and stored at room temperature. Samples (0.2 g) were digested with H_2O_2 and H_2SO_4. N concentrations were determined using a continuous-flow auto-analyzer AA3 (Bran + Luebbe; Norderstedt, Germany). Grain yield was measured at maturity by harvesting 1 m^2 of crop and drying to a moisture content of 14%. Leaf dry matter levels were measured using this material.

2.3. SPAD Measurements

The chlorophyll meter is a spectral instrument, it measures the difference between the transmittance of a red (650 nm) and an infrared (940 nm) light through the leaf, generating a 3-digit SPAD value, which was used to take SPAD readings from the four uppermost fully expanded leaves of 10 randomly selected plants from each plot. Three SPAD values per leaf, including one value around the midpoint of the leaf blade and two values 3 cm apart from the midpoint, were averaged to give the mean SPAD value of the leaf avoiding the midribs. These measurements were taken at each growth stage and averaged [10]. Chlorophyll meter (SPAD) readings were obtained at six growth stages: TI, stem elongation (SE), panicle initiation (PI), BT, HD, flowering (FL), grain filling (GF), and RP, using a SPAD-502 meter (Minolta Camera Co., Osaka, Japan). SPAD readings were obtained from the four, uppermost fully expanded leaves of 10 randomly selected plants in each plot. The normalized difference SPAD index (NDSI) between LFT_i and LFT_j used to evaluate N nutritional status was that for wheat developed by Zhao et al. [29], and the equation for other SPAD-based indices is described in Table 2.

$$NDSI_{Li,j} = (SPAD_i - SPAD_j)/(SPAD_i + SPAD_j) \tag{1}$$

where $SPAD_i$ and $SPAD_j$ are the SPAD readings of leaf positions i and j; i and j vary from 1 to 4, and $i < j$.

Table 2. Equations of soil and plant analysis development (SPAD)-based indices.

Index	Description	Algorithm	Reference
$DSI_{L1\text{-}L3}$	The difference SPAD between 1LFT and 3LFT	$S_{1LFT} - S_{3LFT}$	[24]
$SPAD_{L3\text{-}L4}$	The difference SPAD between 3LFT and 4LFT	$S_{3LFT} - S_{4LFT}$	[18]
$RSI_{L1/L3}$	The relative SPAD index between 1LFT and 3LFT	S_{1LFT}/S_{3LFT}	[23]
$RDSI_{L1,3}$	The relative difference SPAD index between 1LFT and 3LFT	$S_{1LFT}/(S_{1LFT} + S_{3LFT})$	[25]
$NDSI_{L1,3}$	The normalized differences SPAD and index between 1LFT and 3LFT	$(S_{1LFT} - S_{3LFT})/(S_{1LFT} + S_{3LFT})$	[28]
NDSI	The normalized differences SPAD and index between i LFT and j LFT, the range of i, j values is from 1 to 4, i < j	$(S_{iLFT} - S_{jLFT})/(S_{iLFT} + S_{jLFT})$	[29]

Note: "S" means SPAD values; "LFT" represents the fully expanded leaf position form the top; "1-4 LFT" means the first to fourth fully expanded leaf position form the top.

2.4. Data Analysis

2.4.1. Nitrogen Nutrition Index

The NNIs of various rice cultivars at different vegetative growth stages were calculated using the critical N concentration (*Nc*) values obtained from the *Nc* dilution curve of rice developed by Ata-Ul-Karim et al. [36], using the following equation:

$$Nc = 3.53 \times W^{-0.28}\ (12.37 \geq W \geq 1.55\ \text{t ha}^{-1};\ W < 1.55\ \text{t ha}^{-1},\ Nc = 3.05\%) \tag{2}$$

$$NNI = \frac{N_a}{N_c} \tag{3}$$

where *Nc* is the critical rice N, *W* is the weight of the crop, and *Na* is the crop N concentration [36].

2.4.2. Calibration of Dynamic-Critical Curve Models

The critical $NDSI_{L3,4}$ curve of high-yield was in the shape of a sigmoid curve. Different equation models were used to fit the curve, and the Boltzmann model was selected to image the changes of $NDSI_{L3,4}$, based on the R^2 and the relative root mean square error (RRMSE):

$$y = A_2 + \frac{A_1 - A_2}{1 + e^{\frac{x - x_0}{dx}}}, \; Boltzmann\ model \tag{4}$$

where A_1 is the vegetative plateau, A_2 the reproduction plateau, x_0 the x value when $NDSI_{L3,4} = 0$, and dx is a time constant. Three high-yield $NDSI_{L3,4}$ trends were similar as a sigmoid curve. In the $NDSI_{L3,4}$-based high-yield critical curve, the A_1, A_2, x_0, and dx values differed regarding the $NDSI_{L3,4}$ trend.

2.4.3. Statistical Analysis

All SPAD data were normalized using the maximum conversion ratio method. Data from each sampling date and year were subjected to analysis of variance using SPSS ver.20.0 software (IBM, Armonk, New York, NY, USA); this software was also used to compare yields. The least significant difference (LSD) test was employed to compare differences between treatment means. GraphPad Prism 5 software (GraphPad Software, San Diego, CA, USA) was used to fit the critical NDSI curve to different grain yields and to compare the intercept and slope of the regression curve at different growth stages. The R^2 value and the relative root mean square error (RRMSE, %) were used to explore the predictive accuracy of the model:

$$\text{RRMSE} = \sqrt{\frac{\sum_{i=1}^{n}(P_i - Q_i)^2}{n}} \times \frac{100}{\overline{Q_i}} \tag{5}$$

where n is the number of samples, P_i the measured values, Q_i the predicted values, and the average of Q_i. The RRMSE was used to explore agreement between model predictions and measured values.

3. Results

3.1. SPAD Readings of Different Leaves

SPAD readings were measured from the TI to the RP stages under different N treatments. The readings of Japonica rice (27123) increased with increasing N application, and ranged from 34–42 and 40–48 when low, or sufficient N, and excess N were applied, respectively (Figure 1). SPAD reading trends of 1LFT and 2LFT differed greatly from those of 3LFT and 4LFT. The N application rates had no significant effect on the SPAD values of 1LFT and 2LFT, but did affect SPAD values of 3LFT and 4LFT, especially 4LFT. However, a higher N fertilization rate did not significantly change SPAD readings. SPAD reading of 1LFT gradually increased from the TI to the BT and RP stages, in a deformed "S" manner. SPAD reading of 2LFT fell from the TI to the RP stage. SPAD readings of 3LFT and 4LFT were variable from the TI stage to the flowering stage (FL), but then decreased toward the RP stage. Regarding SPAD readings at different N rates, the N0 stage exhibited the lowest readings at every leaf, and the N3 stage the highest. N2 readings were not always greater than those for N1 prior to the 1–3LFT PI stages. Thus, N topdressing fertilizer increased SPAD values and enhanced the stability of the photosynthetic reaction.

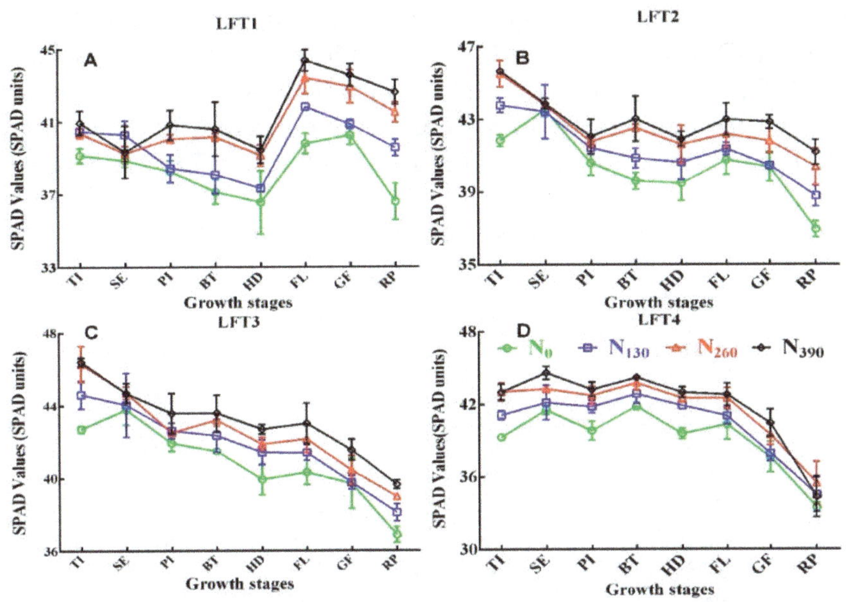

Figure 1. Changes over time in the soil and plant analysis development (SPAD) readings of different leaves ("LFT" represents the fully expanded leaf position form the top; "1-4 LFT" means one to four fully expanded leaf position form the top.; 1LFT, **A**; 2LFT, **B**; 3LFT, **C**; and 4LFT, **D**) of 27123 plants evaluated in 2007 and 2008 at N levels of 120 kg ha^{-1}. The vertical bars are standard error (TI, tillering; SE, stem elongation; PI, panicle initiation; BT, booting; HD, heading; FL, flowering; GF, grain filling; and RP, ripening).

3.2. Differences in the Normalized SPAD Indices

SPAD values were significantly affected by cultivar, growing season, and growth stage. We calculated NDSI values to evaluate leaf performance. Table 3 shows that simple linear SPAD analysis revealed that $NDSI_{L1,3}$, $NDSI_{L1,4}$, and $NDSI_{L3,4}$ did not differ significantly among the seven varieties, showing that SPAD readings were eliminated among variety differences. Furthermore, $NDSI_{L1,3}$, $NDSI_{L2,3}$, and $NDSI_{L3,4}$ values did not differ between the years. Therefore, compared to other SPAD indicators, $NDSI_{L3,4}$ better compared data from different years or varieties. Simple linear analysis of the seven N rates indicated that all the $NDSI_{L1,3}$, $NDSI_{L2,3}$, and $NDSI_{L3,4}$ differed significantly at the 0.01 probability level. For the various growth stages, all the $NDSI_{L1,3}$, $NDSI_{L1,4}$, $NDSI_{L2,3}$, and $NDSI_{L3,4}$ differed significantly at the 0.05 or 0.01 probability levels. Thus, $NDSI_{L1,3}$, $NDSI_{L2,3}$, and $NDSI_{L3,4}$ indices can be used to describe the time course of N nutritional status.

Table 3. Simple grouping linear analysis of SPAD indicator.

SPAD Indicator	Variety			Year			Treatment			Growth Stage			Residual	
	df	MS	*F*-Value	df	MS	*F*-Value	df	MS	*F*-Value	df	MS	*F*-Value	df	MS
$NDSI_{L1,2}$	6	0.00612 **	12.016	2	0.016 *	30.025	3	0.0001 **	0.064	5	0.16 ns	53.7	271	0.0001
$NDSI_{L1,3}$	6	0.013 ns	11.288	2	0.032 ns	27.303	3	0.0001 ns	0.184	5	0.04 *	81.014	271	0.001
$NDSI_{L1,4}$	6	0.034 ns	11.751	2	0.079 **	27.068	3	0.003 ns	0.863	5	0.1 **	81	271	0.002
$NDSI_{L2,3}$	6	0.002 *	1.706	2	0.003 ns	2.84	3	0.016 *	17.755	5	0.009 **	10.771	271	0.001
$NDSI_{L2,4}$	6	0.016 **	8.448	2	0.035 *	17.932	3	0.006 ns	2.62	5	0.048 ns	44.103	271	0.002
$NDSI_{L3,4}$	6	0.004 ns	6.65	2	0.009 ns	13.945	3	0.006 **	10.29	5	0.017 **	60.636	271	0.0001

"df" is degrees of freedom; "MS" is mean square; "ns" means non-significance; "*" indicates significant difference at 0.05 probability level; "**" indicates significant difference at 0.01 probability level. Variety include Japonica rice "9915, 27123, Wuxiangjing14 Wuyunjing19, 27123, Wuxiangjing14, Wuyunjing19, Wuyunjing24, Wuxiangjing19", and Indica rice "Y-Liangyou8"; years are 2007, 2008, and 2013; treatment is the N rates (N1–N6); growth stage form tillering to flowering.

3.3. The Relationship between the $NDSI_{Li,j}$ and Leaf N Concentration

Linear relationships were evident between $NDSI_{Li,j}$ and LNC, irrespective of growth stage or year (2007, 2008, and 2013; Table 4). $NDSI_{L3,4}$ exhibited less variability on regression analysis ($R^2 > 0.82$,

$p < 0.01$); lower coefficients of determination (R^2) were apparent for the other leaves ($0.77 > R^2 > 0.06$). The standard deviation (SD) ranged from 0.15–0.54; the lowest SD was that of the $NDSI_{L3,4}$ and the highest were those of $NDSI_{L1,4}$ and $NDSI_{L2,3}$. $NDSI_{L3,4}$ was thus an ideal index for diagnosis of LNC reliability. Thus, we developed an $NDSI_{L3,4}$-based model to determine LNC (Figure 2A); the variability was 84%, proving that the model afforded good retrieval accuracy (RRMSE = 0.0683; Figure 3a).

Table 4. Quantitative relationships between the SPAD readings of different rice leaves and leaf nitrogen concentrations.

Year	SPAD Index	Quantitative Relationship	R^2	SD
2007	$SPAD_{L3\text{-}L4}$	$LNC = 2.243 \times e^{-0.053SPAD}$	0.21 ns	0.37
	$RSI_{L1/L3}$	$LNC = 2.12 \times e^{-2.66RSI}$	0.56 *	0.27
	$DSI_{L1\text{-}L3}$	$LNC = 11.909 \times e^{-1.75DSI}$	0.53 *	0.33
	$RDSI_{L1,3}$	$LNC = 2.054 \times e^{-0.049RDSI}$	0.35 *	0.29
	$NDSI_{L1,2}$	$LNC = 2.193 \times e^{-11.38NDSI}$	0.77 **	0.31
	$NDSI_{L1,3}$	$LNC = 2.137 \times e^{-5.072NDSI}$	0.61 *	0.33
	$NDSI_{L1,4}$	$LNC = 2.205 \times e^{-2.229NDSI}$	0.36 *	0.46
	$NDSI_{L2,3}$	$LNC = 2.16 \times e^{-7.467NDSI}$	0.36 *	0.23
	$NDSI_{L2,4}$	$LNC = 2.257 \times e^{-2.681NDSI}$	0.19 ns	0.25
	$NDSI_{L3,4}$	$LNC = 2.249 \times e^{-10.58NDSI}$	0.83 **	0.23
2008	$SPAD_{L3\text{-}L4}$	$LNC = 2.195 \times e^{-0.042SPAD}$	0.11 ns	0.27
	$RSI_{L1/L3}$	$LNC = 1.863 \times e^{-6.747RSI}$	0.58 *	0.31
	$DSI_{L1\text{-}L3}$	$LNC = 4.131 \times e^{-0.688DSI}$	0.56 *	0.33
	$RDSI_{L1,3}$	$LNC = 1.862 \times e^{-0.163RDSI}$	0.59 *	0.24
	$NDSI_{L1,2}$	$LNC = 1.896 \times e^{-14.87NDSI}$	0.61 **	0.21
	$NDSI_{L1,3}$	$LNC = 1.898 \times e^{-14.84NDSI}$	0.61 **	0.21
	$NDSI_{L1,4}$	$LNC = 2.092 \times e^{-3.476NDSI}$	0.18 ns	0.48
	$NDSI_{L2,3}$	$LNC = 2.019 \times e^{-0.4133NDSI}$	-	0.54
	$NDSI_{L2,4}$	$LNC = 2.201 \times e^{-1.895NDSI}$	0.06 ns	0.49
	$NDSI_{L3,4}$	$LNC = 2.208 \times e^{-14.71NDSI}$	0.81 **	0.16
2013	$SPAD_{L3\text{-}L4}$	$LNC = 1.974 \times e^{-0.045SPAD}$	0.16 ns	0.41
	$RSI_{L1/L3}$	$LNC = 1.979 \times e^{-0.037RSI}$	0.49 *	0.34
	$DSI_{L1\text{-}L3}$	$LNC = 8.39 \times e^{-1.43DSI}$	0.17 ns	0.52
	$RDSI_{L1,3}$	$LNC = 1.998 \times e^{-1.434RDSI}$	0.57 *	0.36
	$NDSI_{L1,2}$	$LNC = 2.167 \times e^{-3.556NDSI}$	0.56 **	0.34
	$NDSI_{L1,3}$	$LNC = 2.219 \times e^{-2.401NDSI}$	0.57 **	0.29
	$NDSI_{L1,4}$	$LNC = 2.324 \times e^{-1.706NDSI}$	0.18 ns	0.39
	$NDSI_{L2,3}$	$LNC = 2.363 \times e^{-2.673NDSI}$	0.07 ns	0.15
	$NDSI_{L2,4}$	$LNC = 2.385 \times e^{-0.121NDSI}$	-	0.24
	$NDSI_{L3,4}$	$LNC = 2.246 \times e^{-10.56NDSI}$	0.82 **	0.18

"RSI" means relative SPAD index; "DSI" represents difference SPAD index; "RDSI" means relative difference SPAD index; "NDSI" is the normalized difference SPAD index; "SD" is standard deviation value; "LNC" means leaf nitrogen concentration; "NNI" is nitrogen nutrition index; "ns" and "-" mean non-significance; * indicates significant difference at 0.05 probability level; ** indicates significant difference at 0.01 probability level.

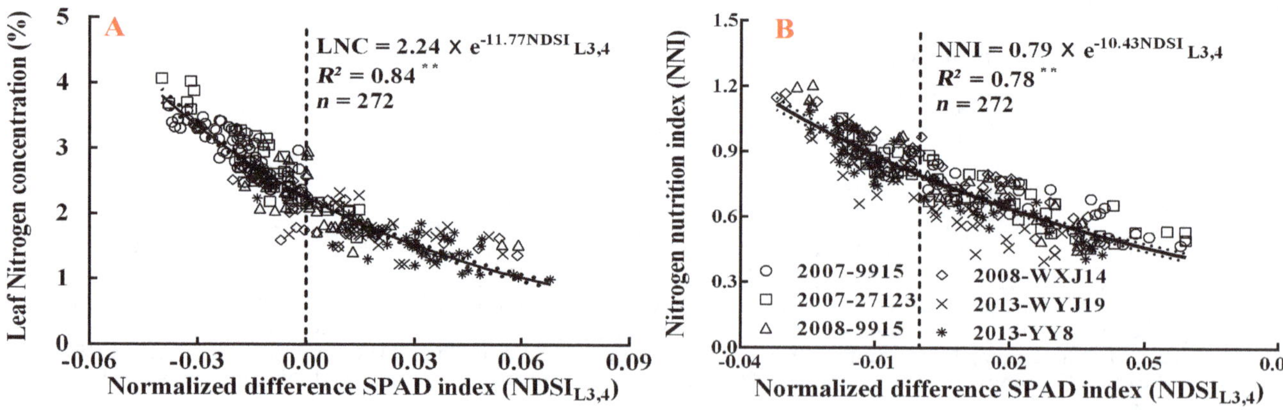

Figure 2. Regression fits between the leaf nitrogen concentration (LNC, **A**), nitrogen nutrition index (NNI, **B**), and $NDSI_{L3,4}$. The experimental years are shown as 2007, 2008, or 2013; 9915, 27123, and WXJ14 are the Wuxiangjing 14, WYJ19 means Wuyunjing19, and YY8 is Yongyou8. "**" means significant difference at 0.01 probability level.

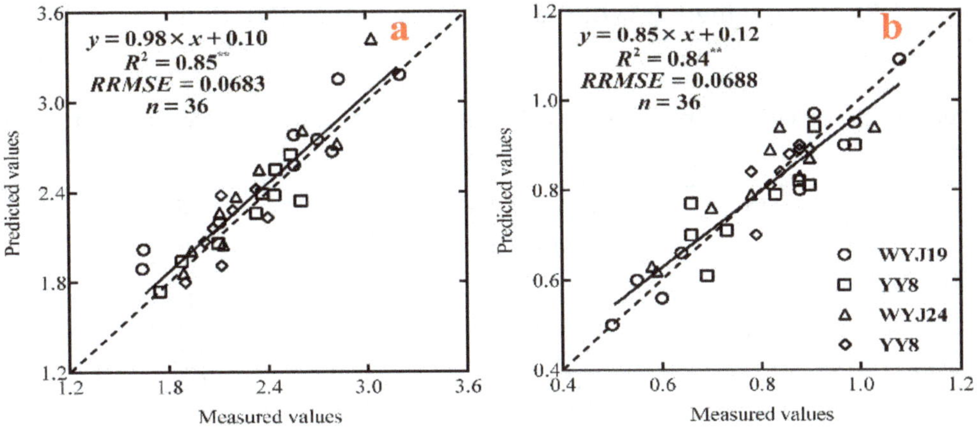

Figure 3. The relationship between measured and predicted leaf nitrogen concentration (LNC, **a**), nitrogen nutrition index (NNI, **b**) of four rice cultivars from the time of stem elongation (SE) to heading (HD) (WYJ19, Wuyungjing19; YY8, Yongyou8; WXJ14, Wuxiangjing14; and WYJ24, Wuyungjing 24; Japonica). The solid line is the linear regression line and the dotted line a line inclined at 45° to the axis. ** indicates significant difference at 0.01 probability level.

In Table 5, using the model equation of $NDSI_{Li,j}$ (LNC $= a \times e^{b \times NDSI_{Li,j}}$), the least significant difference (LSD) test was employed to measure differences in the "a" and "b" parameters among years or varieties. Regarding years, "a" differed significantly among the $NDSI_{L1,2}$ values, and "b" differed significantly among the $NDSI_{L2,4}$ values of the various varieties. In Table 5, "a" did not vary among varieties, and "b" did not change with the year. Thus, $NDSI_{L3,4}$ was an optimal indicator of rice N status.

Table 5. Simple linear regression; fitted curves between the SPAD values of different leaves and rice nitrogen indicators.

Nitrogen Indicator	Parameter	Impact Factor	Mean Square (MS)					
			$NDSI_{L1,2}$	$NDSI_{L1,3}$	$NDSI_{L1,4}$	$NDSI_{L2,3}$	$NDSI_{L2,4}$	$NDSI_{L3,4}$
LNC	a	Year	0.27 *	0.27 ns	0.15 ns	0.0355 ns	0.09 ns	0.001 ns
		variety	0.104 ns	0.035 ns	0.021 ns	0.0178 ns	0.26 ns	0.0252 ns
	b	Year	42.67 ns	42.88 ns	0.83 ns	12.97 ns	1.72 ns	5.71 ns
		variety	36.58 ns	41.24 ns	0.61 ns	8.73 ns	1.22*	3.07 ns
	Residual		0.051	0.061	0.021	0.035	0.014	0.023
NNI	a	Year	0.02 ns	0.01 ns	0.01 ns	-	-	0.02 ns
		variety	0.21 ns	0.04 ns	0.09 *	0.17 ns	0.06 ns	0.51 ns
	b	Year	5.51 ns	5.58 ns	0.54 ns	8.45 ns	5.91 ns	5.62 ns
		variety	1.39 ns	6.71 ns	0.91 ns	4.27 ns	6.53 ns	4.98 ns
	Residual		0.27	0.39	0.17	0.32	0.21	0.04

"LNC" means leaf nitrogen concentration; "NNI" is nitrogen nutrition index; "NDSI" represents normalized difference SPAD index; "ns" and "-" mean non-significance; * indicates significant difference at 0.05 probability level; ** indicates significant difference at 0.01 probability level.

3.4. Relationships between the $NDSI_{Li,j}$ and N Nutrition Index

The NNI is a widely used diagnostic indicator; when NNI = 1, N nutrition is optimal; NNI >1 and NNI <1 indicate excess and deficient N nutrition, respectively. We found a non-linear relationship between the $NDSI_{Li,j}$ and the NNI. Table 6 shows that the relationships between $NDSI_{Li,j}$ values of the differences among $LFT_{L1,2}$, $LFT_{L1,3}$, $LFT_{L3,4}$, and NNI were more stable than those of differences in the other $LFT_{Li,j}$ values across both the growth stages and the cultivars ($0.23 < R^2 < 0.84$ vs. $0.07 < R^2 < 0.16$). Regarding the SDs, these ranged from 0.15–0.54; the $NDSI_{L3,4}$ value was the lowest except in 2013; those of the $NDSI_{L1,4}$ or $NDSI_{L2,3}$ were the highest. The SDs were little affected by the LNC model chosen. Compared to the other $NDSI_{Li,j}$ values, the $NDSI_{L3,4}$ was more closely related to the NNI in both of the earlier years. We created a diagnostic model using the $NDSI_{L3,4}$ values

(Figure 2B). The $NDSI_{L3,4}$ explained 78% of the variability in the NNI; thus, the $NDSI_{L3,4}$ predicted N status (RRMSE = 0.0688; Figure 3b).

Table 6. Quantitative relationships between SPAD readings of different rice leaves and the nitrogen nutrition index.

Year	SPAD Index	Quantitative Relationship	R^2	SD
2007	$SPAD_{L3\text{-}L4}$	$NNI = 0.780 \times e^{-0.028SPAD}$	0.16 ns	0.49
	$RSI_{L1/L3}$	$NNI = 0.739 \times e^{-0.012RSI}$	0.35 *	0.38
	$DSI_{L1\text{-}L3}$	$NNI = 1.017 \times e^{-0.322DSI}$	0.26 ns	0.32
	$RDSI_{L1,3}$	$NNI = 0.746 \times e^{-0.627RDSI}$	0.56 *	0.27
	$NDSI_{L1,2}$	$NNI = 0.776 \times e^{-5.84NDSI}$	0.76 **	0.31
	$NDSI_{L1,3}$	$NNI = 0.759 \times e^{-2.35NDSI}$	0.53 *	0.33
	$NDSI_{L1,4}$	$NNI = 0.766 \times e^{-1.06NDSI}$	0.15 ns	0.46
	$NDSI_{L2,3}$	$NNI = 0.755 \times e^{-2.89NDSI}$	0.61 *	0.23
	$NDSI_{L2,4}$	$NNI = 0.773 \times e^{-1.32NDSI}$	0.14 ns	0.25
	$NDSI_{L3,4}$	$NNI = 0.809 \times e^{-8.85NDSI}$	0.83 **	0.23
2008	$SPAD_{L3\text{-}L4}$	$NNI = 0.764 \times e^{-0.36SPAD}$	0.14 ns	0.51
	$RSI_{L1/L3}$	$NNI = 0.713 \times e^{-2.91RSI}$	0.21 ns	0.34
	$DSI_{L1\text{-}L3}$	$NNI = 0.498 \times e^{0.41DSI}$	0.43 *	0.40
	$RDSI_{L1,3}$	$NNI = 0.707 \times e^{-0.07RDSI}$	0.67 *	0.26
	$NDSI_{L1,2}$	$NNI = 0.709 \times e^{-5.99NDSI}$	0.71 *	0.21
	$NDSI_{L1,3}$	$NNI = 0.710 \times e^{-6.05NDSI}$	0.62 *	0.21
	$NDSI_{L1,4}$	$NNI = 0.749 \times e^{0.50NDSI}$	-	0.48
	$NDSI_{L2,3}$	$NNI = 0.746 \times e^{-2.11NDSI}$	0.68 *	0.54
	$NDSI_{L2,4}$	$NNI = 0.742 \times e^{1.48NDSI}$	0.07 ns	0.49
	$NDSI_{L3,4}$	$NNI = 0.821 \times e^{-11.96NDSI}$	0.85 **	0.16
2013	$SPAD_{L3\text{-}L4}$	$NNI = 0.653 \times e^{-0.02SPAD}$	0.15 ns	0.39
	$RSI_{L1/L3}$	$NNI = 0.610 \times e^{-0.03RSI}$	0.32 *	0.28
	$DSI_{L1\text{-}L3}$	$NNI = 2.492 \times e^{-1.40DSI}$	0.51 *	0.36
	$RDSI_{L1,3}$	$NNI = 0.615 \times e^{-1.40RDSI}$	0.60 *	0.26
	$NDSI_{L1,2}$	$NNI = 0.708 \times e^{-1.85NDSI}$	0.57 *	0.34
	$NDSI_{L1,3}$	$NNI = 0.711 \times e^{-1.52NDSI}$	0.41 *	0.29
	$NDSI_{L1,4}$	$NNI = 0.721 \times e^{-1.96NDSI}$	0.25 *	0.39
	$NDSI_{L2,3}$	$NNI = 0.733 \times e^{-2.96NDSI}$	0.58 *	0.15
	$NDSI_{L2,4}$	$NNI = 0.764 \times e^{-3.36NDSI}$	0.16 ns	0.24
	$NDSI_{L3,4}$	$NNI = 0.734 \times e^{-13.51NDSI}$	0.84 **	0.18

"RSI" means relative SPAD index; "DSI" represents difference SPAD index; "RDSI" means relative difference SPAD index; "NDSI" is the normalized difference SPAD index; "SD" is standard deviation value; "LNC" means leaf nitrogen concentration; "NNI" is nitrogen nutrition index; "ns" and "-" mean non-significance; * indicates significant difference at 0.05 probability level; ** indicates significant difference at 0.01 probability level.

Table 5 also shows a simple exponential function regression exercise, grouping the fitted curves between SPAD values at different positions and the rice NNI. The exponential regression equation is ($NNI = a \times e^{b \times NDSI_{Li,j}}$). N either "a" nor "b" was influenced by year or variety, except for the "a" of $NDSI_{L1,4}$ (which varied by variety). The $NDSI_{L3,4}$ residual was the smallest of all leaves. Thus, Table 5 shows that $NDSI_{L3,4}$ was the optimal indicator of rice N status.

3.5. The Relationship between $NDSI_{L3,4}$ and Grain Yield

Grain yield was positively associated with the $NDSI_{L3,4}$ (Table 7), especially from the SE to HD stages. In Table 7, the linear regressions between grain yield and $NDSI_{L3,4}$ values at varying N addition rates for all growth stages are plotted; we used data from Experiments 1, 2, and 4. The coefficients of determination of the SE-to-HD stages tended to be higher than those of the TI and FL stages. Thus, the $NDSI_{L3,4}$ value was related to grain yield during the SE-to-BT stages, accounting for 53% of the variation (Figure 4). The plateau/linear relationship showed that grain yield decreased linearly with the $NDSI_{L3,4}$ value when that value was >0.001 for paddy rice; at which time the yield approached 10.28 $t \cdot ha^{-1}$.

Table 7. Linear regressions between various normalized SPAD indices and grain yield.

Year	Variety	Growth Stage					
		TI	SE	PI	BT	HD	FL
2007	9915	0.26 *	0.73 **	0.79 **	0.68 *	0.68 *	0.32 *
	27123	0.21 *	0.48 *	0.62 *	0.75 **	0.51 *	0.43 *
2008	27123	0.20 *	0.68 *	0.73 **	0.72 **	0.65 *	0.40 *
	WXJ14	0.26 *	0.75 **	0.79 **	0.72 **	0.54 *	0.37 *
2013	WYJ19	0.22 *	0.57 *	0.63 *	0.61 *	0.52 *	0.34 *
	YY8	0.19 ns	0.74 **	0.62 *	0.47 *	0.62 *	0.42 *

"ns" and "-" mean non-significant difference; * indicates significant difference at 0.05 probability level; ** indicates significant difference at 0.01 probability level. "TI" is tillering stage; "SE" is stem elongation stage; "PI" is panicle initiation stage; "BT" is booting stage; "HD" is heading stage; "FL" is flowering stage; "9915, 27123, WXJ14, WYJ19, WYJ24, YY8, WXJ19" are different rice cultivars; "WXJ14" is Wuxiangjing14; "WYJ19" is Wuyunjing19; "WYJ24" is Wuyunjing24; "YY8" is Yongyou8; "WXJ19" is Wuxiangjing19.

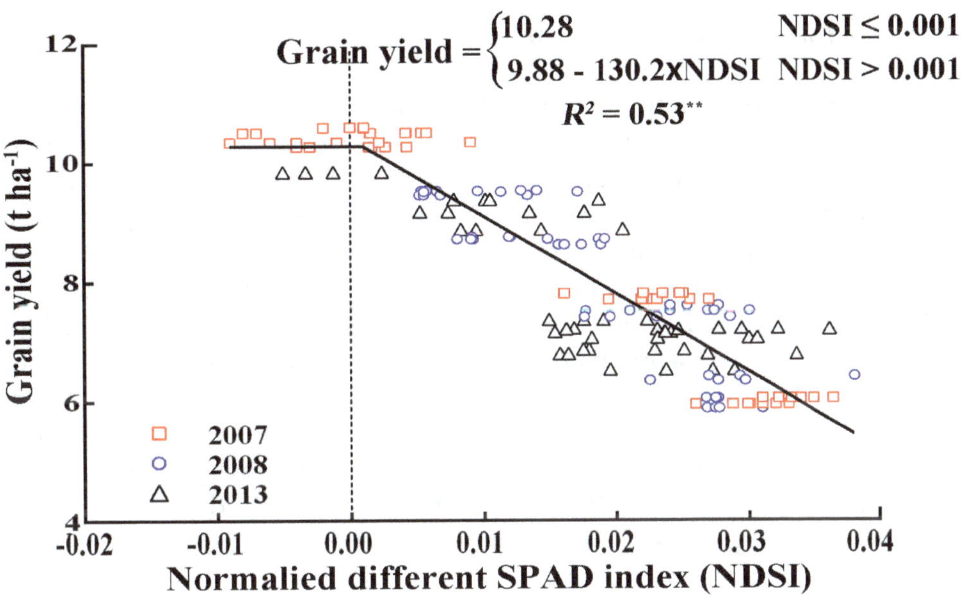

Figure 4. Correlations between $NDSI_{L3,4}$ values and grain yields at critical growth stages (from stem elongation to heading; cultivars: 9915, 27123 (2007); wuxiangjing14, 27123; 2013: wuyunjing19, wuxiangjing19 (2008)).

Table 8 shows results of the differential function models fitted. The results showed that most models had relatively high R^2 ($0.88 < R^2 < 0.99$), low RRMSE values (16% < RRMSE < 30%), and high F-value ($113 < F < 266$; the higher F-value means model fitting). The equation in the shape of the sigmoid curve obtained the higher R^2 and low RRMSE values. The Boltzmann model performed the best among all regression models. Therefore, given $y = A_1 + \frac{A_1 - A_2}{1+e^{\frac{x-x_0}{dx}}}$ (Boltzmann model) during SE-to-HD stages is critical in terms of high yield. We had many data points from SE to GF; we covered all the critical growth stages. Our studied varieties constitute distinct subspecies; all are high-yielding. Next, we created a dynamic model based on a high-yield (10.28 $t{\cdot}ha^{-1}$) $NDSI_{L3,4}$ value (Figure 4). In the $NDSI_{L3,4}$-based high-yield critical curve, the A_1, A_2, x_0, and dx values differed in terms of the $NDSI_{L3,4}$ trend. In the high-yield critical curve, the day of the year (DOY) 205 to 220 (SE) was the vegetative plateau; vegetative and reproductive growth was in describe from DOY220 to DOY250 (thus from PI to BT), and DOY250 was the start of the reproductive stage. When stress develops, N is the pivotal factor limiting grain filling; we found that the $NDSI_{L3,4}$ reliably indicated rice N nutrition.

Table 8. Summary of fitting results of normalize difference SPAD index (NDSI) dynamic model under different fitting equation.

Regression Model	Equation	R^2	RRMSE (%)	*F*-Value
linear model	$y = a + b \times x$	0.899	23.5	213.8
Boltzmann model	$y = A_2 + (A_1 - A_2)/(1 + e^{((x - x0)/dx)})$	0.975	16.8	265.9
Polynomial model	$y = A + B \times x + C \times x^2 + D \times x^3 + \ldots$	0.927	25.2	113.91
Exponential model	$y = a - b \times c^x$	0.894	28.5	99.9
Bradley model	$y = a \times \ln(-b \times \ln(x))$	0.889	28.95	142.5
Power model	$y = a \times x^b$	0.453	48.3	15.3
Nelder model	$y = (x + a)/(b_0 + b_1 \times (x + a) + b_2 \times (x + a)^2)$	0.929	24.4	116.9
DoseResp model	$y = A_1 + (A_2 - A_1)/(1 + 10^{((LOGx_0 - x) \times p)})$	0.931	23.4	119.8
Hyperbola model	-	-	-	-

Note: "RRMSE" indicates relative root mean square error; "a, b, A, A_1, A_2, B, b_0, b_1, C, and D" are parameters of the equation; 'LOG' means the base-10 logarithm.

3.6. Use of the $NDSI_{L3,4}$ Curve for N Management

Estimation of the in-season N requirement (NR) is essential for the management of N topdressing during paddy rice production. However, we found that topdressing at critical growth stages (the SE–HD stages) did not support N management at every growth stage. Therefore, we developed an $NDSI_{L3,4}$ change curve for high-yield, topdressing N management at critical growth stages. We used this curve to first determine $NDSI_{L3,4}$ values (Figure 4) for the in-season N nutritional status of Japonica rice and a high-yield target, and then calculated ΔNNI values by evaluating quantitative relationships between the $NDSI_{L3,4}$ and NNI (Figure 2). Finally, the N fertilizer requirement for Japonica rice was modeled as suggested by Ata-Ul-Karim et al. [2]. This fertilization decision support method precisely estimates crop growth, grain yield, and the NR time-course (an N management strategy). However, the tool cannot be implemented in the early (active) TI stage. Moreover, our N fertilizer topdressing management strategy needs to be tested in other cultivars and rice growing regions to test its reliability for predicting grain yield and crop N status. This section may be divided by subheadings. It should provide a concise and precise description of the experimental results, their interpretation as well as the experimental conclusions that can be drawn.

4. Discussion

The present study analyzed the N nutrient status of rice leaves, determining SPAD values in four field experiments distributed in the Yangtze River Reaches. We examined N distribution by evaluating color differences between upper and lower leaves, and via differential diagnosis of rice N status. A previous study found that chlorophyll content reflected N nutritional status, but was significantly influenced by variety, site, and year of experimentation [31]. A previous study indicated that data normalization prior to modeling eliminated differences caused by variety, soil types, and management strategies, etc. [30]. Therefore, $NDSI_{Li,j}$ indicators can be used to correct traditional SPAD indicators. Most previous studies used threshold SPAD indicators during critical growth periods to diagnose N nutritional status [18]. However, most of the monitoring methods such as saturation index (SI) or NNI were mainly based on the single test of each growth stage [38]. These methods were mostly related to accessing the relationship with poor mechanism, using indicators with the complex calculation methods [39]. In addition, the existence of deviation in the identification of growth stage, environmental factors, time selection also have a greater impact on the single test. However, the single measured value was treated as each stage in the calculation, which results in a greater deviation from the actual value [40]. This minimizes field management but does not consider cropping duration. Thus, a dynamic model incorporating cropping duration, not only the critical growth stages, is better. Such models optimally display nutritional status over time, best-supporting careful agriculture.

We found a better relationship between $NDSI_{L3,4}$ and the N indicators (pooled data from five varieties in Exp. 1, 2, and 4) than between any other combination of $NDSI_{Li,j}$, and the N data, probably due to the non-linear regressions between $NDSI_{Li,j}$ and the N indicators varied over three years at the different sites. We found considerable differences between the linear regression coefficients of SE, PI, and BT stages. When the 3-year data (2007, 2008, and 2013) were pooled, R^2 decreased considerably because of significant regression slopes during different growth periods for the various varieties at different sites in different years. Such results showed that both the growth stage and leaf position significantly influenced $NDSI_{Li,j}$ values. The relationships between 4LFT and NSI4 were a linear/plateau in nature, from the SE to the BT periods. Prost and Jeuffroy indicated that the 1LFT showed poor defined relationship with N indicators, this might be due to the time difference in maturity of the first expanded leaves, due to these differences and wide variation in 1LFT [41]. Wang et al. reported that the lower leaf (3,4LFT) responded more to nitrogen supply than upper leaf (1,2LFT), and also suggested that lower leaf (4LFT) could be the ideal sample leaf for diagnosis of plant nitrogen nutrition [18]. Yuan et al. also proved that lower leaves afforded more reliable estimations of crop N status [17].

At the same target yield, different varieties require very different N topdressing. A previous study revealed that rice grain yield was positively associated with the SPAD readings of 4LFT, but the intercepts of the response curves of grain yield (as a function of SPAD value) differed markedly for the two varieties studied (Xiushui 63 and Hang 43) [27]. It has been reported that use of the NSI4 increased grain yield and N use efficiency compared to 4LFT. This may vary by cultivar and site conditions [17]. The cited authors identified the most sensitive stages at which to measure N nutrition and grain yield, and then created optimal fertilization prescriptions. However, such methods cannot diagnose N status in real time. Attempts have been made to estimate grain yield based on dynamic LAI models [42]. The LAI had the highest, positive indirect effects on grain yield, as measured by kernel number per spike, but it was difficult to describe the physiological condition of the crop directly. Wang et al. reported that SPAD measured the difference between 4LFT and 3LFT [43]. Color differences between these leaves could be used to determine N concentrations [21]. However, all models used to diagnosis N nutrition status or grain yield operate only in the critical growth period of rice [15,17]. If crop parameters are to be monitored, nutritional status should be diagnosed and regulated by simulating the dynamic changes of SPAD indicators appropriate for paddy rice. To date, few studies have sought to establish dynamic models based on spectral indices [35].

Further research on variations in such indices and establishment of index-based dynamic models is necessary to monitor the nutrition of late crops and for dynamic diagnosis [31,44]. We attempted to solve this problem by establishing $NDSI_{L3,4}$ change curves for a high-yield target, thus slightly higher than that of the high-yield rice cultivars grown Yangtze River Reaches [31]. In Figure 4 we established the relationship between $NDSI_{L3,4}$ and grain yield, and comprehensive comparisons of the simulated and observed values at critical growth stages revealed that the slope was 1.02, the $R^2 = 0.74$, and RRMSE = 0.106 (Figure 5).

We developed a dynamic $NDSI_{L3,4}$ model based on a high-yield target (Figure 6) to display the critical change trends. Using the high-yield dynamic model, it is possible to guide N fertilizer rice topdressing efficiently during the critical N growth period (from the SE to the HD). Some authors have used a spectral index or a nitrogen indicator to manage N fertilizer topdressing at critical growth stages [2,45]. N-deficient fields subject to intensive N-regulation become N-sufficient fields after application of local agricultural practices, increasing potential production. Our model allows detection of not only deficient N nutrition, but also excess N nutrition; there is no requirement for N-saturation. However, more data are needed for assessing the reliability of the model in different rice production areas.

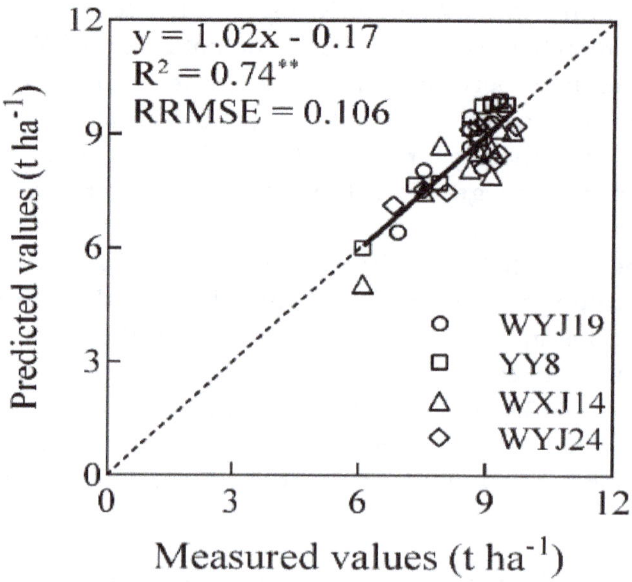

Figure 5. Relationships between measured and predicted grain yields of four rice cultivars in 2009 (WYJ19, Wuyungjing19; YY8, Yongyou8; WXJ14, Wuxiangjing14; and WYJ24, Wuyungjing24; Japonica).'*x*' represents measured values, '*y*' means predicted values. The solid line is the linear regression line and the dotted line a line inclined at 45° to the axis.

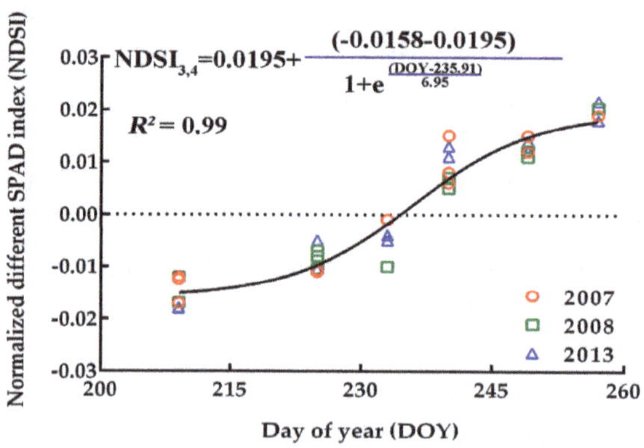

Figure 6. Critical $NDSI_{L3,4}$ data points used to define the changes in the $NDSI_{L3,4}$ curve when data of high-yield targets were pooled. The solid line is the critical $NDSI_{L3,4}$ change curve ($NDSI_{3,4} = 0.0195 + \frac{-0.0158-0.0195}{1+e^{\frac{DOY-235.91}{6.95}}}$; the A_1, A_2, x_0, and dx values differ in terms of their effect on $NDSI_{L3,4}$ trends) of high-yield target rice in the Yangtze river valley.

5. Conclusions

4LFT SPAD values measured using a chlorophyll meter correlated significantly with rice LNC and NNI values. The $NDSI_{L3,4}$ difference between the third and fourth LFT [$NDSI_{L3,4} = SPAD_3 - SPAD_4/(SPAD_3 + SPAD_4)$] could be used to improve LNC and NNI estimations compared to those afforded by isolated SPAD readings and the differences between other leaf positions. We have developed two universal $NDSI_{L3,4}$ based diagnostic models (from the TI to the RP). Both models can be used for effective diagnosis of the LNC ($R^2 > 0.81$, $p < 0.001$) and NNI ($R^2 > 0.83$, $p < 0.01$) with a reasonable distribution of residuals (LNC: RRMSE = 0.0683; NNI: RRMSE = 0.0688; $p < 0.01$). To optimize N topdressing for Japonica rice, we first established the relationship between the $NDSI_{L3,4}$ and grain yield to predict the yield during the SE-to-HD stages. Secondly, a new, critically dynamic $NDSI_{L3,4}$ model was developed based on the previous experiments. Thirdly, the ΔNNI was estimated using both a dynamic model and the NNI–$NDSI_{L3,4}$ model. Finally, the N requirement was determined using the NNI model developed by Ata-Ul-Karim et al. [3]. However, parameters of the newly developed model may require adjustment under varied conditions caused by different climatic features, etc. The robustness and sensitivity of the model should be further tested using data from different rice production region.

Author Contributions: K.Z., Y.T., and X.L. conceived and designed the experiments, S.T.A.-U.-K., J.L., and S.L. performed experiments, Q.C., Y.Z., K.Z., and X.L. analyzed the data, K.Z., X.L., and Y.T. wrote the paper, W.C. and B.K. provided advice and edited the manuscript.

Acknowledgments: The authors also thank the anonymous reviews for their constructive comments and suggestions. The first author also thanks Xiaofang Duan for her support and love all the time.

Abbreviations

TI	Tillering
RP	Ripening period
BT	Booting
HD	Heading
SE	Stem elongation
PI	Panicle initiation
FL	Flowering
GF	Grain filling
SPAD	Soil and plant analysis development
NDSI	Normalized Different SPAD values
RSI	Relative SPAD index
DSI	Difference SPAD index
RDSI	Relative difference SPAD index
NNI	Nitrogen Nutrition Index
LNC	Leaf Nitrogen concentration
LAI	Leaf Area Index
DW	Dry matter weight
SSNM	Site-Specific Nitrogen Management
NUE	Nitrogen Use Efficiency
DAT	Day after Transplanting
RRMSE	Relative root mean square error
LSD	Least significant difference
LFT	The position on upper fully expanded Leaf From the rice Top

References

1. Zhao, B.; Ata-Ul-Karim, S.T.; Duan, A.; Liu, Z.; Wang, X.; Xiao, J.; Liu, Z.; Qin, A.; Ning, D.; Zhang, W. Determination of critical nitrogen concentration and dilution curve based on leaf area index for summer maize. *Field Crop. Res.* **2018**, *228*, 195–203. [CrossRef]
2. Ata-Ul-Karim, S.T.; Liu, X.; Lu, Z.; Zheng, H.; Cao, W.; Zhu, Y. Estimation of nitrogen fertilizer requirement for rice crop using critical nitrogen dilution curve. *Field Crop. Res.* **2017**, *201*, 32–40. [CrossRef]
3. Ata-Ul-Karim, S.T.; Zhu, Y.; Liu, X.; Cao, Q.; Tian, Y.; Cao, W. Comparison of different critical nitrogen dilution curves for nitrogen diagnosis in rice. *Sci. Rep.* **2017**, *7*, 42679. [CrossRef] [PubMed]
4. Ataulkarim, S.T.; Liu, X.; Lu, Z.; Yuan, Z.; Yan, Z.; Cao, W. In-season estimation of rice grain yield using critical nitrogen dilution curve. *Field Crop. Res.* **2016**, *195*, 1–8.
5. Ge, H.X.; Zhang, H.S.; Zhang, H.; Cai, X.H.; Song, Y.; Kang, L. The characteristics of methane flux from an irrigated rice farm in East China measured using the eddy covariance method. *Agric. For. Meteorol.* **2018**, *249*, 228–238. [CrossRef]

6. Yousaf, M.; Li, X.; Zhang, Z.; Ren, T.; Cong, R.; Ata-Ul-Karim, S.T.; Shah, F.; Shah, A.N.; Lu, J. Nitrogen fertilizer management for enhancing crop productivity and nitrogen use efficiency in a rice-oilseed rape rotation system in China. *Front. Plant Sci.* **2016**, *7*, 1496. [CrossRef]
7. Ata-Ul-Karim, S.T.; Cao, Q.; Zhu, Y.; Tang, L.; Rehmani, M.I.; Cao, W. Non-destructive assessment of plant nitrogen parameters using leaf chlorophyll measurements in rice. *Front. Plant Sci.* **2016**, *7*, 1829. [CrossRef]
8. Wang, W.; Yao, X.; Yao, X.F.; Tian, Y.C.; Liu, X.J.; Ni, J.; Cao, W.X.; Zhu, Y. Estimating leaf nitrogen concentration with three-band vegetation indices in rice and wheat. *Field Crop. Res.* **2012**, *129*, 90–98. [CrossRef]
9. Markwell, J.; Osterman, J.C.; Mitchell, J.L. Calibration of the Minolta SPAD-502 leaf chlorophyll meter. *Photosynth. Res.* **1995**, *46*, 467–472. [CrossRef]
10. Yuan, Z.; Qiang, C.; Ke, Z.; Ata-Ul-Karim, S.T.; Tian, Y.; Yan, Z.; Cao, W.; Liu, X. Optimal leaf positions for spad meter measurement in rice. *Front. Plant Sci.* **2016**, *7*, 719. [CrossRef]
11. Muñoz-Huerta, R.F.; Guevara-Gonzalez, R.G.; Contreras-Medina, L.M.; Torres-Pacheco, I.; Prado-Olivarez, J.; Ocampo-Velazquez, R.V. A review of methods for sensing the nitrogen status in plants: Advantages, disadvantages and recent advances. *Sensors* **2013**, *13*, 10823. [CrossRef] [PubMed]
12. Peng, S.; García, F.V.; Laza, R.C.; Cassman, K.G. Adjustment for specific leaf weight improves chlorophyll meter's estimate of rice leaf nitrogen concentration. *Agron. J.* **1993**, *85*, 987–990. [CrossRef]
13. Wang, Y.; Wang, D.; Shi, P.; Omasa, K. Estimating rice chlorophyll content and leaf nitrogen concentration with a digital still color camera under natural light. *Plant Methods* **2014**, *10*, 36. [CrossRef] [PubMed]
14. Gitelson, A.A.; Viña, A.; Ciganda, V.; Rundquist, D.C.; Arkebauer, T.J. Remote estimation of canopy chlorophyll content in crops. *Geophys. Res. Lett.* **2005**, *32*, 93–114. [CrossRef]
15. Zhang, K.; Ge, X.; Liu, X.; Zhang, Z.; Liang, Y.; Tian, Y.; Cao, Q.; Cao, W.; Zhu, Y.; Liu, X. Evaluation of the chlorophyll meter and GreenSeeker for the assessment of rice nitrogen status. In Proceedings of the 11th European Conference on Precision Agriculture (ECPA 2017), Edinburgh, UK, 16–20 July 2017; Volume 8, pp. 359–363.
16. Sudduth, K.A.; Kitchen, N.R.; Drummond, S.T. Nadir and oblique canopy reflectance sensing for n application in corn. *Liccosec* **2011**, *7*, 162–172.
17. Yuan, Z.; Ata-Ul-Karim, S.T.; Cao, Q.; Lu, Z.; Cao, W.; Zhu, Y.; Liu, X. Indicators for diagnosing nitrogen status of rice based on chlorophyll meter readings. *Field Crop. Res.* **2016**, *185*, 12–20. [CrossRef]
18. Wang, S.; Zhu, Y.; Jiang, H.; Cao, W. Positional differences in nitrogen and sugar concentrations of upper leaves relate to plant N status in rice under different N rates. *Field Crop. Res.* **2006**, *96*, 224–234. [CrossRef]
19. Ziadi, N.; Bélanger, G.; Claessens, A.; Lefebvre, L.; Tremblay, N.; Cambouris, A.N.; Nolin, M.C.; Parent, L.É. Plant-based diagnostic tools for evaluating wheat nitrogen status. *Crop Sci.* **2010**, *50*, 2580–2590. [CrossRef]
20. Hussain, F.; Bronson, K.F.; Singh, Y.; Singh, B.; Peng, S. Use of chlorophyll meter sufficiency indices for nitrogen management of irrigated rice in Asia. *Agron. J.* **2000**, *92*, 875–879.
21. Zhao, Q.Z.; Ding, Y.F.; Wang, Q.S.; Huang, P.S.; Ling, Q.H. Relationship between leaf color and nitrogen uptake of rice. *Sci. Agric. Sin.* **2006**, *39*, 916–921.
22. Wang, S.H.; Cao, W.X.; Wang, Q.S.; Ding, Y.F.; Huang, P.S.; Ling, Q.H. Positional distribution of leaf color and diagnosis of nitrogen nutrition in rice plant. *Sci. Agric. Sin.* **2002**, *192*, 45–51.
23. Shen, Z.Q.; Wang, K.; Zhu, J.Y. Preliminary study on diagnosis of the nitrogen status of two rice varieties using the chlorophyll meter. *Bull. Sci. Technol.* **2002**, *18*, 174–176.
24. Lin, F.F.; Qiu, L.F.; Deng, J.S.; Shi, Y.Y.; Chen, L.S.; Ke, W. Investigation of SPAD meter-based indices for estimating rice nitrogen status. *Comput. Electron. Agric.* **2010**, *71*, S60–S65. [CrossRef]
25. Ata-Ul-Karim, S.T.; Cao, Q.; Zhu, Y.; Rehmani, M.I.A.; Cao, W.; Tang, L. In-season assessment of rice protein and amylose content using critical nitrogen dilution curve. *Eur. J. Agron.* **2017**, *90*, 139–151. [CrossRef]
26. Greenwood, D.J.; Lemaire, G.; Gosse, G.; Cruz, P.; Draycott, A.; Neeteson, J.J. Decline in percentage n of c3 and c4 crops with increasing plant mass. *Ann. Bot.* **1990**, *66*, 425–436. [CrossRef]
27. Hu, Y.; Yang, J.P.; Lv, Y.M.; He, J.J. SPAD values and nitrogen nutrition index for the evaluation of rice nitrogen status. *Plant Prod. Sci.* **2014**, *17*, 81–92.
28. Noura, Z.; Marianne, B.; Gilles, B.; Annie, C.; Nicolas, T.; Athynan, C.; Michelc, N.; Léonétienne, P. Chlorophyll measurements and nitrogen nutrition index for the evaluation of corn nitrogen status. *Agron. J.* **2010**, *100*, 2275–2279.

29. Zhao, B.; Liu, Z.; Ata-Ul-Karim, S.T.; Xiao, J.; Liu, Z.; Qi, A.; Ning, D.; Nan, J.; Duan, A. Rapid and nondestructive estimation of the nitrogen nutrition index in winter barley using chlorophyll measurements. *Field Crop. Res.* **2016**, *185*, 59–68. [CrossRef]
30. Liu, X.; Zhang, K.; Zhang, Z.; Cao, Q.; Lv, Z.; Yuan, Z.; Tian, Y.; Cao, W.; Zhu, Y. Canopy chlorophyll density based index for estimating nitrogen status and predicting grain yield in rice. *Front. Plant Sci.* **2017**, *8*, 1829. [CrossRef]
31. Liu, X.; Ferguson, R.B.; Zheng, H.; Cao, Q.; Tian, Y.; Cao, W.; Zhu, Y. Using an active-optical sensor to develop an optimal NDVI dynamic model for high-yield rice production (Yangtze, China). *Sensors* **2017**, *17*, 672. [CrossRef]
32. Debaeke, P.; Rouet, P.; Justes, E. Relationship between the normalized spad index and the nitrogen nutrition index: Application to Durum Wheat. *J. Plant Nutr.* **2006**, *29*, 75–92. [CrossRef]
33. Hou, Y.H.; Chen, C.Y.; Guo, Z.Q.; Hou, L.B.; Dong, Z.Q.; Zhao, M. Establishment of dry matter accumulation dymamic simulation model and analysis of growth charateristc for high-yielding population of spring maize. *J. Maize Sci.* **2008**, *16*, 90–95.
34. Islam, M.R.; Haque, K.M.S.; Akter, N.; Karim, M.A. Leaf chlorophyll dynamics in wheat based on SPAD meter reading and its relationship with grain yield. *Sci. Agric.* **2014**, *4*, 13–18.
35. Zheng, H.; Cheng, T.; Yao, X.; Deng, X.; Tian, Y.; Cao, W.; Zhu, Y. Detection of rice phenology through time series analysis of ground-based spectral index data. *Field Crop. Res.* **2016**, *198*, 131–139. [CrossRef]
36. Ata-Ul-Karim, S.T.; Xia, Y.; Liu, X.; Cao, W.; Yan, Z. Development of critical nitrogen dilution curve of Japonica rice in Yangtze River Reaches. *Field Crop. Res.* **2013**, *149*, 149–158. [CrossRef]
37. Ata-Ul-Karim, S.T.; Zhu, Y.; Yao, X.; Cao, W. Determination of critical nitrogen dilution curve based on leaf area index in rice. *Field Crop. Res.* **2014**, *167*, 76–85. [CrossRef]
38. Bijay-Singh; Varinderpal-Singh; Yadvinder-Singh; Thind, H.S.; Kumar, A.; Gupta, R.K.; Kaul, A.; Vashistha, M. Fixed-time adjustable dose site-specific fertilizer nitrogen management in transplanted irrigated rice (*Oryza sativa* L.) in South Asia. *Field Crop. Res.* **2012**, *126*, 63–69. [CrossRef]
39. Liu, K.; Yazhen, L.I.; Huiwen, H.U. Estimating the effect of urease inhibitor on rice yield based on NDVI at key growth stages. *Front. Agric. Sci. Eng.* **2014**, *1*, 150–157. [CrossRef]
40. Beck, P.S.A.; Atzberger, C.; Høgda, K.A.; Johansen, B.; Skidmore, A.K. Improved monitoring of vegetation dynamics at very high latitudes: A new method using MODIS NDVI. *Remote Sens. Environ.* **2006**, *100*, 321–334. [CrossRef]
41. Prost, L.; Jeuffroy, M.H. Replacing the nitrogen nutrition index by the chlorophyll meter to assess wheat N status. *Agron. Sustain. Dev.* **2007**, *27*, 321–330. [CrossRef]
42. Dente, L.; Satalino, G.; Mattia, F.; Rinaldi, M. Assimilation of leaf area index derived from ASAR and MERIS data into CERES-Wheat model to map wheat yield. *Remote Sens. Environ.* **2008**, *112*, 1395–1407. [CrossRef]
43. Wang, S.H.; Zhi-Jun, J.I.; Liu, S.H.; Ding, Y.F.; Cao, W.X. Relationships between balance of nitrogen supply-demand and nitrogen translocation and senescence of different position leaves on rice. *J. Integr. Agric.* **2003**, *2*, 747–751.
44. Escobar-Gutiérrez, A.J.; Combe, L. Senescence in field-grown maize: From flowering to harvest. *Field Crop. Res.* **2012**, *134*, 47–58. [CrossRef]
45. Xue, L.; Yang, L. Recommendations for nitrogen fertilizer topdressing rates in rice using canopy reflectance spectra. *Biosyst. Eng.* **2008**, *100*, 524–534. [CrossRef]

5

UAV-Borne LiDAR Crop Point Cloud Enhancement using Grasshopper Optimization and Point Cloud Up-Sampling Network

Jian Chen [1], Zichao Zhang [1], Kai Zhang [1], Shubo Wang [1] and Yu Han [2,*]

1 College of Engineering, China Agricultural University, Beijing 100083, China; jchen@cau.edu.cn (J.C.); zhangzc1@cau.edu.cn (Z.Z.); sy20193071234@cau.edu.cn (K.Z.); wangshubo@cau.edu.cn (S.W.)

2 College of Water Resources & Civil Engineering, China Agricultural University, Beijing 100083, China

* Correspondence: yhan@cau.edu.cn

Abstract: Because of low accuracy and density of crop point clouds obtained by the Unmanned Aerial Vehicle (UAV)-borne Light Detection and Ranging (LiDAR) scanning system of UAV, an integrated navigation and positioning optimization method based on the grasshopper optimization algorithm (GOA) and a point cloud density enhancement method were proposed. Firstly, a global positioning system (GPS)/inertial navigation system (INS) integrated navigation and positioning information fusion method based on a Kalman filter was constructed. Then, the GOA was employed to find the optimal solution by iterating the system noise variance matrix Q and measurement noise variance matrix R of Kalman filter. By feeding the optimal solution into the Kalman filter, the error variances of longitude were reduced to 0.00046 from 0.0091, and the error variances of latitude were reduced to 0.00034 from 0.0047. Based on the integrated navigation, an UAV-borne LiDAR scanning system was built for obtaining the crop point. During offline processing, the crop point cloud was filtered and transformed into WGS-84, the density clustering algorithm improved by the particle swarm optimization (PSO) algorithm was employed to the clustering segment. After the clustering segment, the pre-trained Point Cloud Up-Sampling Network (PU-net) was used for density enhancement of point cloud data and to carry out three-dimensional reconstruction. The features of the crop point cloud were kept under the processing of reconstruction model; meanwhile, the density of the crop point cloud was quadrupled.

Keywords: UAV-borne LiDAR scanning system; grasshopper optimization algorithm; GPS/INS integrated navigation; point cloud up-sampling network (PU-net); clustering segmentation; 3-dimensional reconstruction

1. Introduction

Light Detection and Ranging (LiDAR) is an active sensing technology that can quickly acquire spatial information of a target or environment [1,2]. Compared with traditional remote sensing methods, LiDAR has shown great advantages in accuracy and stability [3]. In agriculture, the LiDAR system is one of the most accurate methods to measure regional vegetation structural characteristics and biophysical parameters [4]. The Unmanned Aerial Vehicle (UAV)-borne LiDAR scanning system is one of the most common platforms for crop point cloud obtaining. The crop point clouds captured by UAV-borne LiDAR scanning system are sparse and unordered [5]. The navigation and positioning accuracy of the UAV-borne LiDAR scanning system influences the accuracy of the point cloud data greatly. Density is also an important indicator to measure the quality of point cloud data. Higher point cloud density represents richer information of the target or environment [6]. Ensuring the accuracy and density characteristics of the UAV-borne LiDAR point cloud data while maintaining the cost is crucial for the continuous and efficient execution of LiDAR detection tasks.

For integrated navigation, the most widely used in the world is the global positioning system (GPS)/inertial navigation system (INS) integrated navigation positioning system [7]. The Kalman filter is the most common method of data fusion [8]. Gao et al. [9] developed an INS/GPS/LiDAR integrated navigation system through an extended Kalman filter (EKF) using a hybrid scan matching algorithm. Shabani et al. [10] explored the characteristics, advantages, and disadvantages of direct Kalman filtering and indirect Kalman filtering commonly used in GPS/INS integrated navigation, and a novel asynchronous direct Kalman filter was proposed. Zhang et al. [11] developed a low-cost global navigation satellite system (GNSS) and INS integrated navigation method and an adaptive Kalman filter for overcoming the measurement noise of GNSS, which was proposed based on a supervised machine learning model. Yu et al. [12] used a single-frequency real-time kinematic carrier phase (RTK) technology to deduce and obtain information including speed and orientation under GPS/BeiDou Navigation Satellite System (BDS) for polishing the inaccurate INS information. Norouz et al. [13] proposed an improved Kalman filter algorithm based on small probability distribution of the state variable between the measurement updates of the multi-rate GPS/INS-coupled navigation filter, which greatly reduced the amount of calculation, and the algorithm performed well by numerically simulating. All of the above integrated navigation systems suffered from the low accuracy of integrated navigation, except Yu et al. [12]. However, RTK-based high accuracy integrated navigation systems cannot be widely popularized due to the price gap. During the experiments of data fusion using the Kalman filter, the longitude and latitude errors were sharply affected by the parameter of the Kalman filter, especially system noise sequence Q and measurement noise sequence R, and the same situation also occurred in the realization of Ref. [10,11]. For promoting the accuracy of our integrated navigation-based scanning system, the meta-heuristic method was employed for finding the optimum parameters of integrated navigation in our UAV scanning system. For multi-solution problems, genetic algorithms (GA) [14], particle swarm optimization (PSO) [15], and ant colony optimization (ACO) [16] were the most popular, with swarm intelligence optimization proposed recently such as the bat algorithm (BA) [17] and grasshopper optimization algorithm (GOA) [18]. All of the multi-solution meta-heuristic methods have their own merits for different problems [19]. In order to reduce longitude and latitude errors of integrated navigation, the meta-heuristic methods were employed to find the optimum system noise sequence Q and measurement noise sequence R.

After obtaining the crop point cloud data by UAV scanning system, the raw point cloud data consisted of crop points and other points, and manual classification was extremely inefficient for different point cloud data. Therefore, the clustering segmentation method was employed to extract crop data from raw point cloud data. The density-based clustering method, density-based spatial clustering of applications with noise (DBSCAN), has attracted a lot of attention; besides the improvement of the algorithm itself, finding out the key parameters of clustering was also a new direction of optimization [20]. The same as integrated navigation, the meta-heuristic method was employed to optimize the clustering segmentation method. According to the experimental results of the density distribution characteristics of the plane point cloud under different conditions, Kedzierski et al. [21] found that the point cloud density could be preliminarily estimated based on distance and angle, but the actual point cloud density should take into account the influence of object shape and environment. Rupink et al. [22] proposed the formula and calculation steps for calculating the point cloud density. Huang et al. [23] proposed an edge-sensing point set resampling method, gradually approaching the edges and corners. However, the quality of the results largely depends on the accuracy of the normal of a given point and the fine-tuning of parameters. Based on Ref. [21,23], dense and accurate point clouds are very useful for crop point cloud recognition and 3D reconstruction. The high-cost multi-line LiDAR can be simulated by low-cost single-line LiDAR point clouds through the point cloud up-sampling method without destroying the features.

In view of the current problems and inspired by existing methods, in order to improve the quality of point cloud data, a combined navigation optimization algorithm based on GOA and a point cloud density enhancement algorithm based on Point Cloud Up-Sampling Network (PU-net) were proposed. The novel contribution of this article was as follows:

(1) In order to obtain the optimal solution of the initial parameters of the Kalman filter quickly and accurately, GOA took the sum of the longitude and latitude error variances as the fitness function to find the optimal solution to iterate the system noise variance matrix Q and the measurement noise variance matrix R. Based on the Q and R matrices from GOA, the positioning error of the UAV-borne LiDAR scanning system was greatly reduced; meanwhile, the computing burden was decreased, and the accuracy of point cloud data acquisition was improved;
(2) The density clustering algorithm was improved by the linear decreasing particle swarm optimization (PSO) algorithm. In order to solve the problem of sparse density, in this paper, a pre-trained PU-net was used for enhancing the point cloud density of the point cloud data obtained by UAV-borne LiDAR. Experiments have verified that the enhanced point cloud data retain the features of the raw point cloud data.

2. Framework of the Proposed Method

For UAV-borne LIDAR point cloud data enhancement, the integrated navigation algorithm based on GOA and the point cloud density enhancement algorithm based on neural network were shown in Figure 1. The specific implementation steps were as follows:

(1) Firstly, the GPS/INS integrated navigation information fusion framework was built by the Kalman filter. Then, the GOA was used to optimize the initial parameters of the Kalman filter, and the optimal solution was returned to the Kalman filter to estimate the output state of the GPS/INS integrated navigation system;
(2) After data acquisition, the point cloud data was transformed into the WGS-84 coordinate system from the LiDAR coordinate system. Then the point cloud data was filtered for removing the irrelevant points. Finally, the two-dimensional point cloud data was transformed into a three-dimensional point cloud according to the flight trajectory of UAV;
(3) The PSO algorithm was employed to optimize the density clustering algorithm for crop point cloud segment. Then the crop point cloud data was fed into the trained network to obtain the crop point cloud data after the point cloud density was enhanced. Finally, the point cloud data was reconstructed in three dimensions.

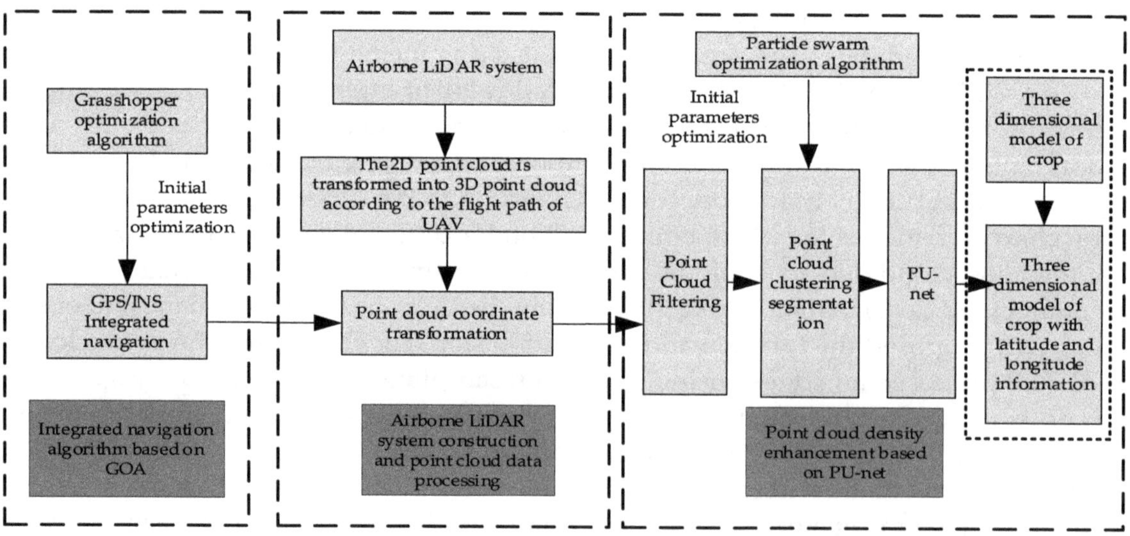

Figure 1. Framework of the proposed method. There are 3 steps: Integrated navigation algorithm based on grasshopper optimization algorithm (GOA), Unmanned Aerial Vehicle (UAV)-borne Light

Detection and Ranging (LiDAR) system construction and point cloud data processing, and point cloud density enhancement based on the Point Cloud Up-Sampling Network of the whole work of this paper.

3. Integrated Navigation Enhancement Based on GOA

In GPS/INS integrated navigation, the Kalman filter was usually used for fusing information of GPS and INS. The better the filter of integrated navigation was, the better it can integrate the advantages of each sensor. Since the GOA is not affected by the nonlinearity or scale, compared with other global optimization algorithms, it can find more effective optimal solutions with faster convergence speed [24].

3.1. Framework of GPS/INS Integrated Navigation

GPS and INS have complementary advantages. By combining GPS and INS in an appropriate way, their own drawbacks could be overcome, and the navigation accuracy would significantly improve. In order to improve the robustness of the system and reduce the complexity of the system, this paper adopted the loose combination mode to fuse the information of GPS and INS, as shown in Figure 2.

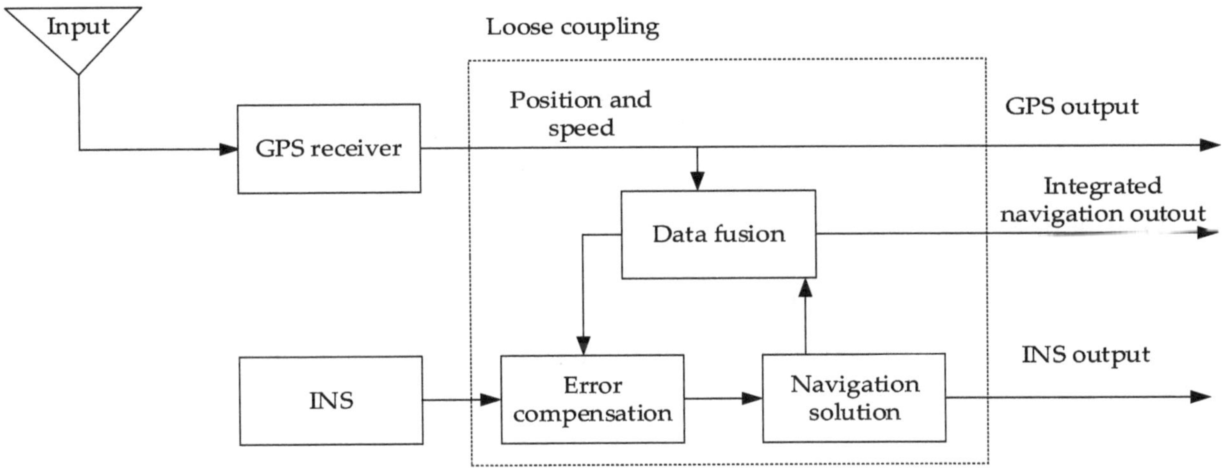

Figure 2. Framework of global positioning system (GPS)/ inertial navigation system (INS) integrated navigation.

The state variables of the integrated navigation filter were composed of the random drift error, velocity error, and position error of INS gyroscope, as well as errors caused by the attitude angle and accelerometer. The equation of the state was as follows:

$$\hat{X} = FX + Gw \tag{1}$$

The state variables of the INS integrated navigation system include longitude error, latitude error, velocity error in east and north directions, attitude angle error in east, north, and sky directions, random drift, and accelerometer bias error. The 13-dimensional equation of the state was as follows:

$$\hat{X}_I = F_I X_I + G_I \omega_I \tag{2}$$

The clock error of the receiver equivalent distance δt_u and equivalent range error of the receiver clock frequency error δt_{ru} were the GPS state errors; the equation of the state can be described as follows:

$$\hat{X}_G = F_G X_G + G_G w_G \tag{3}$$

The measurement equation of the position velocity combination was expressed as follows:

$$\begin{bmatrix} L_{INS} \\ \lambda_{INS} \\ h_{INS} \end{bmatrix} = \begin{bmatrix} L_t + \delta L \\ \lambda_t + \delta\lambda \\ h_t + \delta h \end{bmatrix} \quad (4)$$

where L_{INS}, λ_{INS}, and h_{INS} were the longitude, latitude, and altitude calculated by inertial navigation. L_t, λ_t, and h_t were the real longitude, latitude, and height of the carrier. δL, $\delta\lambda$, and δh were the longitude and latitude error and altitude error calculated by INS.

The position measurement information of the GPS receiver was expressed by subtracting the corresponding error from the real value:

$$\begin{bmatrix} L_{GPS} \\ \lambda_{GPS} \\ h_{GPS} \end{bmatrix} = \begin{bmatrix} L_t - \frac{N_n}{R} \\ \lambda_t - \frac{N_e}{R\cos L} \\ h_t - N_u \end{bmatrix} \quad (5)$$

where N_n, N_e, and N_u were the position error of the GNSS receiver along the east, north, and sky directions. Then the external measurement position error can be defined as:

$$Z_{pos} = \begin{bmatrix} \lambda_{INS} - \lambda_{GPS} \\ L_{INS} - L_{GPS} \\ h_{INS} - h_{GPS} \end{bmatrix} \quad (6)$$

The position measurement equation was as follows:

$$Z_{POS} = H_{POS}(t)X(t) + V_{POS}(t) \quad (7)$$

The velocity measurement information equation of INS can be expressed as follows:

$$\begin{bmatrix} v_{eINS} \\ v_{nINS} \\ v_{uINS} \end{bmatrix} = \begin{bmatrix} v_e + \delta v_e \\ v_n + \delta v_n \\ v_u + \delta v_u \end{bmatrix} \quad (8)$$

where v_{eINS}, v_{nINS}, and v_{uINS} were the east speed, north speed, and sky speed calculated by the inertial navigation system. v_e, v_n, and v_u were the velocity of the object in the east, north, and sky directions.

The velocity information provided by GPS can be expressed by the difference between the real value and the corresponding error in the geographical coordinate system:

$$\begin{bmatrix} v_{eGPS} \\ v_{nGPS} \\ v_{uGPS} \end{bmatrix} = \begin{bmatrix} v_e - M_e \\ v_n - M_n \\ v_u - M_u \end{bmatrix} \quad (9)$$

The speed error of the external measurement can be defined as:

$$Z_{vel} = \begin{bmatrix} v_{eINS} - v_{eGPS} \\ v_{nINS} - v_{nGPS} \\ v_{uINS} - v_{uGPS} \end{bmatrix} \quad (10)$$

The velocity measurement equation was as follows:

$$Z_{vel} = H_{vel}(t)X(t) + V_{vel}(t) \quad (11)$$

The integrated measurement equation of the GPS/INS integrated navigation system was obtained as follows:

$$Z(t) = \begin{bmatrix} \lambda_{INS} - \lambda_{GPS} \\ L_{INS} - L_{INS} \\ h_{INS} - h_{INS} \\ v_{eINS} - v_{eGPS} \\ v_{nINS} - v_{nGPS} \\ v_{uINS} - v_{uINS} \end{bmatrix} = H(t)X(t) + V(t) \tag{12}$$

3.2. Grasshopper Optimization Algorithm (GOA)

GOA is a swarm intelligence optimization algorithm designed by Australian scholar Shahrzad [18] in 2017, simulating the predatory migration behaviors of the grasshopper population in nature. The main characteristics of larval stage grasshoppers were slow movement and small stride. In contrast, long distances and sudden movements were essential features of adult grasshoppers. Searching for food sources was another important feature of the grasshopper swarm. The mathematical model of grasshopper swarm behavior was as follows:

$$X_i = \sum_{\substack{j=1 \\ j \neq i}}^{N} s(|x_j - x_i|)\frac{x_j - x_i}{d_{ij}} - g\hat{e}_g + u\hat{e}_w \tag{13}$$

where X_i represented the position of i_{th} grasshopper. N represents the amount of the grasshopper swarm. s was interaction force between grasshoppers. d_{ij} was the distance between the i_{th} grasshopper and j_{th} grasshopper. The function of the interaction force (attraction and repulsion shown in Figure 3.) s between grasshoppers was defined as follows:

$$s(d) = fe^{\frac{-d}{l}} - e^{-d} \tag{14}$$

where d was the distance between two grasshoppers. f represented the interaction force (attraction and repulsion), l represented the unit of interaction force length. The modified form of Equation (10) was as follows:

$$X_i^{dim} = c * r\left(\sum_{\substack{j=1 \\ j \neq i}}^{N} c\frac{ub_{dim} - lb_{dim}}{2}A\frac{x_j - x_i}{d_{ij}}\right) + \hat{T}_{dim} \tag{15}$$

$$A = s\left(\left|x_j^d - x_i^d\right|\right) \tag{16}$$

where ub_{dim} was the upper boundary of the dim_{th} dimension, lb_{dim} was the lower boundary, and $\hat{T}_{dim}$ was the target value of the dim_{th} dimension. c represented a self-adaption ratio which is defined as follows:

$$c = c_{max} - l\frac{c_{max} - c_{min}}{L} \tag{17}$$

where c_{max} was the max value, while c_{min} was the minimum value; l represented the current iteration, while L is the max iteration.

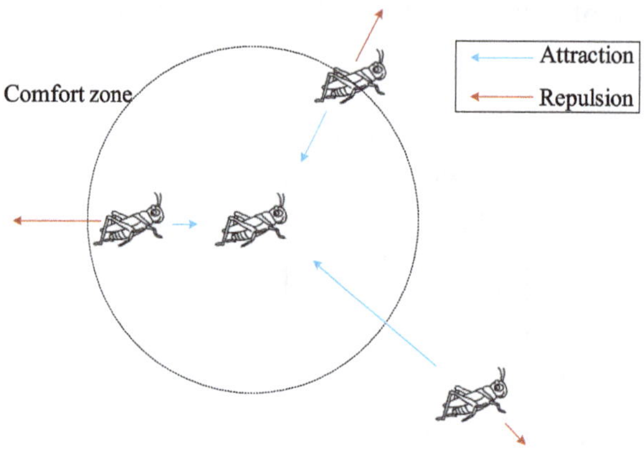

Figure 3. Interaction among grasshoppers.

3.3. *Kalman Filter Optimization Using GOA*

GOA was employed to optimize the initial parameters Q and R of the Kalman filter. Q was the variance matrix of the system noise sequence, which affected the filtering performance and parameter estimation accuracy of the Kalman filter algorithm; R was the variance matrix of the measurement noise sequence, which was related to the correction speed of filtering and the stability of the filtering process. Through the optimization of the initial parameters, the optimal position of the grasshopper was as close to the real value as possible, thus the accuracy of the Kalman filter algorithm was improved greatly. The sum of the variances of longitude and latitude errors was selected as the fitness function. The pipelines of the Kalman filter optimization was shown in Figure 4.

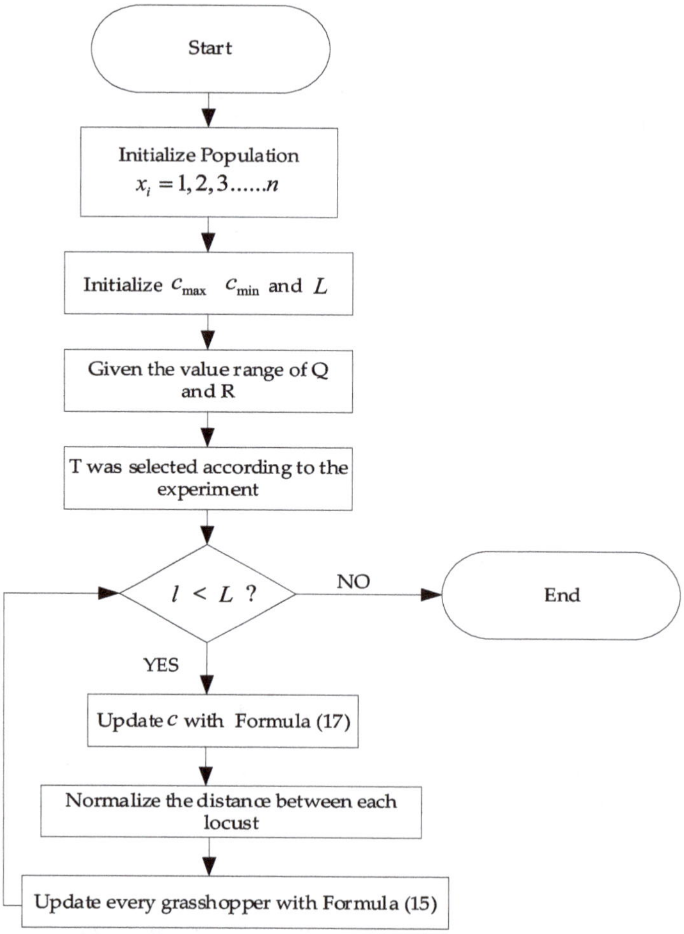

Figure 4. The pipelines of Kalman filter optimization.

4. UAV-Borne LiDAR System Construction and Point Cloud Data Processing

The UAV-borne LiDAR system was mainly composed of the LiDAR sensor, GPS/INS integrated navigation module, and UAV carrier. After the point cloud data were collected by the UAV-borne LiDAR scanning system, it was necessary to transform the point cloud data from the LiDAR coordinate system to the WGS-84 coordinate system.

4.1. UAV-Borne LiDAR System

In view of the application of LiDAR in the agricultural field, this paper selected RPLiDAR-a2 of SLAMTEC®, which was a single-line two-dimensional LiDAR, which greatly reduced the cost of equipment. The WitMotion® GPS/ Inertial Measurement Unit (IMU) was selected to collect the navigation and positioning information of UAV. In order to supply electricity to the LiDAR and the IMU, a small pad was used to combine the system. DJI® M100 UAV was selected as the carrier, which was a lightweight UAV. Finally, the UAV-borne LiDAR scanning system was completed, as shown in Figures 5 and 6.

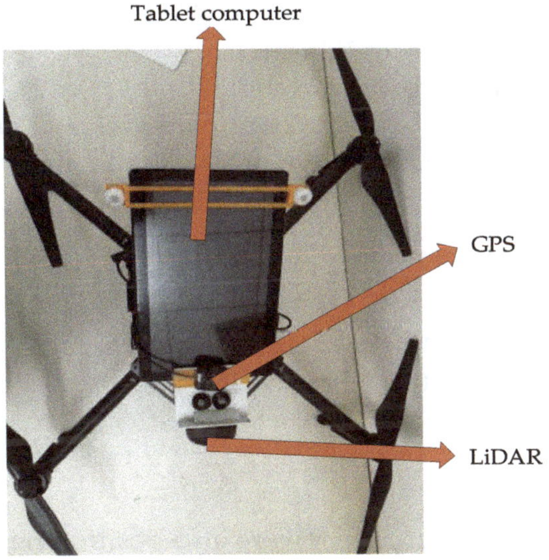

Figure 5. Top view of UAV-borne LiDAR scanning system.

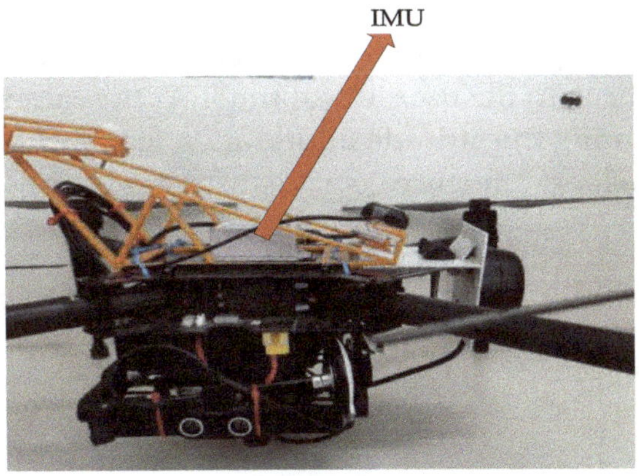

Figure 6. Side view of UAV-borne LiDAR system.

Due to the single scanning line of RPLiDAR-a2, the collected point cloud data was a two-dimensional point cloud, which only contained the data of Y and Z coordinates. In order to generate three-dimensional point cloud data, according to the light trajectory and speed of UAV, the data of the X-axis were inversely solved according to IMU.

4.2. Coordinate Transformation of Point Cloud Data

It was assumed that the coordinate of the laser foot point in the instantaneous laser coordinate system was $(x_{SL}, y_{SL}, z_{SL})^T$:

$$\begin{bmatrix} x_{SL} \\ y_{SL} \\ z_{SL} \end{bmatrix} = \begin{bmatrix} 0 \\ 0 \\ \rho \end{bmatrix} \quad (18)$$

There was an angle θ between the instantaneous laser coordinate system and the laser scanning reference coordinate system. The coordinates of the raw point cloud data in the laser scanning reference coordinate system were $(x_L, y_L, z_L)^T$:

$$\begin{bmatrix} x_L \\ y_L \\ z_L \end{bmatrix} = R_L \begin{bmatrix} x_{SL} \\ y_{SL} \\ z_{SL} \end{bmatrix} \quad (19)$$

where R_L represented the rotation matrix of scan angle θ.

Generally, the error between the different coordinate axes was analyzed and unified to be the placement error angle α, β, and γ. The deviation of the coordinate origin and reference center was treated as error $t_L = (\Delta x_I^L, \Delta y_I^L, \Delta z_I^L)^T$. The coordinates of the raw point cloud in the inertial platform reference coordinate system were $(x_I, y_I, z_I)^T$:

$$\begin{bmatrix} x_I \\ y_I \\ z_I \end{bmatrix} = R_M \begin{bmatrix} x_L \\ y_L \\ z_L \end{bmatrix} + \begin{bmatrix} \Delta x_I^L \\ \Delta y_I^L \\ \Delta z_I^L \end{bmatrix} \quad (20)$$

where $R_M = R(\gamma) \times R(\beta) \times R(\alpha)$.

In the integrated navigation system, there were also positioning errors between GPS and INS. During the installation process, there was a certain deviation between the center of GPS and the reference center of inertial components. The deviation was defined as $t_G = (\Delta x_I^G, \Delta y_I^G, \Delta z_I^G)^T$; it can be measured offline. INS measured the attitude information of the carrier in real time, including roll angle (R), pitch angle (P), and heading angle (H), and got the velocity and position information of the carrier through integration and feed back to the user in real time. At the same time, the coordinate rotation matrix was obtained by calculating the attitude information feedback from IMU. The coordinates of the raw point cloud data in the local reference coordinate system were $(x_{LH}, y_{LH}, z_{LH})^T$:

$$\begin{bmatrix} x_{LH} \\ y_{LH} \\ z_{LH} \end{bmatrix} = R_N \begin{bmatrix} x_I \\ y_I \\ z_I \end{bmatrix} - \begin{bmatrix} \Delta x_I^G \\ \Delta y_I^G \\ \Delta z_I^G \end{bmatrix} \quad (21)$$

where $R_N = R(H) \times R(P) \times R(R)$.

Finally, point cloud data should be transformed to WGS-84 coordinate system. The rotation matrix R_w was a function of the latitude and longitude of the region. The point cloud could be transformed into the WGS-84 coordinate system by rotating the matrix:

$$\begin{bmatrix} x_{84} \\ y_{84} \\ z_{84} \end{bmatrix} = R_W \begin{bmatrix} x_{LH} \\ y_{LH} \\ z_{LH} \end{bmatrix} + \begin{bmatrix} x_{84} \\ y_{84} \\ z_{84} \end{bmatrix}_{\text{Antenna phase center}} \quad (22)$$

Based on the above analysis, the coordinates of the raw point cloud in WGS-84 coordinate system were as follows:

$$\begin{bmatrix} x_{84} \\ y_{84} \\ z_{84} \end{bmatrix} = R_W R_N \left[R_M R_L \begin{bmatrix} 0 \\ 0 \\ \rho \end{bmatrix} + \begin{bmatrix} \Delta x_I^L \\ \Delta y_I^L \\ \Delta z_I^L \end{bmatrix} - \begin{bmatrix} \Delta x_I^G \\ \Delta y_I^G \\ \Delta z_I^G \end{bmatrix} \right] + \begin{bmatrix} x_{84} \\ y_{84} \\ z_{84} \end{bmatrix}_{\text{Antenna phase center}} \tag{23}$$

In addition, in order to get the data containing the latitude and longitude information, it was necessary to convert the three-dimensional rectangular coordinates (XYZ) into geodetic coordinates (λLH). The conversion formula was as follows:

$$L = \arctan(Y/X) \tag{24}$$

$$\lambda = \arctan\left[Z/\sqrt{X^2+Y^2} * \left(1 - e^2N(N+H)\right)^{-1}\right] \tag{25}$$

$$H = \sqrt{X^2+Y^2}/\cos\lambda - N \tag{26}$$

where X, Y, and Z were three-dimensional rectangular coordinate components, L, λ, and H were earth longitude, latitude, and height. e represented the first eccentricity of the ellipsoid. $N = a/\sqrt{1 - e^2\sin^2\lambda}$ was the radius of curvature in the prime vertical, and the a was the semimajor axis of the ellipsoid.

5. Point Cloud Density Enhancement Based on PU-Net

Point cloud density enhancement aims to increase the number of point clouds without destroying the characteristics of the original point cloud. By enhancing the density of point cloud, the richness and accuracy of 3D reconstruction of point cloud data can be improved without increasing equipment cost.

5.1. Point Cloud Preprocessing

After the coordinate transformation, it was necessary to remove irrelevant noise points from the original point cloud data. In this paper, the statistical filtering method was selected for filtering. Statistical filtering calculated the distance between each point and its surrounding neighborhood points: According to neighborhood points and the standard deviation threshold, the points which did not meet the requirements were identified as outlier points.

5.2. Point Cloud Density Clustering Segmentation Method Based on PSO Algorithm

After filtering in the previous section, the point cloud data obtained included ground and crop points. In order to separate the ground and crop, the density clustering algorithm was used for separating, and the linear decreasing PSO algorithm was employed to improve the density clustering algorithm.

DBSCAN was a density-based clustering algorithm proposed by ester et al. It was designed to achieve clustering and segmentation of arbitrary data in spatial and non-spatial high-dimensional databases. By calculating the distance between each point and its neighborhood, the points with similar density were classified into the same class.

ε-Neighborhood of arbitrary point p is defined as follows:

$$N_{eps} = \{q \in D/\text{dist}(p,q) < Eps\} \tag{27}$$

where D represented the target's database, point p was the core point if the ε-neighborhood contained at least the minimum number of points. The core point was defined as follows:

$$N_{eps}(P) > MinPts \tag{28}$$

where Eps and MinPts were the neighborhood radius and minimum number of points in the ε-neighborhood of the core point. Through the calculation of all the points, the density of the points meeting the set conditions was similar, and they were classified as the same kind of points. All points were calculated until the termination condition was met. Because the selection of these two initial parameters mainly depended on human experience and continuous experiments, in order to improve the efficiency, this paper employed the linear decreasing PSO algorithm to iteratively optimize the two initial parameters to determine the optimal solution of the initial parameters.

The PSO algorithm was a swarm intelligent bionic optimization algorithm: Every point in PSO was stamped by a position P_i, the optimal value of swarm was P_g, every point's velocity was V_i, and each particle's velocity and position update formula was shown as follows:

$$V_i^{t+1} = \omega V_i^t + c_1 r_1 (P_i - X_i^t) + c_2 r_2 (P_g - X_i^t) \quad (29)$$

$$X_i^{t+1} = X_i^t + V_i^{t+1} \quad (30)$$

where t represented the iterations, V_i^{t+1} was the velocity of the $t+1$ iteration, ω represented the inertia weight. c_1, $c_2 \in [0,1]$ were the generating functions of a random number, and X_i^{t+1} was the $t+1$ iteration position.

In the initial stage of the algorithm, the inertia weight ω was given a larger value to enhance the global exploration ability of the algorithm. With the continuous exploration process, the value of ω could be linearly reduced to enhance the local search and optimization ability of the algorithm, so that particles could gradually approach the optimal solution in the neighborhood of the optimal value. The value of ω was expressed as follows:

$$\omega = \omega_{min} + \frac{(\omega_{max} - \omega_{min}) \times (max_{iter} - t)}{max_{iter}} \quad (31)$$

where ω_{max} represented the max inertia coefficient, while ω_{min} represented the minimum. max_{iter} was the max iteration.

PSO optimization pipelines are shown in Figure 7:

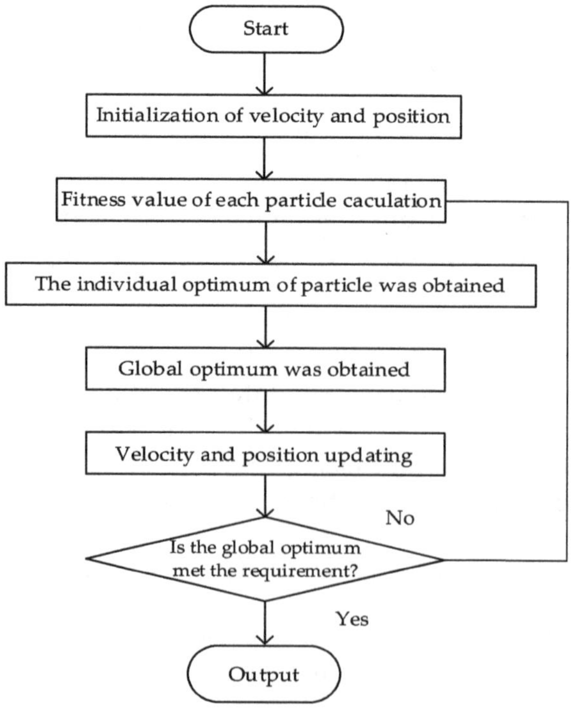

Figure 7. Flow chart of particle swarm optimization (PSO) algorithm.

5.3. Point Cloud Density

The point cloud data acquired by UAV-borne LiDAR was 3-dimensional, which was usually expressed by plane density. The point cloud density in national standards was defined as the normal direction of elevation density [22]:

$$\rho = \frac{N - \sum_{i=0}^{k} N_i}{A - \sum_{i=0}^{k} A_i} \tag{32}$$

where ρ represented the point cloud density, and N represented the total amount of points. Furthermore, N_i was the i_{th} field number of points, and k was the number of scanning area. A was the measure of the scanning area.

The point cloud density calculated by the above formula was the average point cloud density in the scanning area. In order to further quantify the point density distribution in the scanning area, the reciprocal of the area affected by a single point was taken as the density of the point:

$$d' = 1/A' \tag{33}$$

The main factors affecting the density of the point cloud were the LiDAR design parameters, system characteristics, carrier, and scanning area.

The distance between points directly reflected the density of point clouds. For the pulse laser ranging sensor, the distance from the heading point to the edge point determined the density distribution of point cloud. The design of aerial photography parameters was referred from the distance between heading point and edge point. The parameters of the sensor, such as scanning frequency, scanning bandwidth, laser emission frequency, field of view angle, flight height and flight speed, and the distance between point clouds, were considered to design the LiDAR system.

5.4. Density Enhancement Using Point Cloud Up-Sampling Network (PU-Net)

Point cloud up-sampling was essentially similar to the problem of image super-resolution [25]. Yu et al. [26] proposed a data-driven point cloud sampling neural network (PU-net), which was applied to patch level. At the same time, a joint loss function was used for guiding the up-sampled points to retain a uniformly distributed object surface. The main idea of PU-net was capturing the features of each point in every level, then multi-convolution blocks were extended in feature hidden layers, and finally, the hidden layers were divided into multi-level features and up-sampled into the point cloud. The pre-trained PU-net was employed for enhancing the density of the crop point cloud in this paper. There were 4 steps of PU-net for up-sampling of point cloud: Slice extraction, point feature embedding, feature extending, and point coordinate reconstruction. It is noteworthy that a series of sub-pixel convolution layers were employed to extend the features of neighborhood points in the feature extending stage:

$$f' = \mathrm{Reshape}([\mathrm{conv}_1^2(\mathrm{conv}_1^1(f)), \ldots, \mathrm{conv}_r^2(\mathrm{conv}_r^1(f))]) \tag{34}$$

where $\mathrm{conv}_i^1(*)$ and $\mathrm{conv}_i^2(*)$ were the convolution layer with 1*1 kernel size for adjusting the channels of patch feature f. The output, combined feature f′, was born with properties of spatial characteristic which could retain the information of raw data. After coordinate reconstruction, the density elevation in last section was employed to show the enhancement effect directly.

6. Results

In order to verify the stability of the UAV-borne LiDAR scanning system, the experiments were carried out in integrated navigation testing, point cloud density clustering, and point cloud density enhancement.

6.1. GOA-Based Integrated Navigation

The integrated navigation experiment was carried out in MATLAB2017 (a). It was supposed that the initial longitude of UAV $L = 40^{\circ}$, the initial latitude $\lambda = 116^{\circ}$, and the velocity of UAV was 1m/s. The filtering period was T = 1s. The initial value of filtering covariance was as follows:

$$p(0) = \mathrm{diag}[(0.00254^{\circ})^2, (0.00446^{\circ})^2, (0.1\mathrm{m/s})^2, (0.1\mathrm{m/s})^2, (1^{\circ})^2, (1^{\circ})^2, (1^{\circ})^2, (1^{\circ})^2, (1^{\circ})^2, (1^{\circ})^2, (10^{-5})\mathrm{g}, (10^{-5})\mathrm{g}, (10^{-5})\mathrm{g}] \quad (35)$$

The initial value of integrated navigation was:

$$X(0) = [0\mathrm{m}, 0\mathrm{m}, 0.5\mathrm{m/s}, 0.5\mathrm{m/s}, 0.08^{\circ}, 0.08^{\circ}, 0.08^{\circ}, 0.16^{\circ}, 0.1^{\circ}/\mathrm{h}, 0.1^{\circ}/\mathrm{h}, 0.1^{\circ}/\mathrm{h}, 0.1^{\circ}/\mathrm{h}, (10^{-5})\mathrm{g}, (10^{-5})\mathrm{g}, (10^{-5})\mathrm{g}] \quad (36)$$

The initial parameters Q and R obtained when the sum of longitude and latitude errors were used as the fitness function feeding into the system. The GOA based integrated navigation error diagram was shown in Figures 8 and 9, PSO based integrated navigation error diagram was shown in Figures 10 and 11.

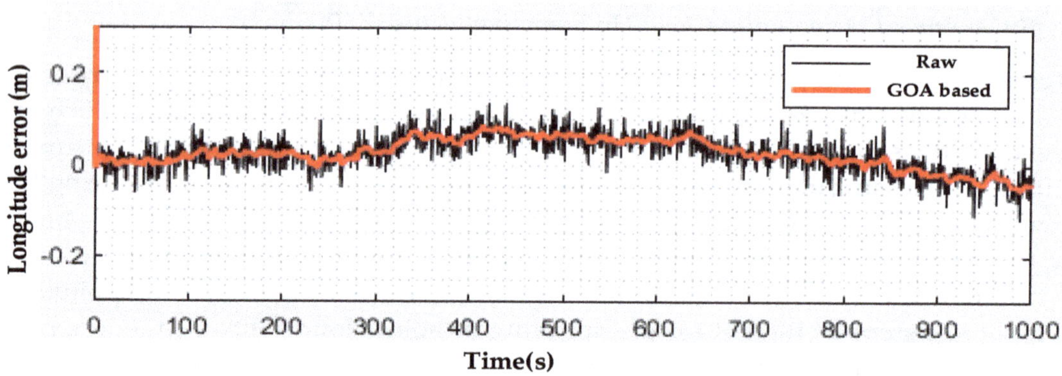

Figure 8. Longitude error of GOA-based integrated navigation.

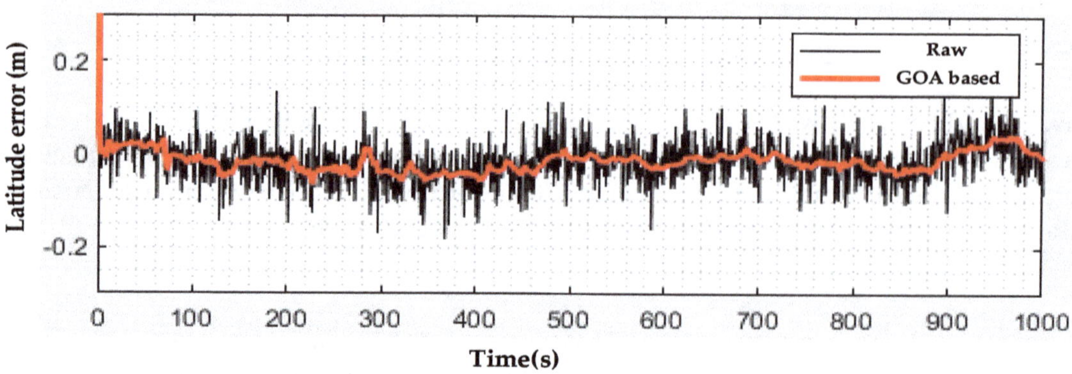

Figure 9. Latitude error of GOA-based integrated navigation.

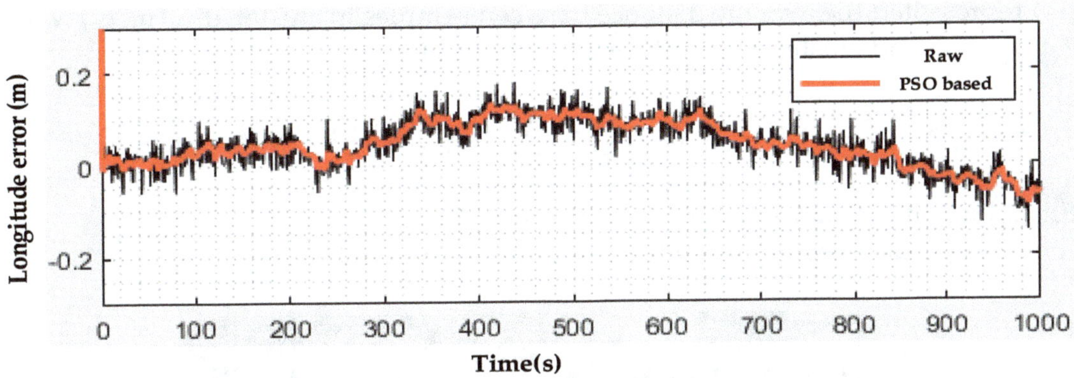

Figure 10. Longitude error of PSO-based integrated navigation.

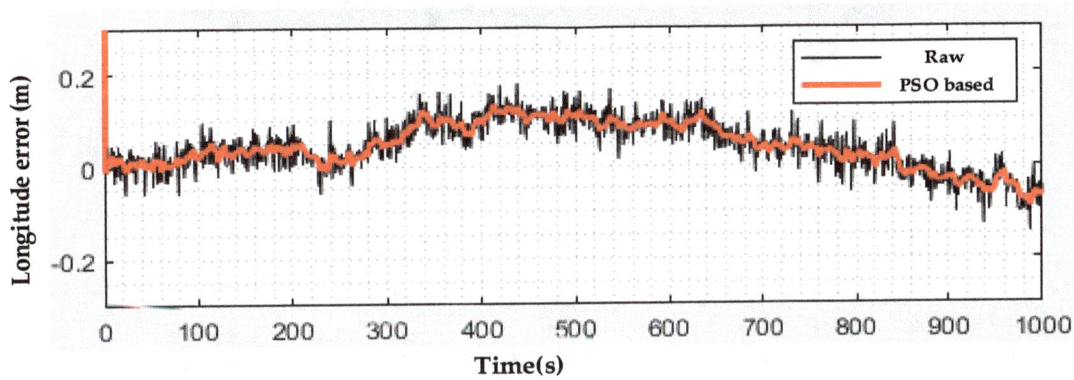

Figure 11. Latitude error of PSO-based integrated navigation.

As shown in Table 1, both the longitude L variance error and latitude λ variance error were reduced greatly compared with raw integrated navigation. It is worth noting that the GOA-based integrated navigation showed better performance than the PSO-based integrated navigation; it is suspected that the ability to find the optimal of GOA was better than for PSO in the integrated navigation task. For more precise optimal Q and R, the GOA-based integrated navigation was employed for the UAV-borne LiDAR scanning system.

Table 1. Error of integrated navigation with different enhancement methods.

	Longitude L Variance Error	Latitude λ Variance Error
GOA-based integrated navigation	0.00046	0.00034
PSO-based integrated navigation	0.00087	0.00079
Raw integrated navigation	0.0091	0.0047

6.2. *PSO-Based Point Cloud Clustering Segmentation*

The meta-heuristic algorithm was employed to optimizing the clustering segmentation key elements: Eps and MinPts. The Davies–Bouldin index (DBI) was employed as the fitness function for the joint optimization of Eps and MinPts:

$$\mathrm{DBI} = \frac{1}{k}\sum_{i=1}^{k} \max_{j \neq i}\left(\frac{\mathrm{avg}(C_i) + \mathrm{avg}(C_j)}{d_{cen}(\mu_i, \mu_j)}\right) \tag{37}$$

where avg(∗) represented the average distance between samples in cluster, $d_{cen}(\mu_i, \mu_j)$ was the center distance of cluster μ_i, and cluster μ_j. avg(*) was defined as follows:

$$avg(C) = \frac{2}{|C|(|C|-1)} \sum_{1 \le i \le j \le |C|} dist(x_i, x_j) \quad (38)$$

where $dist(x_i, x_j)$ was the distance of sample x_i and sample x_j.

Crop point cloud data collection environment was shown in Figure 12.

Figure 12. Crop point cloud raw data collection environment.

A series of crop point cloud raw data was collected from the UAV-borne LiDAR scanning system; the meta-heuristic-based DBSCAN algorithms were initialized by crop raw data and iterated 200 times to find the optimal Eps and MinPts. The fitness iteration curve is shown in Figure 13.

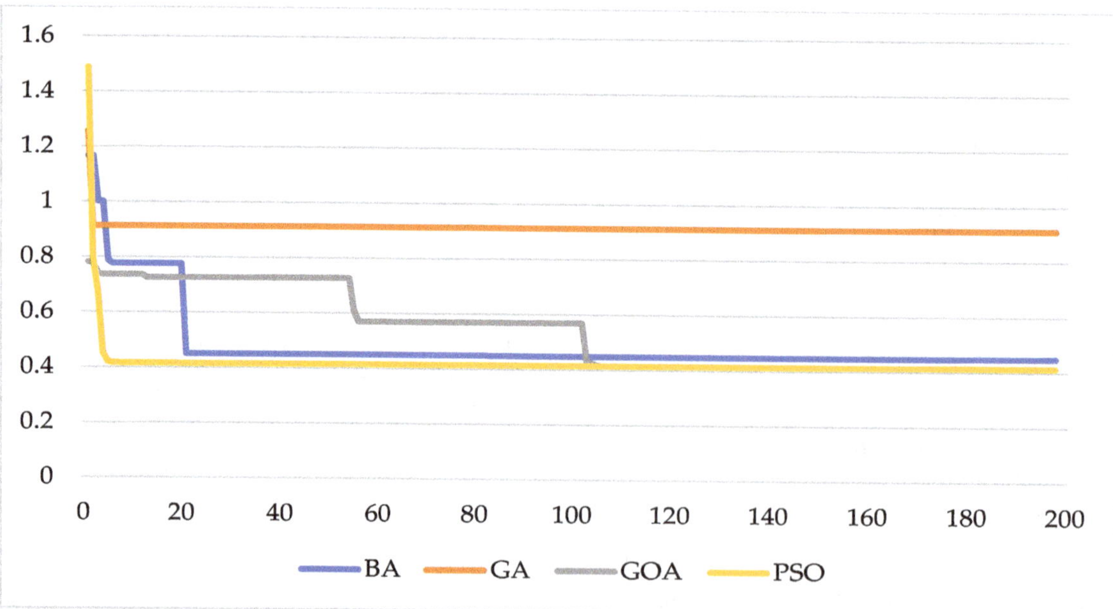

Figure 13. Iteration curve of fitness of the meta-heuristic-based density-based spatial clustering of applications with noise (DBSCAN) algorithms.

For fair comparing with different meta-heuristic algorithms, the BA-based DBSCAN algorithm [17], GA-based DBSCAN algorithm [14], GOA-based DBSCAN algorithm [18], and PSO-based DBSCAN algorithm [15] were initialized with 20 population-sizes, 2 dimensions, and 200 iterations. In particular, the GA-based DBSCAN algorithm was initialized with a 0.7 crossover rate, 0.5 select rate, and 0.01 variation rate. During several times of testing, Figure 13 shows that the GOA and PSO found the best fitness 0.416032. It is suspected that global optimization was 0.416032 under further testing with different fine tuning. Under the same fitness 0.416032, the Eps = 0.8147 and MinPts = 13. In the clustering segmentation, the BA-based DBSCAN algorithm, GA-based DBSCAN algorithm, GOA-based DBSCAN algorithm, PSO-based DBSCAN algorithm, and the K-means clustering were employed for comparison; the K-means was the default tuning. The 200 iterations of time consumption of the BA-based DBSCAN algorithm, GA-based DBSCAN algorithm, GOA-based DBSCAN algorithm, and PSO-based DBSCAN algorithm were 9669 s, 910 s, 1610 s, and 749 s, respectively. The comparison clustering results are shown in Figure 14a–e. Shown in Figure 14f, the crop point cloud is reconstructed in 3-dimensions.

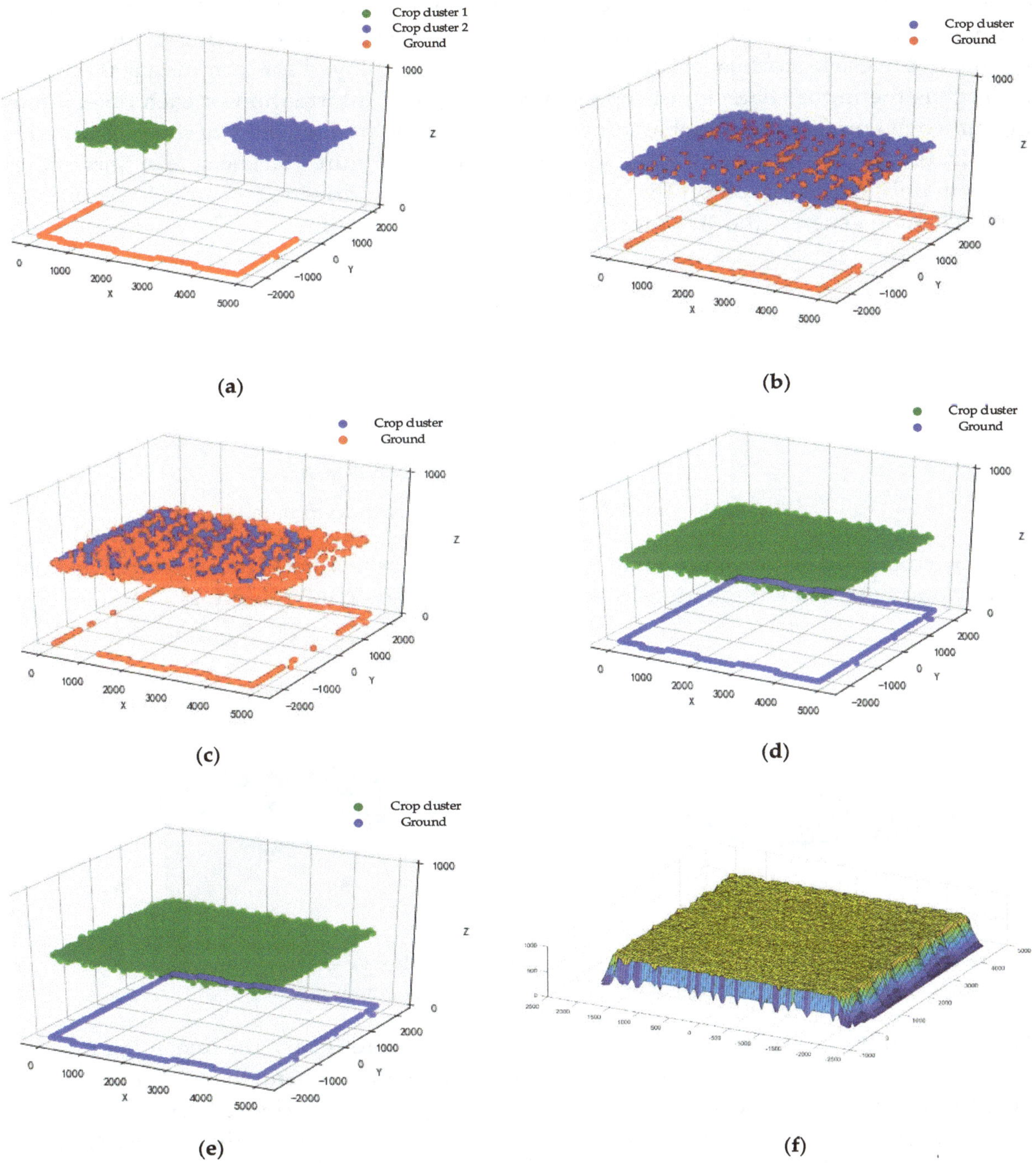

Figure 14. Clustering segmentation result of crop and ground. (**a**) K-means clustering; (**b**) BA-based DBSCAN (Eps = 0.2145, MinPts = 18); (**c**) GA-based DBSCAN (Eps = 0.1006, MinPts = 8); (**d**) PSO-based DBSCAN (Eps = 0.8147, MinPts = 13); (**e**) GOA-based DBSCAN (Eps = 0.8147, MinPts = 13); (**f**) 3D reconstruction of crop field point cloud.

As shown in Figure 14a–c, 3 kinds of clustering results were given from the default tuning of K-means, and the ground cluster could not be separated from the crop cluster using the BA-based DBSCAN algorithm or GA-based DBSCAN algorithm. Under the same fitness 0.416032, the Eps = 0.8147 and MinPts = 13 were obtained using the GOA-based DBSCAN algorithm and PSO-based DBSCAN algorithm. Considering that PSO converges faster and takes less time in iterations, the PSO-based DBSCAN algorithm was employed for crop point cloud clustering segmentation.

6.3. Density Enhancement with PU-Net

After the clustering segmentation of crop point cloud, the crop point cloud was enhanced by a pre-trained PU-net. The PU-net was trained with 60 sets of variation of point clouds from "Visionair". Datasets were divided into train sets and validation sets by 60% and 40%. In the fine-tune training period, the Monte Carlo stochastic sampling method was employed for sampling 5000 points from every object as the input. Each input represented the global information of each object; fine-tune training aimed at enhancing the global information collection performance of the pretrained model. The visualization of raw crop point cloud data was shown in Figure 15. The sparse part of the point cloud was mainly concentrated on the X-axis, which was the direction of the UAV-borne LiDAR scanning system to generate the 3D point cloud by moving forward.

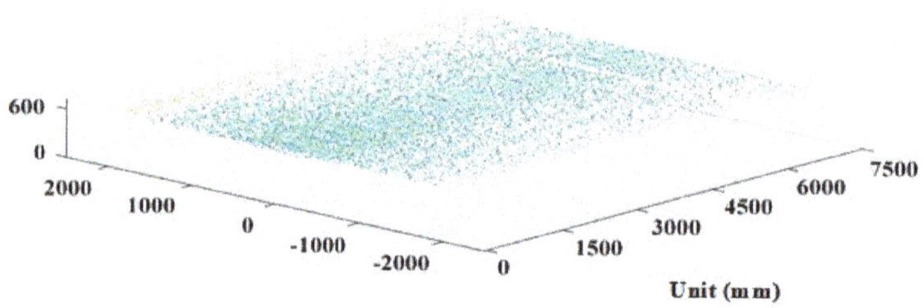

Figure 15. Visualization of raw crop point cloud data.

The raw crop point cloud was fed into the fine-tune trained PU-net. The 3D reconstruction of the crop point cloud with and without density enhancement is shown in Figure 16.

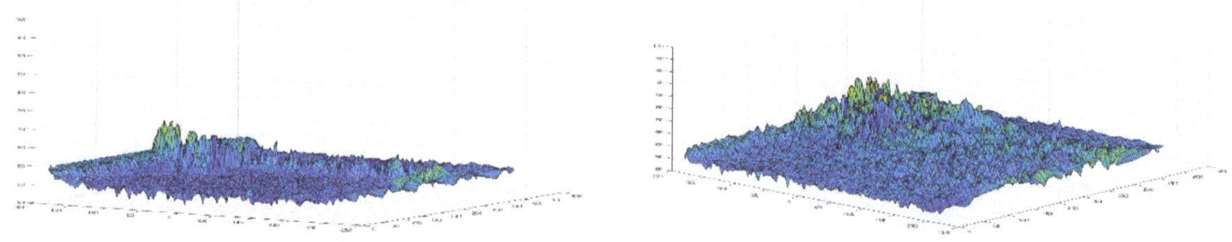

(a) (b)

Figure 16. 3D reconstruction of crop point cloud with and without density enhancement. (**a**) The raw crop point cloud; (**b**) The crop point cloud with density enhancement.

The total number of points of the raw crop point cloud were 10,000, the density of the raw crop point cloud was 1250 points per square meter. After density enhancement, the total number of points of the crop point cloud were 40,000, the density of the crop point cloud was 5000 points per square meter, and the number and density of the point cloud had been significantly increased. Compared with the raw crop point cloud, the enhanced crop point cloud contained more information, and also retained the shape and contour characteristics of the raw data. In order to further verify the reliability and effectiveness of the method, 50 sets of density enhancement tests were carried out and 5 sets of results were sampled randomly for statistical tests. The results were shown in Table 2.

Table 2. The statistical parameters of the enhanced point cloud and the raw data.

	Mean Value	Variance	Standard Deviation
Raw crop point cloud	558.64	171.60	13.10
1st set of the enhanced crop point cloud	568.84	298.48	17.27
2nd set of the enhanced crop point cloud	578.68	289.34	17.01
3rd set of the enhanced crop point cloud	554.32	278.56	16.69
4th set of the enhanced crop point cloud	565.44	325.08	18.03
5th set of the enhanced crop point cloud	563.56	317.51	17.82

7. Discussion

For integrated navigation testing, because the GOA and PSO showed better performance in fitness convergence velocity in the local field refer to Ref. [19]. Although this was an unfavorable feature for global optimization, for time varying and unpredictable problems, such as the integrated navigation system in this paper, better performance in the fitness convergence velocity in the local field tended to find the suboptimal solution quickly, which would reduce the error sharply. Shown as the GOA based integrated navigation error diagram (Figures 8 and 9) and the PSO based integrated navigation error diagram (Figures 10 and 11), both the longitude L variance error and latitude λ variance error were reduced greatly compared with raw integrated navigation. It is worth noting that the GOA-based integrated navigation showed better performance than the PSO-based integrated navigation; it is suspected that the ability to find the optimal of GOA was better than for PSO in the integrated navigation task. For more precise optimal Q and R, the GOA-based integrated navigation was employed for the UAV-borne LiDAR scanning system.

For point cloud density clustering segmentation, default-turning K-means clustering divided crop into two clusters and separates ground points clearly, but parts of data points were ignored for clustering. Different meta-heuristic-based DBSCAN algorithms shown different optimization capabilities according to Figure 13. GA algorithm was the weakest in point cloud density clustering, different variation rates and 800 iterations more were set in further fine-turning, the GA-based DBSCAN algorithms still could not find the lower fitness value which was close to 0.416032. The BA-based DBSCAN algorithm shown the better optimization capability but lower computing efficiency, 9669 s of time consumption make it worse for practical application. Both GOA-based DBSCAN algorithm and PSO-based DBSCAN algorithm found the same fitness 0.416032. It is suspected that global optimization was 0.416032 under further testing with different fine tuning. Considering that PSO converges faster and takes less time in iterations, the PSO-based DBSCAN algorithm was employed for crop point cloud clustering segmentation.

For point cloud density enhancement, the pre-trained PU-net [26] was trained by "Visionair" datasets, there were many objects point cloud dataset for learning features by Convolutional Neural Network (CNN). After point cloud density enhancement, the density of point cloud was greatly enhanced, and Table 2 shown that there was no gap in mean value between the raw crop point cloud and 5 sets of the enhanced crop point cloud, which indicated that the enhanced point cloud kept the height distribution of the raw data. A faint change of variance and standard deviation indicated that the 5 sets of the enhanced point cloud also kept the shape and contour of the raw data.

8. Conclusions

This paper developed a UAV-borne LiDAR scanning system with enhanced integrated navigation and dense point cloud optimization. The GOA was employed for optimizing the integrated navigation; the longitude and latitude variances of optimized integrated navigation were 0.00046 and 0.00034, respectively, and the longitude and latitude variances of raw integrated navigation were 0.0091 and 0.0047, respectively. The positioning accuracy of the UAV carrier was improved greatly. The PSO was employed to find the optimal key elements of DBSCN clustering segmentation, and accurate point cloud segmentation was obtained. At last, the fine-tuned PU-net was used for point cloud density enhancement. Compared with the raw point cloud, the density of the enhanced crop point cloud was 4 times more, and the statistical parameters showed that the enhanced crop point cloud retained the shape and contour characteristics of the raw data. In this paper, each meta-heuristic optimization algorithm had its own merits for the different problems, but we still have not found an optimization algorithm that can meet all the processing requirements. In the future, owing to the flexibility of a multi-rotor UAV in remote sensing operations in specific areas, a multi-rotor UAV equipped with a small LiDAR or optical image remote sensing system will be more suitable for remote sensing operations in the agricultural field [27,28].

Author Contributions: Conceptualization, J.C. and Z.Z.; methodology, J.C. and Z.Z.; software, Z.Z. and K.Z.; validation, Z.Z., K.Z., and S.W.; formal analysis, S.W. and Y.H.; investigation, S.W. and Y.H.; data curation, K.Z. and S.W.; writing—original draft preparation, J.C. and Z.Z.; writing—review and editing, K.Z., S.W., and Y.H.; supervision, J.C. and Y.H.; project administration, J.C. and Y.H.; funding acquisition, J.C. and Y.H. All authors have read and agreed to the published version of the manuscript.

References

1. McManamon, P.F. Review of LiDAR: A historic, yet emerging, sensor technology with rich phenomenology. *Opt. Eng.* **2012**, *51*, 060901. [CrossRef]
2. Vázquez-Arellano, M.; Griepentrog, H.W.; Reiser, D.; Paraforos, D.S. 3-D Imaging Systems for Agricultural Applications—A Review. *Sensors* **2016**, *16*, 618. [CrossRef] [PubMed]
3. Lovell, J.L.; Jupp, D.L.B.; Newnham, G.J. Measuring tree stem diameters using intensity profiles from ground-based scanning LiDAR from a fixed viewpoint. *ISPRS J. Photogramm. Remote Sens.* **2011**, *66*, 46–55. [CrossRef]
4. Strahler, A.H.; Jupp, D.L.B.; Woodcock, C.E. Retrieval of forest structural parameters using a ground-based LiDAR instrument (Echidna). *Can. J. Remote Sens.* **2008**, *34*, 426–440. [CrossRef]
5. Zheng, G.; Moskal, L.M.; Kim, S.H. Retrieval of effective leaf area index in heterogeneous forests with terrestrial laser scanning. *IEEE Trans. Geosci. Remote Sens.* **2013**, *51*, 777–786. [CrossRef]
6. Lai, X.D.; Liu, Y.S.; Li, Y.X. Application status and development of point cloud density characteristics of airborne LiDAR. *Geospat. Inf.* **2018**, *16*, 1–5.
7. Quan, W.; Gong, X.; Fang, J.; Li, J. Prospects of INS/CNS/GNSS Integrated Navigation Technology. In *INS/CNS/GNSS Integrated Navigation Technology*; Springer: Berlin/Heidelberg, Germany, 2015.
8. Fu, B.; Liu, J.; Wang, Q. Multi-sensor integrated navigation system for ships based on adaptive Kalman filter. In Proceedings of the 2019 IEEE International Conference on Mechatronics and Automation (ICMA), Tianjin, China, 4–9 August 2019.
9. Gao, Y.; Liu, S.; Atia, M.M.; Noureldin, A. INS/GPS/LiDAR integrated navigation system for urban and indoor environments using hybrid scan matching algorithm. *Sensors* **2015**, *15*, 23286–23302. [CrossRef] [PubMed]
10. Shabani, M.; Gholami, A.; Davari, N. Asynchronous direct Kalman filtering approach for underwater integrated navigation system. *Nonlinear Dyn.* **2015**, *80*, 71–85. [CrossRef]
11. Zhang, G.; Hsu, L.T. Intelligent GNSS/INS integrated navigation system for a commercial UAV flight control system. *Aerosp. Sci. Technol.* **2018**, *80*, 368–380. [CrossRef]
12. Yu, H.; Han, H.; Wang, J.; Xiao, H.; Wang, C. Single-frequency GPS/BDS RTK and INS ambiguity resolution and positioning performance enhanced with positional polynomial fitting constraint. *Remote Sens.* **2020**, *12*, 2374. [CrossRef]
13. Norouz, M.; Ebrahimi, M.; Arbabmir, M. Modified Unscented Kalman Filter for improving the integrated navigation system. In Proceedings of the 25th Iranian Conference on Electrical Engineering, Tehran, Iran, 2–4 May 2017.
14. Goldberg, D. *Genetic Algorithms in Search, Optimization and Machine Learing*; Addison-Wesley: Boston, MA, USA, 1989.
15. Eberhart, R.; Kennedy, J. A new optimizer using particle swarm theory. In Proceedings of the 6th International Symposium on Micro Machine and Human Science, Nagoya, Japan, 4–6 October 1995; pp. 39–43.

16. Colorni, A.; Dorigo, M.; Maniezzo, V. Distributed optimization by ant colonies. In Proceedings of the 1st European Conference on Artificial Life, Paris, France, 11–13 December 1991; pp. 134–142.
17. Yang, X. A new metaheuristic bat-inspired algorithm. In *Proceedings of Nature Inspired Cooperative Strategies for Optimization (NICSO 2010)*; Springer: Berlin/Heidelberg, Germany, 2010; pp. 65–74.
18. Saremi, S.; Mirjalili, S.; Lewis, A. Grasshopper optimisation algorithm: Theory and application. *Adv. Eng. Softw.* **2017**, *105*, 30–47. [CrossRef]
19. Feng, H.; Ni, H.; Zhao, R.; Zhu, X. An enhanced grasshopper optimization algorithm to the Bin packing problem. *J. Control Sci. Eng.* **2020**, *2020*, 3894987. [CrossRef]
20. Chen, Y.; Zhou, L.; Bouguila, N.; Wang, C.; Chen, Y.; Du, J. BLOCK-DBSCAN: Fast clustering for large scale data. *Pattern Recognit.* **2020**, *109*, 107624. [CrossRef]
21. Kedzierski, M.; Fryskowska, A. Methods of laser scanning point clouds integration in precise 3D building modelling. *Measurement* **2015**, *74*, 221–232. [CrossRef]
22. Rupink, B.; Mongus, D.; Zalik, B. Point density evaluation of airborne LiDAR datasets. *J. Univers. Comput. Sci.* **2015**, *21*, 587–603.
23. Huang, H.; Shi, W.U.; Gong, M. Edge-aware point set resampling. *ACM Trans. Graph.* **2013**, *32*, 1–12. [CrossRef]
24. Mafarja, I.A. Evolutionary population dynamics and grasshopper optimization approach for feature selection problems. *Knowl. Based Syst.* **2018**, *145*, 25–45. [CrossRef]
25. Ledig, C.; Theis, L.; Huszar, F. Photo-realistic single image super-resolution using generative adversarial network. In Proceedings of the IEEE Conference on Computer Vision and Pattern Recognition, Honolulu, HI, USA, 21–26 July 2017.
26. Yu, L.; Li, X.; Fu, C.W. PU-net: Point cloud upsampling network. In Proceedings of the IEEE Conference on Computer Vision and Pattern Recognition, Salt Lake City, UT, USA, 19–21 June 2018.
27. Zhang, Z.; Han, Y.; Chen, J.; Wang, S.; Wang, G.; Du, N. Information extraction of ecological canal system based on remote sensing data of unmanned aerial vehicle. *J. Drain. Irrig. Mach. Eng.* **2018**, *36*, 1006–1011.
28. Wang, S.; Han, Y.; Chen, J.; Pan, Y.; Cao, Y.; Meng, H. Weed classification of remote sensing ecological irrigation area by UAV based on deep learning. *J. Drain. Irrig. Mach. Eng.* **2018**, *11*, 1137–1141.

6

Assimilation of Sentinel-2 Leaf Area Index Data into a Physically-Based Crop Growth Model for Yield Estimation

Francesco Novelli [1], Heide Spiegel [2], Taru Sandén [2] and Francesco Vuolo [1,*]

[1] Institute of Surveying, Remote Sensing & Land Information (IVFL), University of Natural Resources and Life Sciences, Vienna (BOKU), Peter Jordan Str. 82, 1190 Vienna, Austria; francesco.novelli@boku.ac.at

[2] Department for Soil Health and Plant Nutrition, Austrian Agency for Health and Food Safety (AGES), Spargelfeldstrasse 191, 1220 Vienna, Austria; adelheid.spiegel@ages.at (H.S.); taru.sanden@ages.at (T.S.)

* Correspondence: francesco.vuolo@boku.ac.at

Abstract: Remote sensing data, crop growth models, and optimization routines constitute a toolset that can be used together to map crop yield over large areas when access to field data is limited. In this study, Leaf Area Index (LAI) data from the Copernicus Sentinel-2 satellite were combined with the Environmental Policy Integrated Climate (EPIC) model to estimate crop yield using a re-calibration data assimilation approach. The experiment was implemented for a winter wheat crop during two growing seasons (2016 and 2017) under four different fertilization management strategies. A number of field measurements were conducted spanning from LAI to biomass and crop yields. LAI showed a good correlation between the Sentinel-2 estimates and the ground measurements using non-destructive method. A correlating fit between satellite LAI curves and EPIC modelled LAI curves was also observed. The assimilation of LAI in EPIC provided an improvement in yield estimation in both years even though in 2017 strong underestimations were observed. The diverging results obtained in the two years indicated that the assimilation framework has to be tested under different environmental conditions before being applied on a larger scale with limited field data.

Keywords: crop growth model; data assimilation; Leaf Area Index; Sentinel-2; EPIC model; yield estimation

---

1. Introduction

Agriculture is facing great challenges to increase food production while decreasing global impact [1]. In this context, better crop management strategies can help to guarantee sustainable food production and security [2]. It is of paramount importance to set up accurate and timely monitoring tools that ensure spatially detailed and accurate information on crop yields and on the required production inputs (e.g., water, fertilizer). Remotely-sensed data can provide spatially and temporally consistent information over large agricultural areas [3] to monitor the seasonal patterns related to the biological lifecycle of the crop [4]. Through the use of some variables estimated by remotely-sensed data, it is possible to obtain crop yield predictions. Thus, different estimation approaches are described, based on remote sensing data. Statistical regression methods are the most commonly adopted for quantifying the expected yield [5]. For instance, the Normalized Difference Vegetation Index (NDVI) is commonly used as an independent variable in regression models to estimate the crop yield. The main limitations of this technique is the need for sufficiently long and consistent time series of both remote sensing data and agricultural statistics to calibrate the empirical models [6]. In order to improve the predictive power of remotely-sensed data, it is possible to add independent meteorological (or bio-climatic) variables into the regression models as shown in several studies [7,8]. The interaction of these two different sources of data provides different types of information even if the bio-climatic indicators, especially if derived from satellites as well, are not totally independent from the vegetation indices [6]. An alternative approach to obtain crop yield is to consider the crop physiological processes using semi-empirical approaches such as the Monteith's efficiency model [9]. The latter computes the daily production of dry matter (DM) integrating incident global solar radiation, usually available from the meteorological stations, and the diagnostic estimation of the status and the activity of the plant provided by remote sensing data [10]. The approach uses only the amount of photosynthetically active solar radiation (PAR) adsorbed by the plant, although it is known that the crop yield depends also on the interaction of other variables, such as fertilizer, soil, cultivar diversity, weather conditions, water availability, pests, and diseases.

A more sophisticated approach to account for the interaction of environmental factors and field management strategies is based on dynamic crop growth models (CGMs) [11]. The main advantage of CGMs is their capacity to take into account such limiting factors (e.g., soil, weather, water, nitrogen) dynamically [12]. CGMs are able to describe the behavior of the real crop by predicting the physiological mechanisms of the crop growth (e.g., phenological development, photosynthesis, dry matter, portioning, and organogenesis) using mechanistic equations [13]. Their applicability is, however, complex at field or regional scales because they are designed to describe the plant growth at the point scale, assuming that the field conditions are uniform [14]. In this context, the use of satellite data to provide spatially integrated information over large areas is widely recognized to support CGMs [11,12,15,16]. The crop phenology, which can be assimilated into the crop growth models, becomes fundamental to control the crop dry matter distribution during the growth process inside the model itself [17].

The integration of remote sensing data and CGM can be achieved in different ways [18,19]:

(1) Forcing strategy, where the remote sensing data are used to replace the crop model simulation data. For each time step of the model simulation (e.g., daily), direct ingestion of the variables into the model is done. It requires continuous data and it is subject to frequent instabilities in the overall model setting.

(2) Updating strategy, which consists of continuously updating the model state variables whenever new observations from the satellite are available. This method assumes that better-simulated state variables at day t will also improve the accuracy of the simulated state variables on succeeding days.

(3) Calibration method, which considers adjusting either the model or initial state variables to obtain an optimal agreement between simulated and observed state variables. Key model's state variables are changed according to the driving variable obtained from the satellite. Generally, a minimization algorithm is used to minimize the difference between the model state variable and the driving variable obtained from the satellite [20,21].

Various implementations exist in the literature. For example, the commonly used EPIC model (Environmental Policy Integrated Climate) [14] shows good accuracy in the estimation of yield by assimilating variables derived from remote sensing data with a calibration approach [22]. The model is suitable for simulating crop yield over a large range of soils, crops, and management scenarios [23]. Satellite-based Leaf Area Index (LAI) is one of the most commonly used variables in model assimilation for yield estimation [21,24,25]. LAI is one of the key crop state biophysical variables required by many CGMs and is able to describe the relationship between soil, plant, and atmosphere system [26]. LAI can be obtained using freely available data from operational satellites such the USGS Landsat-8 or the Copernicus Sentinel-2 missions. The latter is composed of two identical satellites (Sentinel-2 A and B) providing 5 days of revisit time [27] and this increases the possibility to have cloud-free observations in key phenological periods during the crop growing cycle. Sentinel-2 mission provides improved possibilities for retrieving biophysical parameters such as LAI, thanks to the higher spectral, spatial, and temporal resolutions [28]. Recent studies show good correlation between LAI ground measurements and LAI estimates starting from Sentinel-2 reflectance data and using machine learning approaches [29,30]. Pre-calculated LAI maps with an accuracy of around 12% at 10 m spatial resolution can be obtained from the service platform for Sentinel-2 implemented by the University of Natural Resources and Life Science, BOKU [29].

Building on the BOKU Sentinel-2 LAI data, in this study we apply a data assimilation framework using the EPIC model over a winter wheat cereal crop for two growing seasons (2016 and 2017). A re-calibration of the model parameters was done using a minimization function in order to find the minimum difference between the remote sensing LAI data and the simulated LAI data. The main goal is the evaluation of (1) the data assimilation approach and (2) of the capability of the model to estimate crop yield for possible future application at regional scale. For this reason, the yields estimated by EPIC were validated using measured yield data. The validation of the LAI values obtained from Sentinel-2 was also done through the comparison with LAI ground measurements collected during both the growing seasons.

2. Materials and Methods

2.1. Study Area

The winter wheat experimental fields used in this study are located in the Marchfeld region, an agricultural area in the east of Vienna in Lower Austria (Lat. 48.20° N, Long. 16.72° E) (Figure 1). This area was designated as a pilot area for the H2020 FATIMA (Farming Tools for external nutrient Inputs and Water Management) project [31] and field data were collected in this framework. The Marchfeld region covers about 75,000 ha and the most important crop types are: winter wheat, vegetables (e.g., carrots, onions, green peas, asparagus, spinach, green salad), sugar beet, grain maize, and potatoes [31]. The climate is semi-arid with cold winters characterized by frequently strong frosts and limited snow cover, and hot and intermittently dry summers [32]. The average annual temperature is between 9 and 10 °C and the annual precipitation ranges from 500–550 mm with a maximum in early summer, from May to July. During the vegetation period from April to September, the precipitation is between 200 and 440 mm [32]. The presence of wind, blowing at an average speed of 3.5 m/s, encourages the dry conditions in the summer and, additionally, leads to high evapotranspirative demand. Different types of soils with high spatial variability can be found [33] and the general soil conditions are characterized by a humus-rich A horizon and a sandy C horizon [34].

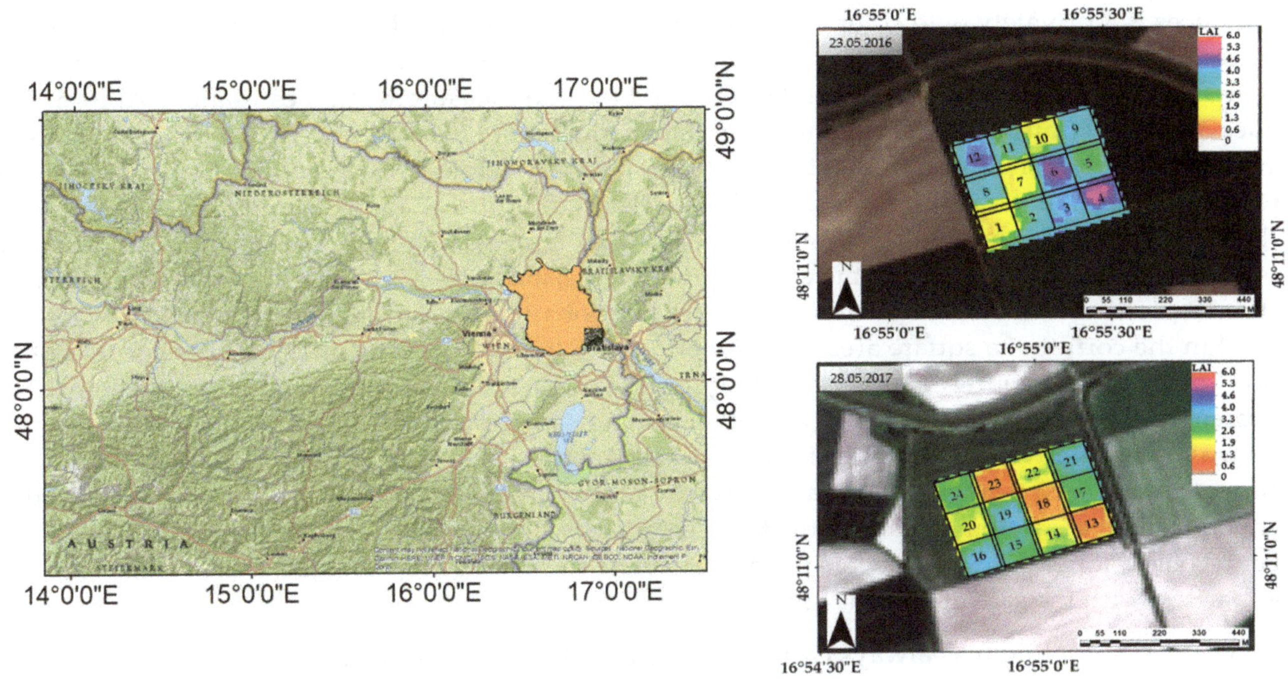

Figure 1. Study area of Marchfeld (left), Lower Austria, and the experimental plots in 2016 (top right) and 2017 (bottom right). It is possible to notice the variability in LAI values as a result of different fertilization rates given during the growing cycle.

2.2. *Experimental Plots*

Field measurements were conducted in 2016 and 2017 in two experimental fields of winter wheat (cultivar Capo) with total field area of 12 hectares (ha), divided in 12 plots, each of 1 ha (100 m × 100 m). Each plot received different fertilization rates with 3 repetitions of four different levels of fertilization (variant) in a range between 0 and 180 kg N/ha (0, 60, 120, 180 kg N/ha) (Figure 1).

The soil characteristics were measured before, during and after the crop growing cycle for each of the 12 plots in both years, the information was provided by the Austrian Agency for Health and Food Safety (AGES) together with the soil management practice (including date of sowing and harvest). The soil texture in the first 30 cm is silty-clay-loam or loam with small difference amongst the plots. For the experimental plots in the year 2016, the range of sand particles was between 10.9% and 33.7% with a median value of 19.65%, while presence of silt was between 40.5% and 57.4% with a median value of 48.65%. In 2017, the presence of sand was between 15.4% and 28.9% with a median value of 22.8%, while the presence of silt was between 43.9% and 50.9% with a median value of 46.3%. In both the experimental fields, the soil was slightly alkaline, indeed the pH in the first 30 cm was around 7.5–7.6 [35]. The before seeding management was characterized by a disk harrow soil processing at twelve centimeters depth and a seedbed preparation at 5 cm in depth. After sowing, the mineral fertilization was applied using calcium ammonium nitrate (N = 27%) at three times (March, April, and May). The same procedure was applied in both years. At harvest, the final yield of dry matter in kg/ha was measured and provided by AGES.

2.3. *Leaf Area Index Measurements*

Ground measurements of LAI were undertaken once a week during both years alternatively one week in the experimental plots and the following week in other randomly selected fields in the region. The 2016 field data collection began in the middle of April and finished in the middle of May, while for 2017 the data collection started at the end of April and ended in the middle of June.

In situ LAI measurements were collected with a non-destructive and indirect method using a Li-Cor LAI-2200C Plant Canopy Analyzer (Li-Cor Biosciences, Lincoln, NB, USA) [36]. Li-Cor

Biosciences is a privately held company based in Lincoln, Nebraska that was founded in 1971 and it is a leader in the design, production, and marketing instruments for biological and environmental research [37]. The LAI-2200C has been largely used for LAI measurement; it estimates LAI from the values of canopy transmittance by identifying the attenuation of the radiation as it passes through the canopy [36]. Therefore, measurements were taken above- (A) and below-canopy (B). The measures were done in elementary sampling units (ESU) within a radius of about 10 m of a geo-referenced point (accuracy ± 3–5 m using GPS signals); each ESU represents a homogenous area (in terms of crop development conditions). In the experimental fields, the ESUs were located in the center of each of the 12 plots. In the random fields that were additionally sampled, the centers of the ESU were placed in the corner of a square area of about 60 × 60 m, measured from an accessible border of the field. This ensures a representative sampling for the whole field and makes sure that the point also represents the Sentinel-2 pixel at 10 m and 20 m ground sampling distances. In both cases, in order to obtain a representative LAI sample within a radius of about 10 m, each ESU was ideally divided into four sectors; for each sector one measurement was done above the canopy and six below it. The latter were performed at random in each of the four sectors to well represent the ESU. A total of 24 readings were recorded for each ESU and were used to calculate the average representative LAI value used for comparison with the satellite data (Figure 2).

The LAI sampling was always carried out early in the morning or later in the afternoon under diffuse light conditions. Previous studies showed that the best results for LAI were measured in diffuse light conditions to avoid direct sunlight and, therefore, minimizing the underestimation of LAI [38,39].

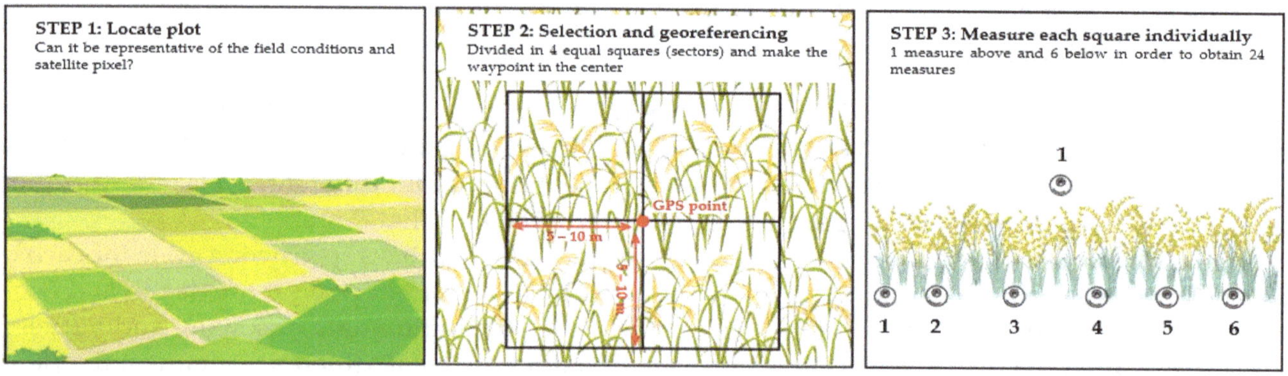

Figure 2. Workflow for collecting in situ LAI measurements using the Li-Cor LAI 2200C Plant Canopy Analyzer. Step 1 is the selection of the field; in Step 2 the GPS point, representative of the center of the ESU, is taken; and in Step 3 the measurements are performed obtaining an LAI value as a mean value of the 24 measurements below the canopy (six for each square).

2.4. Satellite Data

Leaf Area Index

Satellite-based estimates of LAI were obtained using an artificial neural network (ANN) developed at INRA. An ANN consist of a large number of simple processors, called "neurons", linked by weighted connections where each output depends only on the information that is locally available [40]. The ANN is a "black box" approach which has great capacity in predictive modelling but all the characters describing the unknown situation must be presented to the trained ANN and the identification (prediction) is then given [41]. The network was trained using PROSPECT and SAIL radiative transfer models and it uses eight Sentinel-2 spectral bands (B3, B4, B5, B6, B7, B8A, B11, and B12) [42]. On the basis of this approach, LAI was calculated and available as LAI maps at 10 m resolution at the BOKU data service platform for Sentinel-2 [29]. The LAI image data used for the 2016 and 2017 are listed in Table 1. In both years the maximum timespan between satellite acquisitions and ground measurements of LAI was five days. To guarantee that the ground reference points were not covered by clouds or shadows in the image data, the scene classification layer (SCL) was used [43]. For the validation of the LAI retrieval, LAI values were extracted for the central points of each LAI ESU measurements with a buffer of 10 m (an equal expansion in each direction from our GPS point) and used for the comparison with the field data as a mean value inside of the buffer for each ESU. This was done to compensate for the location of the center points, as some did not cover exactly one pixel or were located on a pixel border.

Table 1. List of Sentinel-2 (S2) acquisitions and dates of LAI ground measurements (three in 2016 and four in 2017) used for the LAI validation. The cloud cover indicates the percentage of the land surface covered by clouds in the image data.

	S2 Acquisition Date	Cloud Cover	LAI Ground Measurements Date
2016	April 13	0.0%	April 13–18 [1]
	April 26	18.4%	April 25
	May 06	3.8%	May 02–09
2017	April 21	10.3%	April 25
	May 18	51.9%	May 18
	May 28	0.0%	May 23–29
	June 20	0.0%	June 19

[1] Considering the timespan is possible to use LAI ground measurements from both the days.

For the assimilation of LAI in the EPIC model, LAI values were extracted from the image data using a crop mask of the fields of interest (divided in different plots). The latter involved restricting the analysis to a subset of pixels, rather than using all the pixel of the image [4]. For each polygon representing each plot, considered as an independent field, an inner buffer of 20 m (equal reduction in every direction of the space of our spatial analysis) was created to extract the median value of LAI. The inner buffer is useful in this case to exclude possible outliers in the boundaries of the field. This operation was carried out for each LAI image where the fields of interest were cloud-free and available in the Sentinel-2 data service platform. The date of acquisition and cloud cover of the image data are listed in Table 2.

Table 2. List of Sentinel-2 (S2) LAI acquisitions and cloud cover during both the growing seasons of winter wheat in the area of interest (nine in 2016 and 12 in 2017).

	S2 Acquisition Date	Cloud Cover		S2 Acquisition Date	Cloud Cover
2016	March 14	1.3%	2017	March 29	11.7%
	March 17	3.7%		April 1	0.0%
	April 13	0.0%		April 21	10.3%
	May 6	3.8%		May 11	79.3%
	May 23	35.6%		May 18	52%
	June 2	31.8%		May 28	0.0%
	June 22	4.9%		June 10	43%
	July 2	0.0%		June 20	0.0%
	July 12	4.0%		June 30	29.6%
				July 5	7.7%
				July 12	14.8%
				July 17	12%

2.5. *EPIC Input and Forcing*

2.5.1. EPIC Model

EPIC is a physically-based crop growth model that operates on a daily time-step. One of the most important processes simulated by EPIC is the crop growth, and a wide range of crop types can be simulated using a generic growth routine. The crop growth process includes the interception of solar radiation, conversion of intercepted light to the biomass, partition of biomass into roots, above ground biomass and economic yield, and also simulated root growth [44]. The annual crop growth from planting to harvest depends on the potential heat unit for the crop. The growing period starts when the average daily temperature exceeds the base temperature of the plant and the phenological development of the crop is based on the daily heat unit accumulation (growing degree day approach) [44]. The heat units are used to calculate the heat unit index (ranging between 0 and 1) as a fraction of the potential heat units (PHU) required for the maturation of the crop under study [45].

The potential daily increase of the biomass is estimated using Monteith's approach [9] as a function of intercepted photosynthetic active radiation. The interception of solar radiation depends on the LAI [23], which is described by five model parameters: DMLA, DLAP1, DLAP2, DLAI, and RLAD. The model LAI curve is built considering two simulation stages: from emergence to leaf decline, LAI depends on three input factors DLAP1 and DLAP2, which are two points used for the definition of the optimal leaf development curve and DMLA, namely the maximum potential LAI. The second simulation stage starts from the leaf decline (which depends on DLAI parameter, the fraction of the growing season when leaf area starts to decline) to the end of the growing season. The rate of the LAI is influenced by LAI decline parameter (RLAD). Both simulation stages are also influenced by the heat units and, theoretically, by the estimated stresses caused by water, nutrients, temperature, aeration, and radiation. The potential biomass is also adjusted daily, taking into consideration the minimum of the EPIC plant growth constraints (nutrients, water, temperature, aeration, and radiation) [23]. The crop yield is estimated using the harvest index (HI) approach as a nonlinear function of heat units from zero (at the planting stage) to the optimal value (at maturity) [44]. The HI could be altered by high temperature, low solar radiation, or water stress during critical crop stages [23]. In particular, under water stress conditions, the default parameterization of the model would adjust the HI from the second half of the growing season considering a lower limit of HI. The latter is called water stress yield factor (WSYF) [46], a fraction between 0 and the HI value that represents the lowest HI expected due to the water stress [45]. The water stress effect on HI is controlled by PARM (3), namely the fraction of the growing season where water stress starts reducing HI, which also controls the accumulation of actual and potential evapotranspiration until harvest. At harvest, the ratio between potential and actual evapotranspiration (ET) is used with an S-shaped curve to estimate the final HI [45].

2.5.2. EPIC Input Data

EPIC is designed to simulate field, farm or small watershed, homogenous with respect to climate, soil, land use, and topographic management system parameters [45]. In this study each plot was considered as an independent field. The model enables the preparation of different parameters, operations, and run files to execute the model. The soil characteristics, such as soil hydrologic group, number of the soil layers and layer depth, soil texture, pH, organic carbon concentration, calcium carbonate content of soil, the sum of bases, cation exchange capacity, and bulk density were used to build the soil parameters file. In the latter file, three soil layers were considered with 30 cm of layer's depth. The soil hydraulic parameters were also calculated by default empirical equations provided by EPIC. These equations give as a result the wilting point, field capacity, saturated soil water content, and saturated hydraulic conductivity based on measured values of bulk density, sand content, and silt content [47]. Site files were prepared for every plot while the operation schedule files were prepared for each variant. The climate data (solar radiation, maximum and minimum temperature, precipitation, relative humidity, and wind velocity) were taken from the ZAMG (Zentralanstalt für Meteorologie und Geodynamik) weather station of Gross-Enzersdorf (48°19′ N 16°55′ E) located 30 km away from the experimental fields. Weather files were built using the monthly weather statistic generator WXPM p3020 available from EPIC model website [48], using historical data from 1997.

2.5.3. Model Calibration

Four plots (Figure 1; site 1,2,3,4) with different fertilization variants from the 2016 dataset were used to calibrate two unknown parameters of the EPIC model namely the potential heat unit (PHU) and the potential evapotranspiration (PET), which was estimated using the Penman–Monteith approach. For normal operation, knowing the date of sowing and harvest, it is possible to check the output file of the model HUSC index. If this index is higher than 1.2 or smaller than 1, the PHU must be adjusted in order to obtain a value around 1. In this way it is possible to arrive to the right value of PHU from planting to maturity. The other unknown parameters inside of the model, except those involved in the temporal definition of the LAI curve, were left as the default setting.

2.5.4. Data Assimilation

The temporal evolution of the optimal LAI curve is defined by the five parameters described in Section 2.5.1. First we identify the DMLA (maximum potential LAI) as the maximum value (98th percentile) observed in the image data during the full growing season for each field. Subsequently, the re-calibration of the other parameters (DLAP1, DLAP2, DLAI, and RLAD) that define the LAI curve in EPIC was performed on the basis of the minimization of the root mean square error (RMSE) (Equation (1)) between the LAI simulated by the model and the LAI observed by satellite at the day of the satellite observation. Default values were used for the parameters that were not measured or not available for our specific experiment.

$$RMSE = \sqrt{\frac{1}{n}\sum_{i=1}^{n}(LAI_{sim} - LAI_{obs})^2} \tag{1}$$

The minimization of the cost function (Equation (1)) was done using a MATLAB optimization toolbox called Fmincon that allows to find the minimum of a constrained nonlinear multivariable function. It uses the interior-point approach for solving nonlinearity constrained optimization problems (Equation (2)) where the objective is to find the local minimum of the function x subject to constraints [49]:

$$min\ f(x)\ subject\ to\ h(x) = 0\ g(x) \leq 0 \tag{2}$$

Starting from the original problem definition in Equation (2), the crucial steps of the optimization algorithm are the formulation and solution of the equality constrained barrier sub-problems, with i steps [50] (Equation (3)). Each sub-problem is called barrier sub-problem and could be approximated at the form (for each $\mu > 0$):

$$\min_{x,s} f(x) - \mu \sum_{i=1}^{m} lns_i\ subject\ to\ h(x) = 0\ g(x) + s = 0 \tag{3}$$

where μ is the barrier parameter and s_i is the slack variables that are restricted to be positive by adding ln (s_i) bounded [50]. As μ decrease close to zero, the minimum of f_μ should approach the minimum of f. The added logarithmic term is called a barrier function [50] and this is easier to solve than the original inequality-constrained problem (Equation (2)). For the optimization, we identified the bound constraints (minimum and maximum values for DLAP1, DLAP2, DLAI, and RLAD) by studying the model LAI outputs in correspondence of the four calibration plots (Section 2.5.3, Figure 1; site 1, 2, 3, 4). The bound constraints values found respectively for DLAP1, DLAP2, DLAI, and RLAD are: 05.01, 25.70, 0.50, and 0.50 for the lower bound and 20.10, 60.99, 1.00, and 3.00 for the upper bound constraints.

2.6. Accuracy Assessment

The model performance was estimated using the root mean square error (RMSE), the coefficient of determination (R^2) and relative root mean square error (RRMSE). The R^2 (Equation (4)) defines the proportion of the dependent variable variance that is predictable by the independent one [51]. It represents a normalized measure between 0 and 1 and it is given by the ratio between the "error sum of the square", that quantifies how much the data points Y_i vary around the estimated regression line $\hat{Y}_i$, and the "total sum of square", that quantifies how much the data points Y_i vary around their mean, $\overline{Y}$. In environmental research, a good value of R^2 is above 0.6 [52] even if it depends on the type of estimated variable. For instance, if we consider the range of LAI values, R^2 value larger than 0.9 is considered as excellent performance and between 0.5 and 0.8 a good one [53].

The RMSE is a performance indicator of a model and gives a quantity estimates of the deviation of the model's output from measurements [54]. The RMSE produces results in the same units as that used for measurements [55] and it is calculated as the root square of the mean of the difference between predicted values (V_p) and observed values (V_{obs}). If the RMSE is 0, there is no difference between predicted and measured values and the perfect fit is achieved. The relative RMSE (Equation (5)) is dimensionless and it is obtained by dividing the RMSE by the mean of the observed variables and thus is less sensitive to the magnitude of the variables [53]. The desired result of the RRMSE for an excellent performance of the model is less than 10% and, for a good one, is between 10% and 20% [53]:

$$R^2 = 1 - \frac{\sum_{i=1}^{n} (Y_i - \hat{Y}_i)^2}{\sum_{i=1}^{n} (Y_i - \overline{Y})^2} \tag{4}$$

$$RMSE = \sqrt{\frac{\sum_{i=1}^{n}\left(V_p - V_{obs}\right)^2}{n}} \tag{5}$$

$$RMSE = 100\frac{RMSE}{Mean\ (obs)} \tag{6}$$

3. Results

3.1. LAI Validation

Good agreement and low error between Sentinel-2 LAI for the winter wheat and LAI reference measurements was found in both years (Figure 3). Better RMSE and RRMSE values were obtained in 2017 compared to 2016 (RMSE = 0.46 vs. 0.44) (RRMSE = 19% vs. 17%). In the 2016 year, a slightly lower R^2 value was found compared to 2017 (R^2 = 0.72 vs 0.89). Although the comparison between ground and satellite LAI estimations showed good results in both years, there are some cases in which we can find a small disagreement. In 2016, a general small overestimation by the LAI-2200 compared to satellite observations was observed. In 2017, there are three cases for which Sentinel-2 overestimated LAI remarkably.

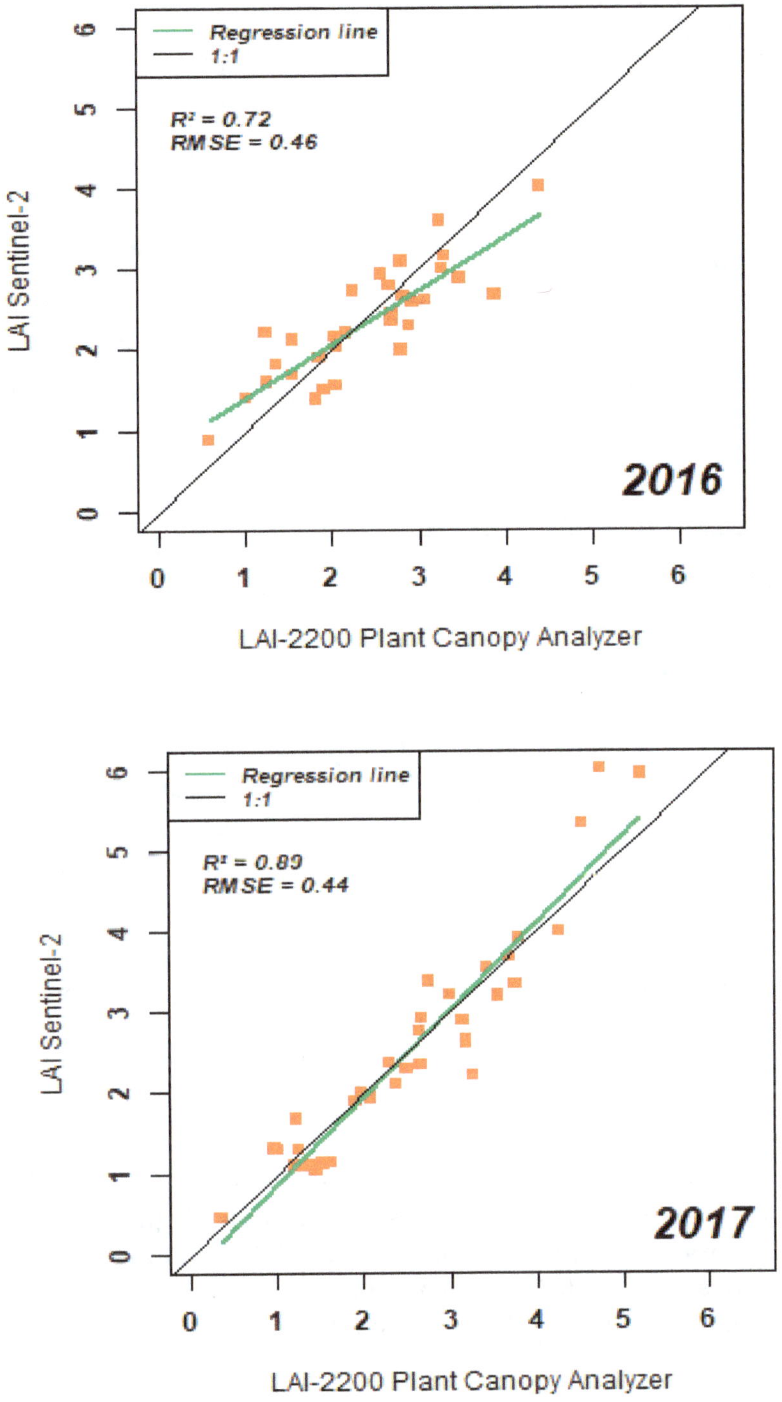

Figure 3. Scatterplot between ground and satellite LAI measurement for Marchfeld in 2016 (n = 31, Feb.–May) and in 2017 (n = 35, Feb.–June) for winter cereals.

3.2. EPIC Calibration and Assessment

Figure 4 shows the EPIC-LAI curves achieved after the assimilation of satellite data and the LAI curves built using the Sentinel-2 observations for the calibration plots in 2016. In this case, the shape of the LAI curves is similar and the only use of the bound constraints on the parameters to optimize seems sufficient to ensure a good fit between the simulated (EPIC) and the observed (Sentinel-2) LAI curves. However, Figure 4D (180 kg N/ha) shows a temporal delay in the crop development trend simulated by EPIC compared to the satellite observations.

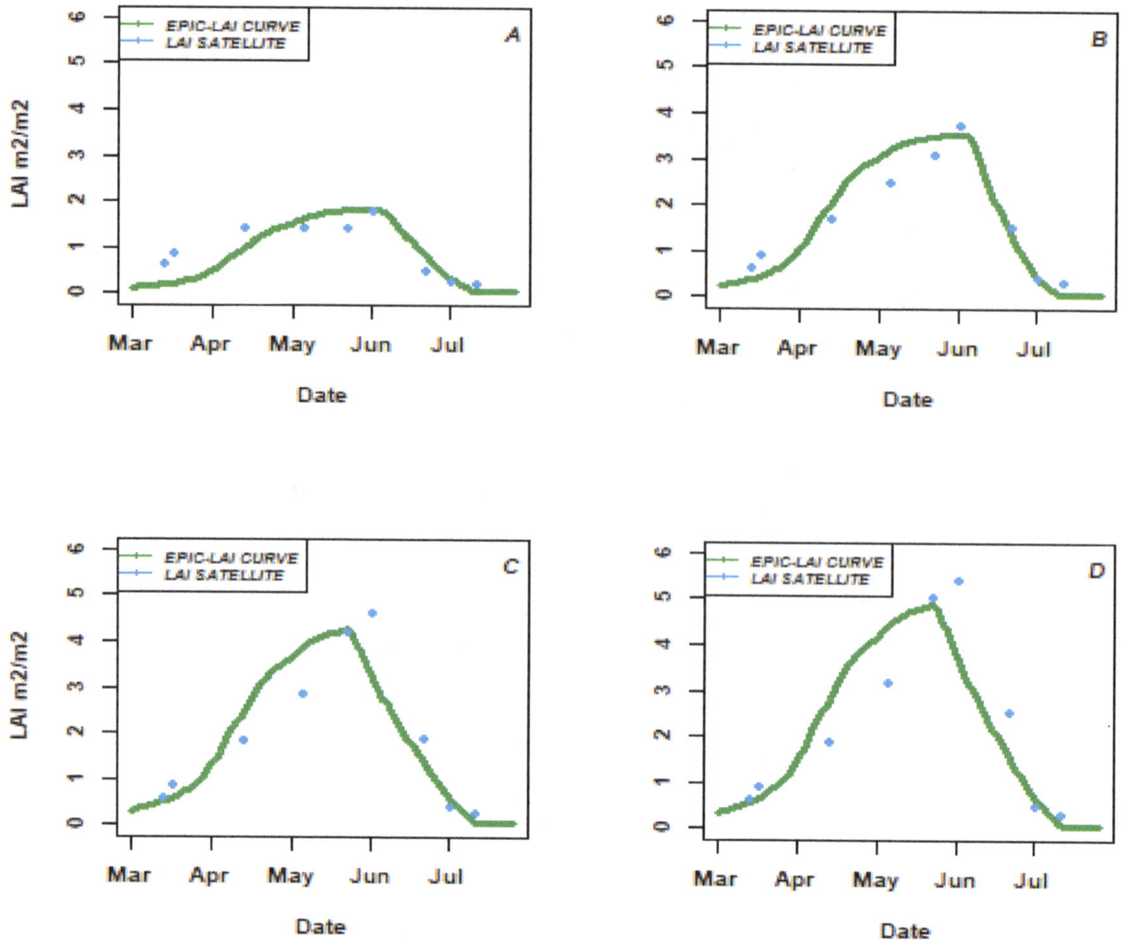

Figure 4. Calibration: the graphs show the different LAI values in four plots for 2016 for different levels of fertilization: 0 kg N/ha (**A**, id: 1-1), 60 kg N/ha (**B**, id: 2-1), 120 kg N/ha (**C**, id: 3-1), and 180 kg N/ha (**D**, id: 4-1). The green line is the daily LAI curve obtained by EPIC model, while the light blue points are the LAI measurements obtained using Sentinel-2 data.

This temporal displacement between the two LAI curves is larger in 2017 compared to 2016 as shown in Figure 5 for the highest fertilization treatment plots (Figure 5C,D), for which an incorrect parameterization appears clear especially for DLAP1, DLAP2, and DLAI. Probably, only use of the bound constraints is not always sufficient to reach a good calibration. It could also be that the model is not able to perfectly represent the satellite curve in all conditions.

3.3. *Yield Estimation in 2016*

Figure 6A shows a relevant good agreement (R^2 = 0.95) and low error (RMSE = 317 kg/ha and RRMSE = 6%) between the yield estimated by the EPIC model (with assimilation of LAI data from Sentinel-2) and the observed yield for the year 2016. The yield estimation notably improves with LAI assimilation compared to the one obtained using the default EPIC parameterization as shown in Figure 6B (R^2 = 0.93, RMSE = 572 kg/ha and RRMSE = 11%). Using the latter calibration, a model overestimation was found in the lower range of yield values (0 and 60 kg/ha of nitrogen during the growing season). Figure 6B also shows that the high yield values are approximately the same in both the plot with 120 and 180 kg N/ha, while using LAI assimilation from satellite, the yield values are better distributed along the regression line. Thus, the assimilation of LAI data seems to improve the sensitivity to the different amount of nitrogen provided during the growing season.

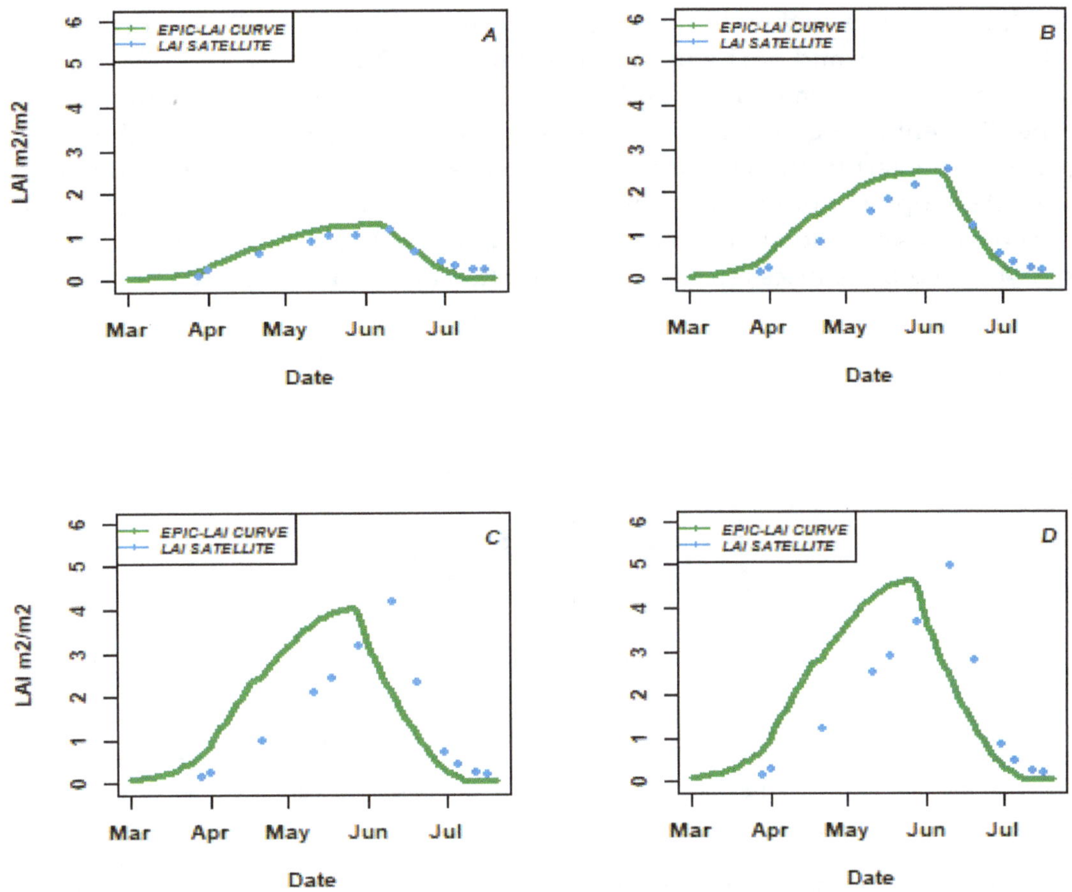

Figure 5. Assessment showing the different LAI values in four 2017 plots for different levels of fertilization: 0 kg N/ha (**A**, id: 1-1), 60 kg N/ha (**B**, id: 2-1), 120 kg N/ha (**C**, id: 3-1), and 180 kg N/ha (**D**, id: 4-1). The green line is the daily LAI curve obtained by EPIC model while the light blue points are the LAI measurements obtained using Sentinel-2 data.

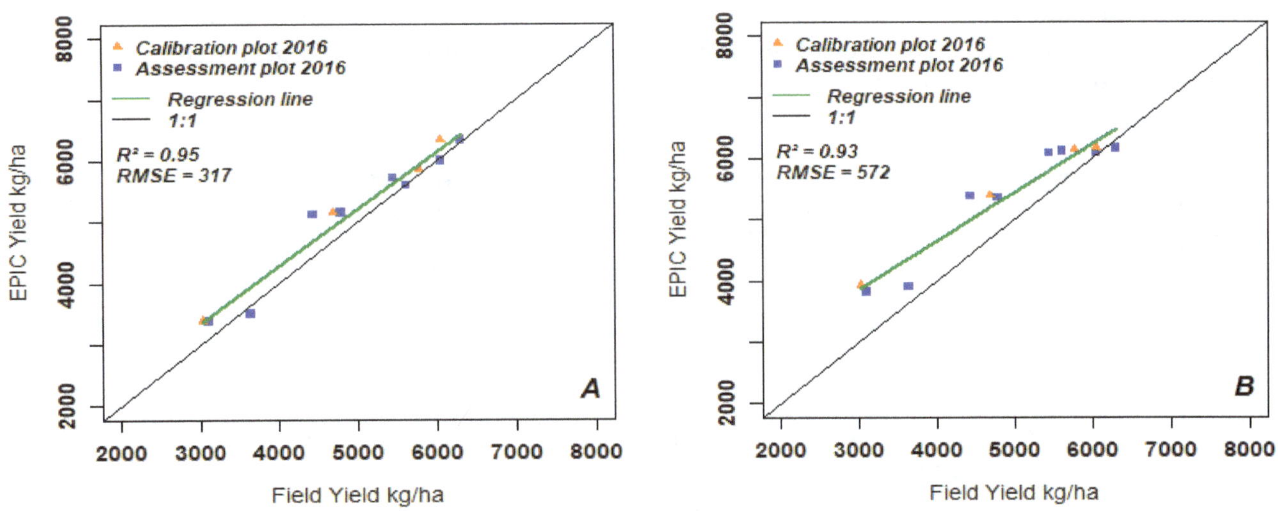

Figure 6. Yield estimation with (**A**) and without (**B**) the assimilation of LAI from Sentinel-2 data. The orange triangles are the plot used for the calibration of *fmincon* and the EPIC unknown parameters, and the blue square symbols represent plots used for the assessment only.

However, we encountered a major flaw for the assimilation of LAI data in EPIC under nitrogen stress conditions. The model lacks a correction feedback on simulated LAI. Only simulated biomass is corrected for the stress induced by limited nitrogen application. The EPIC-LAI curve is influenced mainly by the temperature and not by the variable amount of N.

Figure 7 shows the simulated LAI and biomass curves obtained using default parameters of the model for the different nitrogen fertilization applications. It appears clear that the LAI curve does not change even if the fertilization management changes. Therefore, the EPIC-LAI curve is referred to the optimal development of the LAI, without considering limiting stress conditions for the crop. With this approach, a saturation in the biomass is clearly visible for the high fertilization treatments (120–180 kg N/ha) and this causes errors in the estimation of the yield (Figure 7B).

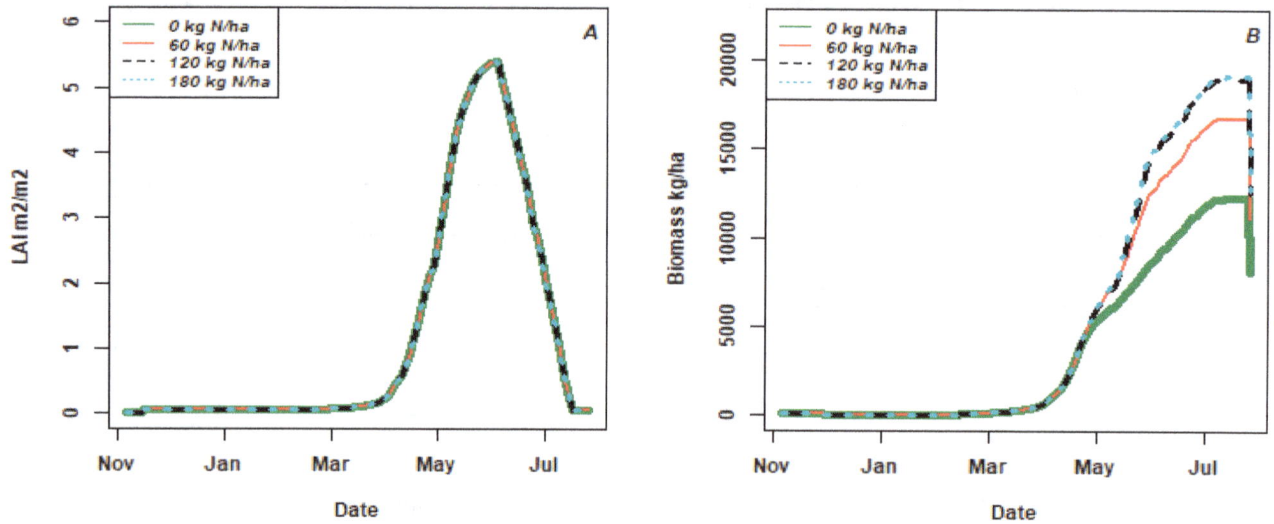

Figure 7. LAI (**A**) and biomass (**B**) curves obtained with default LAI parameters for the four different fertilization management treatments (0, 60, 120, and 180 Kg N/ha) in 2016. The graphs show a variation in the biomass levels but not in LAI.

3.4. Yield Estimation for 2017

Regarding the year 2017, we found a strong underestimation of the model compared to the yield measured in the field with $R^2 = 0.97$, RMSE = 1961, and RRMSE = 55% (Figure 8). A little improvement can be seen by assimilating LAI from the satellite (Figure 8A) compared to a parameterization using default model values (Figure 8B).

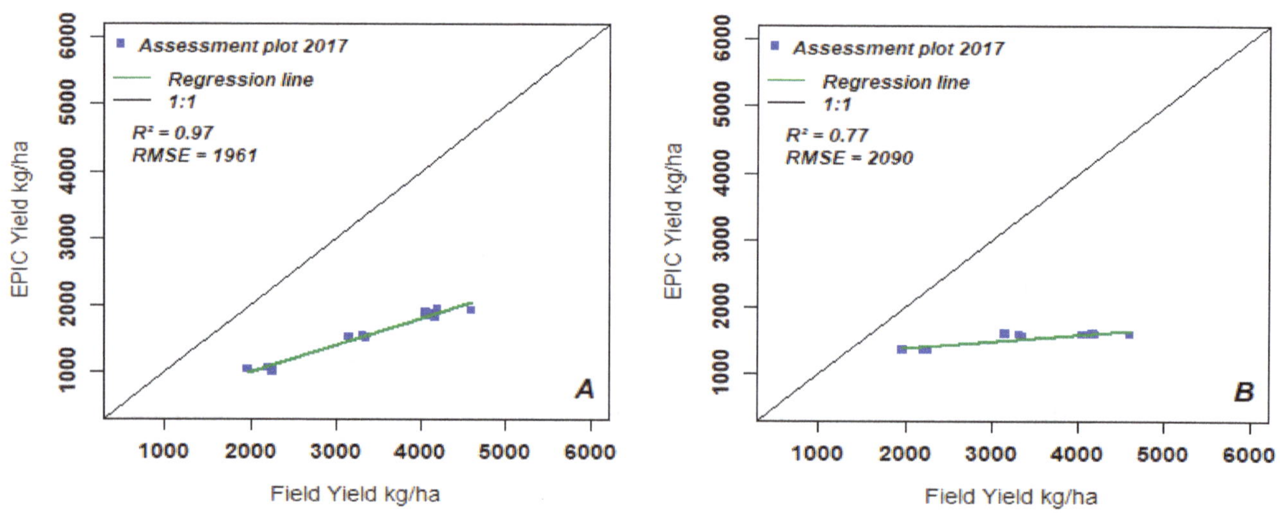

Figure 8. Comparison between (**A**) yield estimated with the assimilation of LAI from Sentinel-2 and (**B**) yield obtained with default EPIC parameters for 2017. The green line represents the regression line between the values.

The output of the model was further analyzed to understand the strong underestimation. One main reason for the underestimation was found to be due to a major decrease in the harvest index (HI) for the year 2017 (0.214–0.218) compared to 2016 (0.416–0.425). The small intra-year variations in the HI values are probably due to the different phenological development of the crop during the growing season related to the different LAI parameterization for each plot. The strong decrease in the HI in 2017 is probably related to water stress induced by limited rainfall. For comparison, Figure 9 shows the amount of precipitation during the growing season 2016 and 2017. It is known that drought stress at any growth stages decreases grain yield, especially in the stem elongation, flowering, and grain falling stages.

It appeared clear that the impact of water stress excessively reduces the HI index. Therefore, we further tuned the HI by modifying the parameter PARM(3) inside of the parameter file. For this parameter a value of 3 was set. Using the letter value, the water stress influences HI only when the heat unit accumulation is equal to three times the potential heat unit accumulation of the crop. In addition, the lower limit of HI, namely the WSYF that was set equal to the HI value. As a result of this adjustment, Figure 10 shows an improvement of the yield estimation (R^2 = 0.97, RMSE = 281, and RRMSE = 8.1%).

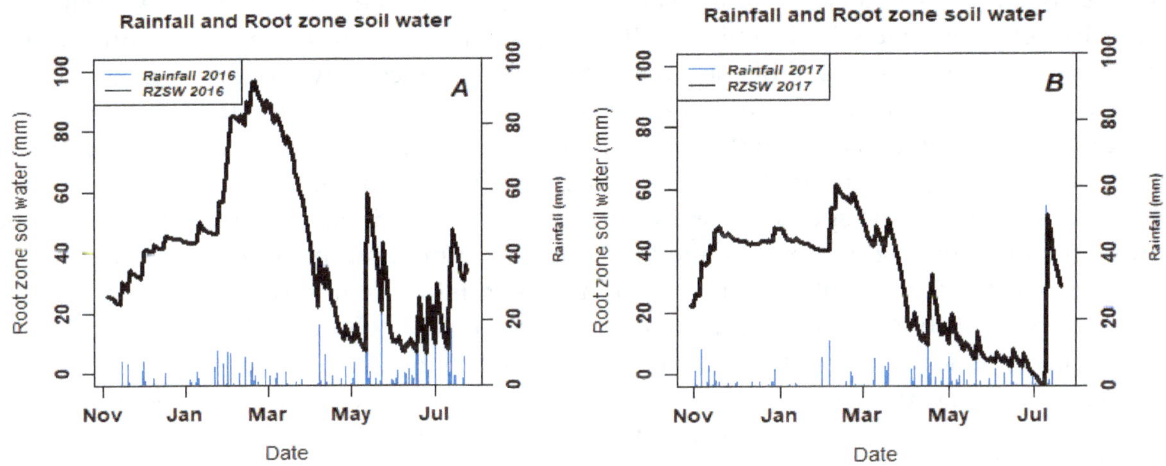

Figure 9. Precipitation (blue bars) and simulated water content in the root zone (black line) in 2016 (**A**) and in 2017 (**B**) for one of the high fertilizer plots (180 kg N /ha).

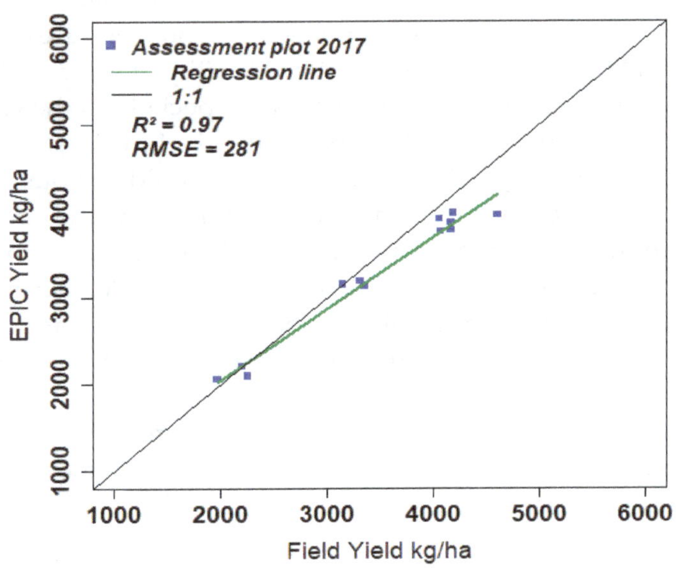

Figure 10. Yield obtained using the modified PARM(3) value that allows to reduce the impact of the water stress on the harvest index.

4. Discussion

Regarding the accuracy of LAI estimation with Sentinel-2 data, our study confirmed a previous study [29] that reported an R^2 of 0.83 and a RMSE of 0.32 m^2/m^2. When assimilated in a crop model, satellite-based LAI improves yield estimates as reported also in other studies that used a similar assimilation strategy [12,56]. The only use of the bound constraints on the model parameters, without including other restriction as nonlinear constraints, seems sufficient to find the minimum of the minimization problem but a good fit between the satellite observations and the LAI curve of the model was found only in the plot where a low amount of fertilizer was provided. However, in both the years, a yield improvement was obtained and this may be due to a forcing of the DMLA parameter using satellite observations. The use of the maximum value of LAI (98th percentile) to set DMLA can be considered as a priori knowledge of the canopy conditions. In this study, an ideal maximum LAI reachable by the crop was found for each field. In this context, the combination of Sentinel-2A and Sentinel-2B satellite data can provide timely and dense sets of observations to precisely characterize the crop growth curves and maximum LAI values.

The combination of the EPIC model and satellite-based LAI was found to be not ideal because there is no nitrogen stress feedback on the LAI in the model. In real field conditions, LAI is significantly influenced by nitrogen levels [57], while, in EPIC, the LAI curve is modeled without taking into consideration the nitrogen stress. Only simulated biomass is affected by the nitrogen constraints.

However, under high level of nitrogen application (above 120 kg/ha in our study), the simulated biomass saturates and, in this case, forcing LAI can help the model to achieve an increased precision. Therefore, the coupling of satellite-based LAI observations and potential EPIC LAI curves might improve the description of the spatial variability, for both LAI and biomass estimation, as confirmed by an improvement in yield estimation obtained in both years under study. Regarding the year specific results: in 2016, a good yield estimation was found assimilating remotely-sensed data even if only a slight improvement in R^2, RMSE, and RRMSE was obtained if compared to the yield estimation without data assimilation. This could be partially attributed to the site-specific parameterization of the model using field observations and, thus, an accurate description of the environment where the crop has been growing up. In 2017, a slight improvement in the yield results was also found especially regarding the correlation (improvement of R^2). However, both a data assimilation approach and a standard EPIC parameterization provided a strong underestimation in yield due to water stress. This is probably caused by three reasons: (1) the rainfall data may not be representative of the field condition and they can affect the model estimation, (2) the default parameters of the model are non-representative for the specific year of simulation; or (3) the impact of water stress conditions are erroneously modelled.

One of the problems that we should consider is the capability of the model to estimate yields under different scenarios and in the presence of high variability, especially in the regional scale application. Previous studies have shown promising results considering the EPIC yield estimation at the regional scale in the presence of high variability [58]. More examples should be obtained using remote sensing data that allows the improvement in the information about the crop and the environment of interest even if, due to a large amount of parameters required by the model and, therefore, the large rate of uncertainty, the regional scale application of EPIC seems difficult. Furthermore, the definition of the model parameter-bound constraints in the minimization function without a site-specific calibration is not always representative of the different geographical areas, and a multi-location calibration appears fundamental, for the application in different sub-areas (such as at the field scale) or at the regional scale. It seems necessary to add additional information in the minimization function. The application of EPIC model at the sub-field scale seems more suitable than the regional application, but specific field and crop data are needed.

5. Conclusions

The use of satellite data together with a crop growth model gives a wide range of possibilities to obtain yield estimates. However, the availability of satellite data and the ability of the model to represent both satellite observations and the characteristics of the crop system has to be evaluated. In this study, the fit between the modelled LAI and the observed LAI curves was reached for both years under investigation (2016 and 2017). However, yield was not always well estimated. This problem could be attributed to the non-capability of the model to represent the environment of study or to issues related to the data used for the model calibration. EPIC was chosen after a literature review showing the potential of the model to simulate grain yield under different environmental and management conditions. In the evaluation of our results, the adoption of a specific calibration of the model parameters, taking into consideration four plots from one growing season only (2016), has to be taken into consideration. Therefore, the cost function minimization approach, together with the EPIC capability to assimilate satellite data, need to be tested under different environmental and management conditions, at different spatial scales, and using general constraints, not specific for the study site. Limitations of the re-calibration approach used in this study are the fact that it does not consider uncertainties in the satellite observations, in the model parameterization and in the model variables' estimations. This approach is also expensive in terms of computing time. For future applications, the updating methods could be considered in order to alleviate the shortcoming of the re-calibration methods. Two classes exist: on the one hand, stochastic methods that are generally implemented using ensemble data assimilation (e.g., the ensemble Kalman filter or particle filter). On the other hand, the variational methods used in other disciplines, as for example the 4D-Var approach, seem to be promising also in the crop growth model assimilation. The 4D-Var methods can simultaneously assimilate the observational data at multiple times in an assimilation window starting with the assumption that the model is perfect. The choice of the best possible model according to the available data and the amount of the data required by the model is one of the key points, together with the quality of the satellite observations and applicability of the data assimilation method that should be considered in the data assimilation framework.

Author Contributions: Conceptualization, F.V.; formal analysis, F.N.; methodology, F.N.; field data collection and pre-processing of site data H.S., T.S.; supervision, F.V.; writing—original draft, F.N.; writing—review and editing, F.V.

Acknowledgments: This research has received funding from the European Union's Horizon 2020 research and innovation programme under grant agreements No 773903, No 774234 and No 633945.

References

1. Mueller, N.D.; Gerber, J.S.; Johnston, M.; Ray, D.K.; Ramankutty, N.; Foley, J.A. Closing yield gaps through nutrient and water management. *Nature* **2012**, *490*, 254–257. [CrossRef]
2. Battude, M.; Al Bitar, A.; Morin, D.; Cros, J.; Huc, M.; Marais Sicre, C.; Le Dantec, V.; Demarez, V. Estimating maize biomass and yield over large areas using high spatial and temporal resolution Sentinel-2 like remote sensing data. *Remote Sens. Environ.* **2016**, *184*, 668–681. [CrossRef]
3. Yuan, W.; Chen, Y.; Xia, J.; Dong, W.; Magliulo, V.; Moors, E.; Olesen, J.E.; Zhang, H. Estimating crop yield using a satellite-based light use efficiency model. *Ecol. Indic.* **2015**, *60*, 702–709. [CrossRef]
4. Atzberger, C. Advances in remote sensing of agriculture: Context description, existing operational monitoring systems and major information needs. *Remote Sens.* **2013**, *5*, 949–981. [CrossRef]
5. Wall, L.; Larocque, D.; Léger, P.M. The early explanatory power of NDVI in crop yield modelling. *Int. J. Remote Sens.* **2008**, *29*, 2211–2225. [CrossRef]
6. Rembold, F.; Atzberger, C.; Savin, I.; Rojas, O. Using low resolution satellite imagery for yield prediction and yield anomaly detection. *Remote Sens.* **2013**, *5*, 1704–1733. [CrossRef]
7. Manjunath, K.R.; Potdar, M.B.; Purohit, N.L. Large area operational wheat yield model development and validation based on spectral and meteorological data. *Int. J. Remote Sens.* **2002**, *23*, 3023–3038. [CrossRef]
8. Balaghi, R.; Tychon, B.; Eerens, H.; Jlibene, M. Empirical regression models using NDVI, rainfall and temperature data for the early prediction of wheat grain yields in Morocco. *Int. J. Appl. Earth Obs. Geoinf.* **2008**, *10*, 438–452. [CrossRef]

9. Monteith, J.L.; Moss, C.J. Climate and the Efficiency of Crop Production in Britain [and Discussion]. *Philos. Trans. R. Soc. B Biol. Sci.* **2006**, *281*, 277–294. [CrossRef]
10. Moulin, S.; Bondeau, A.; Delecolle, R. Combining agricultural crop models and satellite observations: From field to regional scales. *Int. J. Remote Sens.* **1998**, *19*, 1021–1036. [CrossRef]
11. Huang, Y.; Zhu, Y.; Li, W.; Cao, W.; Tian, Y. Assimilating Remotely Sensed Information with the WheatGrow Model Based on the Ensemble Square Root Filter For improving Regional Wheat Yield Forecasts. *Plant Prod. Sci.* **2013**, *16*, 352–364. [CrossRef]
12. Launay, M.; Guerif, M. Assimilating remote sensing data into a crop model to improve predictive performance for spatial applications. *Agric. Ecosyst. Environ.* **2005**, *111*, 321–339. [CrossRef]
13. Gowda, P.T.; Satyareddi, S.A.; Manjunath, S. Crop Growth Modeling: A Review. *Res. Rev. J. Agric. Allied Sci. Crop.* **2013**, *2*, 1–11.
14. Kasampalis, D.; Alexandridis, T.; Deva, C.; Challinor, A.; Moshou, D.; Zalidis, G. Contribution of Remote Sensing on Crop Models: A Review. *J. Imaging* **2018**, *4*, 52. [CrossRef]
15. Prevot, L.; Chauki, H.; Troufleau, D.; Weiss, M.; Baret, F.; Brisson, N. Assimilating optical and radar data into the STICS crop model for wheat. *Agronomie* **2003**, *23*, 297–303. [CrossRef]
16. de Wit, A.J.W.; van Diepen, C.A. Crop model data assimilation with the Ensemble Kalman filter for improving regional crop yield forecasts. *Agric. For. Meteorol.* **2007**, *146*, 38–56. [CrossRef]
17. Jin, X.; Kumar, L.; Li, Z.; Yang, G.; Wang, J. Review article A review of data assimilation of remote sensing and crop models. *Eur. J. Agron.* **2018**, *92*, 141–152. [CrossRef]
18. Dorigo, W.A.; Zurita-milla, R.; De Wit, A.J.W.; Brazile, J. A review on reflective remote sensing and data assimilation techniques for enhanced agroecosystem modeling. *Int. J. Appl. Earth Obs. Geoinf.* **2007**, *9*, 165–193. [CrossRef]
19. Machwitz, M.; Giustarini, L.; Bossung, C.; Frantz, D.; Schlerf, M.; Lilienthal, H.; Wandera, L.; Matgen, P.; Hoffmann, L.; Udelhoven, T. Enhanced biomass prediction by assimilating satellite data into a crop growth model. *Environ. Model. Softw.* **2014**, *62*, 437–453. [CrossRef]
20. Fang, H.; Liang, S.; Hoogenboom, G.; Teasdale, J.; Cavigelli, M. Corn-yield estimation through assimilation of remotely sensed data into the CSM-CERES-Maize model. *Int. J. Remote Sens.* **2008**, *29*, 3011–3032. [CrossRef]
21. Ma, G.; Huang, J.; Wu, W.; Fan, J.; Zou, J.; Wu, S. Assimilation of MODIS-LAI into the WOFOST model for forecasting regional winter wheat yield. *Math. Comput. Model.* **2013**, *58*, 634–643. [CrossRef]
22. Ren, J.; Yu, F.; Qin, J.; Chen, Z.; Tang, H. Integrating remotely sensed LAI with EPIC model based on global optimization algorithm for regional crop yield assessment. In Proceedings of the IEEE International Geoscience and Remote Sensing Symposium, Honolulu, HI, USA, 25–30 July 2010.
23. Tan, G.; Shibasaki, R. Global estimation of crop productivity and the impacts of global warming by GIS and EPIC integration. *Ecol. Model.* **2003**, *168*, 357–370. [CrossRef]
24. Liu, J.; Chen, Z.; Sun, L.; Ren, J.; Jiang, Z.; Chen, J.; Li, H.; Li, Z. Application of Crop Model Data Assimilation with a Particle Filter for Estimating Regional Winter Wheat Yields. *IEEE J. Sel. Top. Appl. Earth Obs. Remote Sens.* **2014**, *7*, 4422–4431.
25. Dente, L.; Satalino, G.; Mattia, F.; Rinaldi, M. Assimilation of leaf area index derived from ASAR and MERIS data into CERES-Wheat model to map wheat yield. *Remote Sens. Environ.* **2008**, *112*, 1395–1407. [CrossRef]
26. Baret, F.; Buis, S. Estimating Canopy Characteristics from Remote Sensing Observations: Review of Methods and Associated Problems. In *Advances in Land Remote Sensing*; Springer: Dordrecht, The Netherlands, 2008; pp. 173–201.
27. European Space Agency (ESA). *SENTINEL-2 User Handbook*; ESA: Paris, France, 2015; pp. 1–64.
28. Richter, K.; Atzberger, C.; Vuolo, F.; Weihs, P.; Urso, G.D. Experimental assessment of the Sentinel-2 band setting for RTM-based LAI retrieval of sugar beet and maize. *Can. J. Remote Sens.* **2009**, *35*, 230–247. [CrossRef]
29. Vuolo, F.; Zóltak, M.; Pipitone, C.; Zappa, L.; Wenng, H.; Immitzer, M.; Weiss, M.; Baret, F.; Atzberger, C. Data service platform for Sentinel-2 surface reflectance and value-added products: System use and examples. *Remote Sens.* **2016**, *8*, 938. [CrossRef]
30. Upreti, D.; Huang, W.; Kong, W.; Pascucci, S.; Pignatti, S.; Zhou, X.; Ye, H.; Casa, R. A Comparison of Hybrid Machine Learning Algorithms for the Retrieval of Wheat Biophysical Variables from Sentinel-2. *Remote Sens.* **2019**, *11*, 481. [CrossRef]

31. Vuolo, F.; Essl, L.; Zappa, L.; Sandén, T.; Spiegel, H. Water and nutrient management: The Austria case study of the FATIMA H2020 project. *Adv. Anim. Biosci.* **2017**, *8*, 400–405. [CrossRef]
32. Fatima-H2020. Marchfeld Pilot Area in Austria. Available online: http://fatima-h2020.eu/pilots/austria-marchfeld/ (accessed on 14 January 2019).
33. Thaler, S.; Eitzinger, J.; Trnka, M.; Dubrovsky, M. Impacts of climate change and alternative adaptation options on winter wheat yield and water productivity in a dry climate in Central Europe. *J. Agric. Sci.* **2012**, *150*, 537–555. [CrossRef]
34. Vuolo, F.; Neugebauer, N.; Bolognesi, S.F.; Atzberger, C.; D'Urso, G. Estimation of leaf area index using DEIMOS-1 data: Application and transferability of a semi-empirical relationship between two agricultural areas. *Remote Sens.* **2013**, *5*, 1274–1291. [CrossRef]
35. USDA. Natural Resources Conservation Service Soil. Available online: https://www.nrcs.usda.gov/wps/portal/nrcs/detail/soils/ref/?cid=nrcs142p2_054253 (accessed on 7 May 2019).
36. LI_COR. *LAI-2200 Plant Canopy Analyzer Instruction Manual*; LI-COR: Lincoln, Nebraska, 2017; p. 262.
37. LI-COR Biosciences—Impacting Lives Through Science. Available online: https://www.licor.com/ (accessed on 7 May 2019).
38. Vuolo, F.; Dash, J.; Curran, P.J.; Lajas, D.; Kwiatkowska, E. Methodologies and uncertainties in the use of the terrestrial chlorophyll index for the sentinel-3 mission. *Remote Sens.* **2012**, *4*, 1112–1133. [CrossRef]
39. González-Sanpedro, M.C.; Le Toan, T.; Moreno, J.; Kergoat, L.; Rubio, E. Seasonal variations of leaf area index of agricultural fields retrieved from Landsat data. *Remote Sens. Environ.* **2008**, *112*, 810–824. [CrossRef]
40. Nelson, D.; Wang, J. Introduction to artificial neural systems. *Neurocomputing* **2003**, *4*, 328–330. [CrossRef]
41. Lek, S.; Guégan, J.F. Artificial neural networks as a tool in ecological modelling, an introduction. *Ecol. Model.* **1999**, *120*, 1–9. [CrossRef]
42. Weiss, M.; Baret, F. *Sentinel2 ToolBox Level2 Products S2ToolBox Level 2 Products: LAI, FAPAR, FCOVER Version 1.1*; INRA-CSE: Avignon, France, 2016; p. 53.
43. Müller-Wilm, U. *Sen2Cor Configuration and User Manual—Ref. S2-PDGS MPC-L2A-SUM-V2.5.5*; Telespazio VEGA: Luton, Bedfordshire, UK, 2018; p. 54.
44. Williams, J.R.; Jones, C.A.; Kiniry, J.R.; Spanel, D.A. The EPIC Crop Growth Model. *Trans. ASAE* **1989**, *32*, 0497–0511. [CrossRef]
45. Williams, J.R.; Dagitz, S.; Magre, M.; Meinardus, A.; Staglich, E.; Taylor, R. *Environmental Policy Integrated Climate Model*; Model User Manual Version 0810; Blackland Research and Extension Center A&M AgriLife: College Station, TX, USA, 2015.
46. Kiniry, J.R.; Williams, J.R.; Major, D.J.; Izaurralde, R.C.; Gassman, P.W.; Morrison, M.; Bergentine, R.; Zentner, R.P. EPIC model parameters for cereal, oilseed, and forage crops in the northern Great Plains region. *Can. J. Plant Sci.* **2011**, *75*, 679–688. [CrossRef]
47. Huang, M.; Gallichand, J.; Dang, T.; Shao, M. An evaluation of EPIC soil water and yield components in the gully region of Loess Plateau, China. *J. Agric. Sci.* **2006**, *144*, 339–348. [CrossRef]
48. EPIC & APEX Model. Model Executables. Available online: https://epicapex.tamu.edu/model-executables/ (accessed on 9 October 2018).
49. Constrained Nonlinear Optimization Algorithms. MATLAB & Simulink. Available online: https://ch.mathworks.com/help/optim/ug/constrained-nonlinear-optimization-algorithms.html#brnpd5f (accessed on 9 November 2018).
50. Byrd, R.H.; Hribar, M.E.; Nocedal, J. An interior point method for large scale nonlinear programming. *SIAM J. Optim.* **1999**, *9*, 877–900. [CrossRef]
51. Backhaus, K.; Erichson, B.; Weiber, R. *Fortgeschrittene Multivariate Analysemethoden eine Anwendungsorientierte Einführung*; Springer-Lehrbuch: Basel, Switzerland, 2013.
52. Soliani, L. Fondamenti di Statistica Applicata All'analisi e Alla Gestione Dell'ambiente. 2001. Available online: https://dokumen.tips/documents/fondamenti-di-statistica-applicata-allanalisi-ambientalepdf25-distribuzioni.html (accessed on 7 May 2019).
53. Richter, K.; Atzberger, C.; Hank, T.B.; Mauser, W. Derivation of biophysical variables from Earth observation data: Validation and statistical measures. *J. Appl. Remote Sens.* **2012**, *6*, 063557. [CrossRef]
54. Bellocchi, G.; Rivington, M.; Donatelli, M.; Matthews, K. Validation of biophysical models: Issues and methodologies. *Agron. Sustain. Dev.* **2010**, *30*, 109–130. [CrossRef]

55. Vanino, S.; Nino, P.; De Michele, C.; Falanga Bolognesi, S.; D'Urso, G.; Di Bene, C.; Pennelli, B.; Vuolo, F.; Farina, R.; Pulighe, G. Capability of Sentinel-2 data for estimating maximum evapotranspiration and irrigation requirements for tomato crop in Central Italy. *Remote Sens. Environ.* **2018**, *215*, 452–470. [CrossRef]
56. Wang, H.; Zhu, Y.; Li, W.; Cao, W.; Tian, Y. Integrating remotely sensed leaf area index and leaf nitrogen accumulation with RiceGrow model based on particle swarm optimization algorithm for rice grain yield assessment. *J. Appl. Remote Sens.* **2014**, *8*, 083674. [CrossRef]
57. Lazauskas, S.; Povilaitis, V.; Antanaitis, Š.; Sakalauskaite, J.; Sakalauskiene, S.; Pšibišauskiene, G.; Auškalniene, O.; Raudonius, S.; Duchovskis, P. Winter wheat leaf area index under low and moderate input management and climate change. *J. Food Agric. Environ.* **2012**, *10*, 588–593.
58. Balkovič, J.; van der Velde, M.; Schmid, E.; Skalský, R.; Khabarov, N.; Obersteiner, M.; Stürmer, B.; Xiong, W. Pan-European crop modelling with EPIC: Implementation, up-scaling and regional crop yield validation. *Agric. Syst.* **2013**, *120*, 61–75. [CrossRef]

7

Integrating Modelling and Expert Knowledge for Evaluating Current and Future Scenario of Large Cardamom Crop

Sajana Maharjan, Faisal Mueen Qamer *, Mir Matin, Govinda Joshi and Sanjeev Bhuchar

International Centre for Integrated Mountain Development (ICIMOD), Kathmandu 44700, Nepal
* Correspondence: fqamer@icimod.org

Abstract: Large Cardamom (*Amomum subulatum* Roxb.) is one of the most valuable cash crop of the Himalayan mountain region including Nepal, India, and Bhutan. Nepal is the world's largest producer of the crop while the Taplejung district contributes a 30%–40% share in Nepal's total production. Large cardamom is an herbaceous perennial crop usually grown under the shade of the Uttis tree in very specialized bioclimatic conditions. In recent years, a decline in cardamom production has been observed which is being attributed to climate-related indicators. To understand the current dynamics of this under-canopy herbaceous crop distribution and its future potential under climate change, a combination of modelling, remote sensing, and expert knowledge is applied for the assessment. The results suggest that currently, Uttis tree cover is 10,735 ha in the district, while 50% (5198 ha) of this cover has a large cardamom crop underneath. When existing cultivation is compared with modelled suitable areas, it is observed that the cultivatable area has not yet reached its full potential. In a future climate scenario, the current habitat will be negatively affected, where mid elevations will remain stable while lower and higher elevation will become infeasible for the crop. Future changes are closely related to temperature and precipitation which are steadily changing in Nepal over time.

Keywords: large cardamom; remote sensing; species modelling; habitat assessment; climate change

---

1. Introduction

Large cardamom (*Amomum subulatum* Roxb.) is mostly being cultivated in the Himalayan mountain region of Nepal, India, and Bhutan [1]. There is an increasing demand of the spices from local to global markets which has fascinated farmers in its commercial cultivation [2]. The high-value crop cultivation has substantially improved the livelihood of farmers residing in these mountain regions. However, several studies have highlighted threats of cardamom farming which include problems in disease management, change in climatic conditions, and human activities such as infrastructure development in cardamom growing areas [2,3]. Alongside this, comprehensive information on its current distribution and potentially suitable areas for cultivation is unavailable, which is essential for crop planning and management. Further, for its distribution and current suitability, information on the impacts of climate change is also vital for future planning as climate has a significant impact on the geographical distribution of plant species and also alters habitat conditions [4,5].

Large cardamom is an herbaceous perennial crop usually grown under shade. Uttis (*Alnus nepalensis*) trees provide excellent shade, supply a good amount of litter from twigs and leaves, and nitrogen from the root nodules to understory cardamom when they are young. The crop is generally grown at an altitude of 700 to 2000 m above sea level in humid conditions, with temperature ranges between 4–20 °C and annual precipitation around 2000–2500 mm [6].

Although satellite remote sensing data can be used for accurate, timely, and consistent information on the agricultural productivity at local and regional scales [7,8], detection of understory plants is inhibited due to canopy cover, canopy gap shadowing, and terrain variability [9]. Differentiating signals of understory vegetation from overstory canopy is still challenging due to complex interactions between overstory and understory vegetation [10]. Several studies have been conducted to distinguish understory vegetation using remote sensing approaches. In this regard, evergreen understory vegetation were identified from Landsat images of leaf off-season in deciduous forest [11,12]. Phenological difference between overstory and understory vegetation was also used to detect understory vegetation using Landsat images [13,14]. However, such studies require multi-temporal data which are generally available in coarser spatial resolutions. Leduc et al. [15] have mapped wild leek which grows on forest floors using low flying Unmanned Aerial Vehicle (UAV). Other studies have used data from active sensor, mainly LIDAR data, to map understory plants in boreal forests [16,17]. Combinations of LIDAR data and hyperspectral images were used to map understory invasive species in tropical forests [9], and the use of LIDAR data and high-resolution IKONOS imagery to identify understory plant invasion in urban forests [18]. Nevertheless, there are limitations on the use of these data due to its high cost and low availability, mainly in developing countries [19].

Understory vegetation are usually hard to identify only from remote sensing, thus a substantial efforts have been made to identify these vegetation through the integration of several approaches. For instance, Wang et al. [20] have mapped understory bamboo by integrating neural network and Geographic Information System (GIS) expert system, and Tuanmu et al. [10] have detected understory vegetation using phenology metrics derived from time series of Moderate Resolution Imaging Spectroradiometer (MODIS) data together with a suitability model like the maximum entropy (Maxent) model. The Maxent model is a species distribution model (SDM) that provides an approach for making predictions from existing distribution and a set of predictors [21]. Such models predicting the potential distribution of species are useful for several applications in conservation biology [22,23]. Other approaches, such as object-based image analysis (OBIA) on multispectral and hyperspectral data, were found to be effective in identifying understory plant species [24] due to its characteristics of using contextual relationships together with spectral information. In addition, expert knowledge [25] and other ancillary data such as elevation, slope, and aspect can also be used to improve classification accuracy. Moreover, the combination of participatory mapping with remote sensing technique can further improve the accuracy as the two methods complement and validate each other [26]. Participatory mapping is an effective tool to obtain accurate baseline data of the field [27], which can be integrated in remote sensing analyses to produce more accurate maps [28].

Several species distribution models such as the genetic algorithm for rule-set production (GARP), ecological niche factor analysis (ENFA), bioclimatic modeling (BIOCLIM), CLIMEX (climate change experiment), domain environmental envelope (DOMAIN), and Maxent (maximum entropy) have been used for species distribution prediction. Among these, Maxent is widely used as it can perform better even with small sample sizes compared to other modelling methods [29–31]. In addition, it has other merits, such as: it requires only presence data of the species and environmental variables for the study area, it can use both categorical and continuous data and can incorporate interactions among different variables, it can produce spatially explicit habitat suitability maps, and the importance of each environmental variable on the model can be evaluated using the built-in jackknife test.

This study aims to evaluate existing spatial distribution, potential suitable areas for cultivation under various scenarios, and core distributional shift with future climate condition in the Taplejung district of Nepal. The study uses a combination of expert knowledge and species modelling approach to capture all aspects of habitat conditions in the complex environmental conditions in the mountain region.

2. Materials and Methods

2.1. Study Area

The study area, the Taplejung district, lies in eastern Nepal. The district (27°57′10′′–27°16′5′′ N and 87°26′40′′–88°12′6′′ E) with an area of 3646 km^2 physiographically lies in the mid-hill to high-himal region. It lies at an elevation of 498 m to 8464 m. Forest is the major land cover of the district followed by agriculture, bare land, grass, snow, shrub, water body, and built-up, respectively. The district is the major producer of large cardamom in the country, which is the major source of income for farmers in the district. In Nepal, an estimated 12,000 ha of land in over 40 mid-hills districts are under cardamom cultivation with estimated annual production of 6000 metric tons. While the Taplejung district contribute around 4500 ha of land and yearly production of around 2600 metric tons [32].

The study broadly followed two broad approaches for Cardamom habitat mapping including species modelling- and expert knowledge-based assessment (Figure 1).

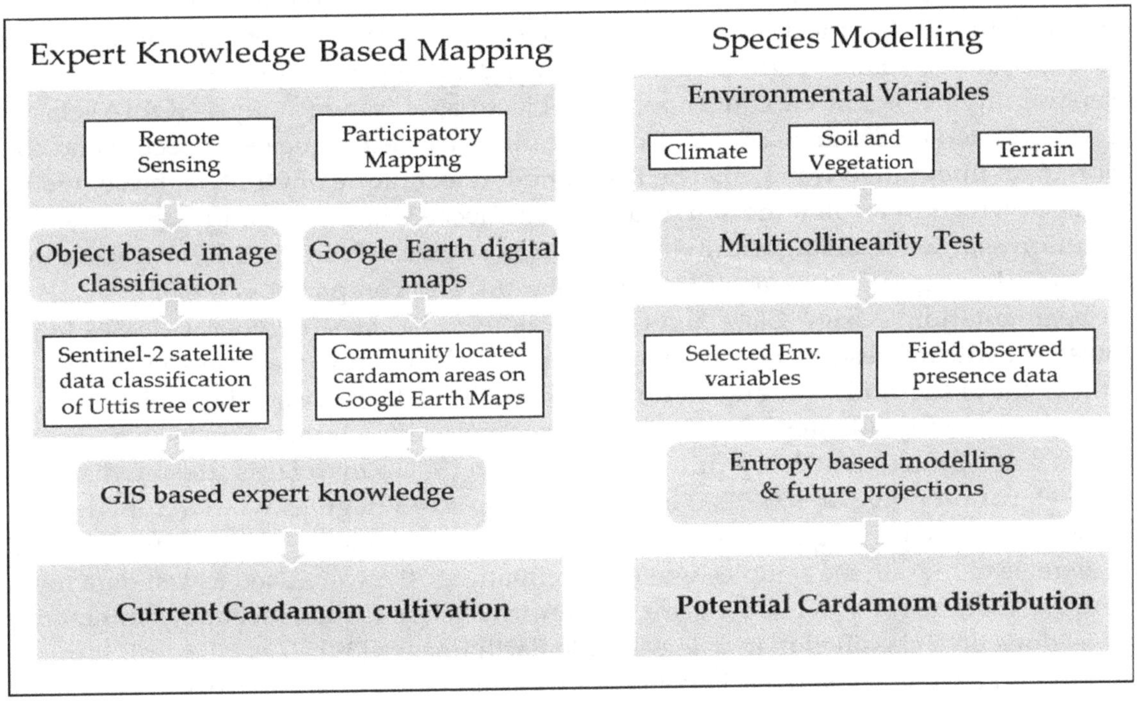

Figure 1. Methodology flow diagram.

2.2. Expert Knowledge-Based Mapping

2.2.1. Participatory Mapping

Participatory mapping with the community in developing countries is found to be effective for understanding local level phenomena [33]. A two-day exercise was conducted with a community representative to map existing large cardamom farming areas on the high-resolution data of Google Earth images available in digital format as well as on the printed maps. During the exercise, large clusters of cardamom cultivated fields were identified. However, detailed delineation of cardamom fields including small patches was not possible. The data was gathered to get existing farming areas on the district which was also cross-checked with outputs obtained from remote sensing data. A part of the pockets identified by the farmers in the participatory exercise was further verified through ground validation exercise. This validation also provided GPS samples of cardamom field points along with other vegetation in the area.

2.2.2. Uttis (*Alnus nepalensis*) Mapping Using High-Resolution Satellite Data

During the participatory mapping and the field activity, it was observed that the Uttis is the major tree species grown for shade to cultivate large cardamom. Since the understory crop could not be directly mapped using remote sensing data, delineation of Uttis cover was consider as proxy to map the large cardamom crop. Our focus was to map Uttis in the entire district, which can provide information on prospective cultivation of large cardamom. Uttis tree cover was mapped by using Sentinel-2A Level 1C product image acquired in February 2016. The image has very high-resolution spectral coverage that includes 12 bands (coastal aerosol, blue green, red, 3 vegetation red edge, NIR, vegetation red edge water vapor, SWIR Cirrus, and 2 SWIR respectively). The spatial resolution of blue green, red and NIR is 10 m. The resolution of the vegetation red edge band is 20 m and for the rest of the bands it is 60 m. These features of the sensor are suited for agricultural monitoring systems [34]. Level 1C product is a Top of Atmosphere (TOA) product for which atmospheric correction had to be done to get reflectance values of the image so that the image can be used for mapping. SNAP (Sentinel toolbox) software (version 5.0.0) was used for atmospheric correction of the image.

Spectral separability of forest types such as coniferous forest, broadleaf forest, Uttis, and shrub was studied before the classification. The mean pixel values of the abovementioned forest types and shrub were plotted against eight bands of Sentinel-2A image to evaluate the potential of image spectral separability before the classification. The object-based image analysis (OBIA) classification approach was adopted for Sentinel-2A image classification. The technique uses spectral and contextual information in an integrative way [25]. The fundamental technique of OBIA is the segmentation of satellite images which overcome the salt and pepper effect [35]. In this study, the chessboard and multi-resolution segmentation algorithm in eCognition software (version 8.7,) was used to develop image objects [36] using scale parameter = 80, shape = 0.1, and compactness = 0.8.

After segmentation, Assign Class algorithm was used to classify general classes (agriculture, conifer forest, broadleaf forest, shrub, water, and snow). This was done to filter the land cover which were not the focus of the study. Several features such as NDVI, brightness, slope, elevation, and field information were used in this step. For the rest of the unclassified image objects, Nearest Neighbor Classification was applied, which is a powerful approach [37] to map Uttis trees. Uttis trees grown between the elevation range of 800–2200 m and slope up to 45 degrees were mapped. This is also the elevation and the slope range where large cardamom is cultivated [38]. Seventy percent (70%) of field data were used to train the samples while the remaining 30% were used as test data for accuracy assessment. Further, in order to reduce error and improve classification accuracy, interactive visual analysis was done on a classified map using Google Earth images [39].

The remote sensing-derived Uttis cover and participatory mapping-based identification of cardamom crop areas were overlaid to get the existing large cardamom farming area. The obtained large cardamom maps in the Taplejung district were further analyzed based on elevation, slope, aspect, and Village Development Committees (VDCs). The analysis was done to understand the pattern of cultivation and environment suitability conditions in the district and the information could be useful in further planning and management of large cardamom farming in the district. Elevation range of the study area was categorized into 9 ranges (below 800 m, 800–1000 m, 1000–1200 m, 1200–1400 m, 1400–1600 m, 1600–1800 m, 1800–2000 m, 2000–2200 m, and above 2200 m). In order to comprehend the farming area slope-wise, the slope was categorized into four gradient levels (0°–15°, 15°–30°, 30°–45°, 45° above).

2.3. *Species Modelling*

2.3.1. Environmental Variables and Species Occurrence Records

The habitat suitability model was primarily developed based on the variables related to climate, soil, terrain, and vegetation type. Initially, 24 variables were selected to model the current distribution pattern. These comprised of 19 bioclimatic variables with 30 arc seconds (~1 km) spatial resolution from WorldClim dataset (http://www.worldclim.org/), 12.5 m resolution digital elevation model (DEM) generated by The Japan Aerospace Exploration Agency (JAXA) using ALOS PALSAR RTC products [40], slope and aspect layers generated from the DEM using spatial analyst tools in ArcGIS 10.6.1 soil pH with the resolution of 250 m was downloaded from https://soilgrids.org/#!/?layer=TAXNWRB_250m&vector=1 and Uttis cover thematic layer. All bioclimatic variables, topographic layers, and pH were resampled into 20 m spatial resolution to make them compatible with the resolution of Uttis cover. This was done in ArcGIS 10.6.1 with the nearest neighbor resampling technique. For future scenarios (year 2050), we initially selected DEM, slope, aspect, and 19 bioclimatic variables for RCP2.6 (the minimum greenhouse gas emission scenario), and RCP8.5 (the maximum greenhouse gas emission scenario) as adopted by the Intergovernmental Panel on Climate Change (IPCC) in its fifth assessment report (AR5). The climatic variables were also resampled into 20 m to make the variables uniform at one resolution.

Multicollinearity between predictor variables was tested as it can lead to inaccurate prediction by excluding significant explanatory variables [41]. The test was conducted calculating Pearson's Correlation Coefficient (r) to assess the cross-correlation and the one variable from any pair of variables with a cross-correlation coefficient value of > ±0.8 was excluded [42]. Variables were chosen based on biological relevancy to the species. For example, pH was highly correlated with temperature seasonality (BIO4) ($r = 0.88$), from this pair BIO4 (temperature seasonality) was removed as pH plays a crucial role in cardamom plantation [2]. In addition, the variation inflation factor (VIF) was also used to check collinearity among the variables in R software (version 3.6). Variables with VIF values greater than 10 were excluded for modelling [43]. Out of 24 variables, 8 environmental variables (Uttis cover, pH, slope, aspect, isothermality, maximum temperature of warmest month, minimum temperature of coldest month, and precipitation of wettest month) were selected for the current scenario. For the future scenario of RCP2.6-2050 and for RCP8.5-2050, 6 variables (slope, aspect, isothermality, maximum temperature of warmest month, minimum temperature of coldest month, and precipitation of wettest month) were selected for modelling. As part of the species presence data, a total of 102 occurrence records (GPS coordinates of points) of large cardamom were collected randomly during the field visit conducted in May 2016.

2.3.2. Spatial Modelling and Statistical Analysis

The maximum entropy modelling approach using Maxent software (version 3.3.3k) was applied in this study for predicting habitat suitability of the large cardamom. Maxent is a machine-learning method with a simple and precise mathematical formulation. It uses a maximum entropy algorithm to produce a model that shows the probability of presence of the species that varies from 0 to 1, i.e., from the lowest to the highest probability [21]. We selected 70% of the data for training and the remaining 30% for testing. Area under the ROC (receiver operating characteristic) curve (AUC) was used to evaluate the model performance which ranges from 0 to 1. The curve is plotted with True Positive Rate (sensitivity) at the vertical axis and False Positive Rate (1-specificity) at the horizontal axis. The jackknife was used to evaluate the importance of the variables on the model. The model used logistic format. The final distribution maps have values ranges from 0 to 1 which were grouped into four classes of suitable habitat viz., unsuitable (<0.2), marginally suitable (0.2–0.4), moderately suitable (0.4–0.6), and highly suitable (>0.6). Further, analysis of existing farming area in comparison to current habitat suitability map was performed to understand the gaps and opportunities. In order to assess the change between current and future suitable areas, we quantified the areas (ha) of classes of habitat suitability under different scenarios across different elevation ranges.

3. Results and Discussions

3.1. Participatory Mapping of Large Cardamom

The spatial location of large cardamom fields in the Taplejung District was recorded on the basis of local people knowledge. Out of 50 VDCs in the district, 49 VDCs were found cultivating (Figure 2) large cardamom. However, the farming area varies from VDC to VDC. Participatory mapping has given a general overview of farming area in the district.

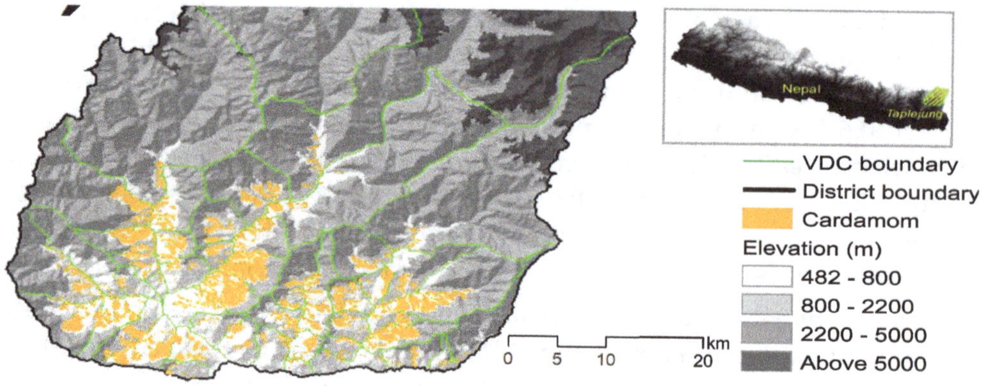

Figure 2. Mapping of large cardamom crop clusters across the Taplejung district based on participatory mapping.

3.2. Uttis (Alnus nepalensis) Cover Mapping and Delineation of Accurate Large Cardamom Map

Spectral separability of major vegetation of the study area using Sentinel-2A image shows that the vegetation are largely separable from each other in NIR and Red Edge bands (Figure 3a). In this study, integration of ancillary data with Sentinel-2A image was applied to map Uttis using OBIA (Figure 3b,c). The classification approach in eCognition has helped to use expert knowledge in differentiating trees within agricultural land and forests, resulting in better classification accuracy. During post classification improvement, small tree patches of Uttis within agricultural land were separated. Shadow in high mountain areas was affecting the classification which was improved during post classification using Google Earth images. There are several studies done on forest type classification [44,45] and land cover classification [39], however, these studies do not include separation of tree species within the forest cover classes.

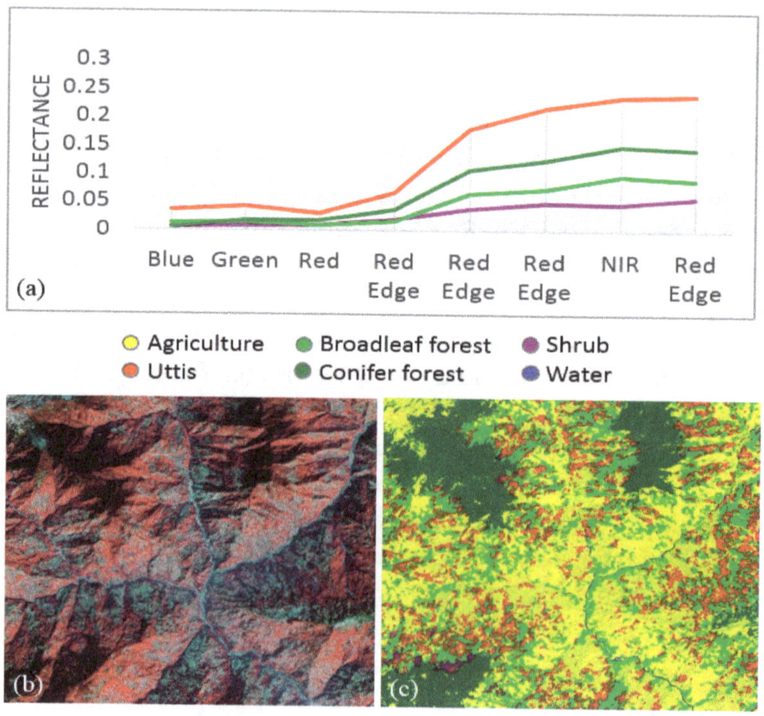

Figure 3. (**a**) Spectral separability of vegetation, (**b**) Sentinel-2A satellite image (False color), (**c**) object-based classification of satellite data.

Uttis cover in the Taplejung district is about 10,735 ha (Figure 4a). Overall, accuracy of the classification was 80% with producer's and user's accuracies at 89% and 84%, respectively (Table 1). The obtained Uttis cover and participatory maps were overplayed to get fine resolution large cardamom farming area (Figure 4b). The overlay produced 5198 ha of existing farming area in the district.

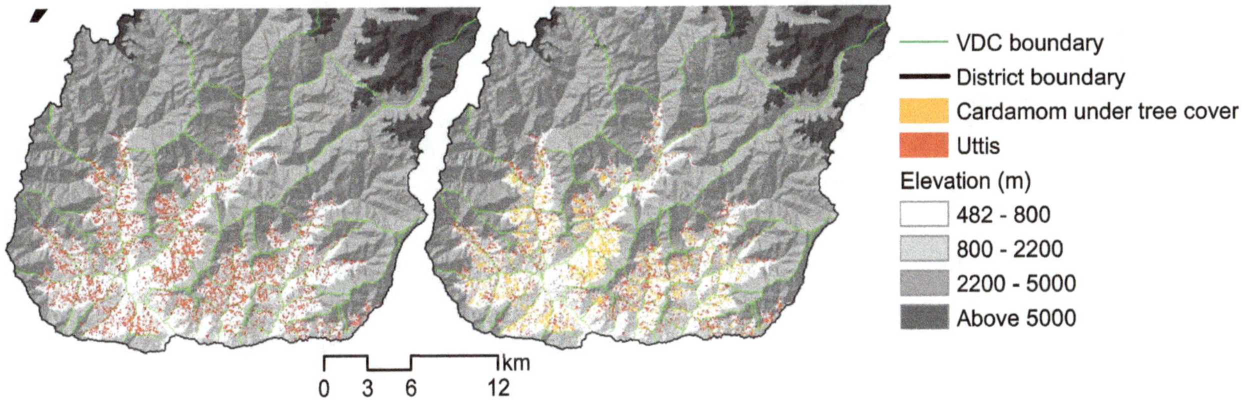

Figure 4. (**a**) Spatial distribution of Uttis tree cover, (**b**) spatial distribution of cardamom under Uttis tree and standalone Uttis tree cover.

Table 1. Contingency matrix for accuracy assessment.

		Observed Vegetation Classes					Grand Total	User's Accuracy
		Agriculture	**Conifer**	**Other Broadleaf**	**Shrubs**	**Uttis**		
Mapped Vegetation Classes	Agriculture	15	0	0	3	1	19	79
	Conifer	0	16	2	0	1	19	84
	Other Broadleaf	0	2	12	0	2	16	75
	Shrubs	2	0	3	16	0	21	76
	Uttis	2	0	3	1	25	31	81
Grand Total:		19	18	20	20	29	106	
Producer's Accuracy:		79	89	60	80	86		
		Overall accuracy: 80%						

The highest cultivation area is found at the elevation range of 1600–1800 m followed by 1800–2000 m, 1400–1600 m, 1200–1400 m, 2000–2200 m, 1000–1200 m, and 800–1000 m, respectively (Figure 5a). Although appropriate elevation for large cardamom is 800–2200 m [38], the farming is also observed below 800 m and above 2200 m. However, there are several species of large cardamom which can be farmed based on altitude [2]. The optimal productivity is possible if large cardamom species is chosen based on elevation.

The largest cultivation area is found in the range of 15°–30° slope while the least is found in the range of 45° and above (Figure 5b). The cultivation area is nearly doubled on the slope of 30°–45° compared to the 0°–15° slope. Several literatures [46,47] stated that the best aspect for cardamom cultivation is North and North-East facing slope. The field data collected for the study showed that cardamom is being cultivated in all aspects. There is no such strict constraint in the selection of aspect for farming. The result shows that the farming area is primarily found in South-West followed by West, North, East, South, North-East, South-East, and North-West (Figure 5c). This indicates that large cardamom can be grown in all kinds of aspects. An evaluation on productivity of large cardamom cultivated at various aspects is essential to identify the best aspect for the crop farming.

Out of 50 VDCs of the Taplejung district, Dhungesanghu, Hangdewa, Hangpang, Phurumbu, Sikecha, and Thukinba are the top six VDCs which have the highest cultivation area (Figure 5d). Among those VDCs, Phurumbu and Hangdewa consist of more than 400 ha of farming area whereas

Dhungesanghu and Hangpang contain approximately 285 ha to 305 ha of cultivation area. Sikecha and Thukimba hold around 200 ha of cardamom field.

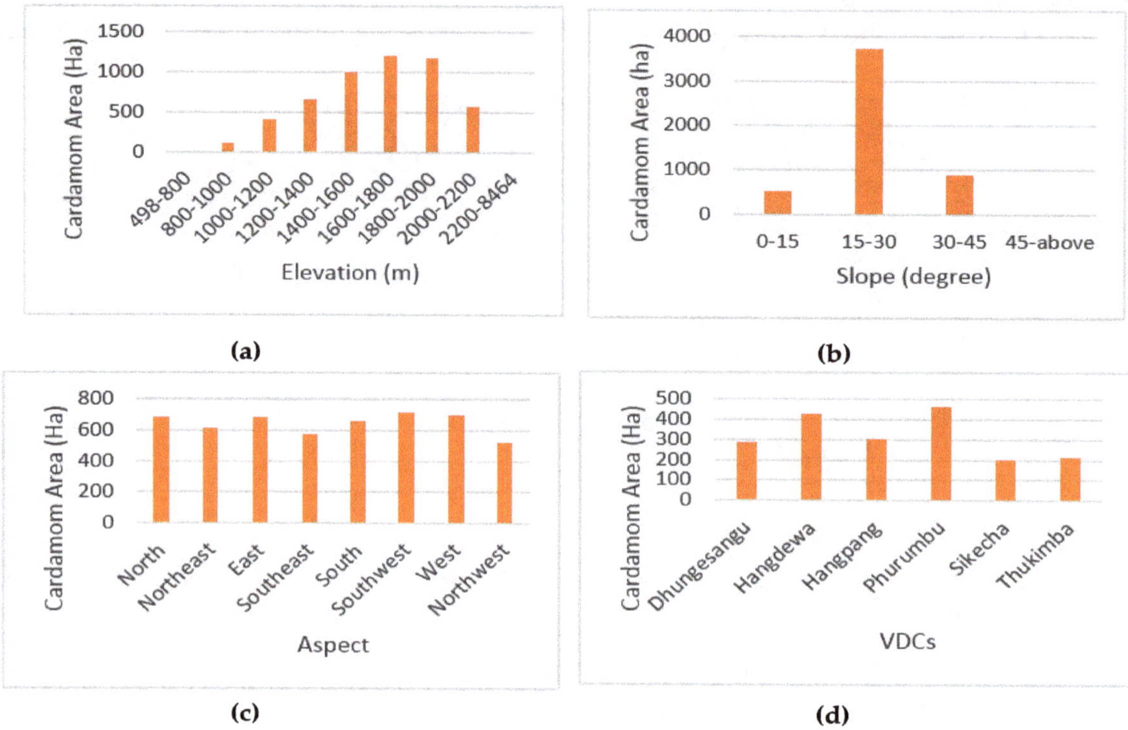

Figure 5. Analysis of large cardamom cultivation area based on elevation, slope, aspect, and VDCs. (**a**). Cardamom cultivation area with elevation; (**b**). Cardamom cultivation area with slope; (**c**). Cardamom cultivation area in various aspects; (**d**). Top 6 VDCs with the highest cultivation area.

3.3. Current Suitable Habitat

The Maxent model provided a comprehensive understanding of the distribution of large cardamom (Figure 6a). Currently, the highly suitable area in the district is about 13,679 ha and the moderately suitable habitat is about 27,778 ha. The most suitable habitat for large cardamom was predicted in the southern part of Taplejung and its distribution is almost continuous. Suitability decreases with an increase in altitude. The maxent-predicted model had high accuracy with an AUC value of 0.941 for training data and 0.934 for test data. Jackknife results showed that Bio6 (minimum temperature of coldest month) among the eight variables considered for the model had the highest predictive power. The Bio5 (maximum temperature of warmest month) is the second most important variable followed by Bio13 (precipitation of wettest month), pH, Uttis cover, and Bio3 (Isothermality) (Figure 6b).

SDM are influenced by several factors, such as data quality [48,49] and decisions taken during the model fitting [50], sample size [51], multicollinearity [52], and selection of independent variables [53]. Despite these, SDM are increasingly used for various purposes, such as species conservation planning [42] and risk analysis [50]. In this study, we have dealt with some of these issues, such as multicollinearity by removing highly correlated variables, selection of important variables for the species, and considered default settings in Maxent as it provided the best model. Maxent performs best among other modelling methods and even performs better with small sample sizes compared to other modelling methods [30,54,55].

3.4. Current Cardamom Cultivation and Habitat Suitability Analysis

Almost 33% of the current farming area is found in the highly suitable class, 44% of the area is found in the moderately suitable class, and nearly 23% of the area is in the marginally suitable class. This indicates that the cultivation has not yet reached its full potential range. Therefore, the farmers need sensitization on potential areas for the farming which would eventually improve productivity (Figure 7).

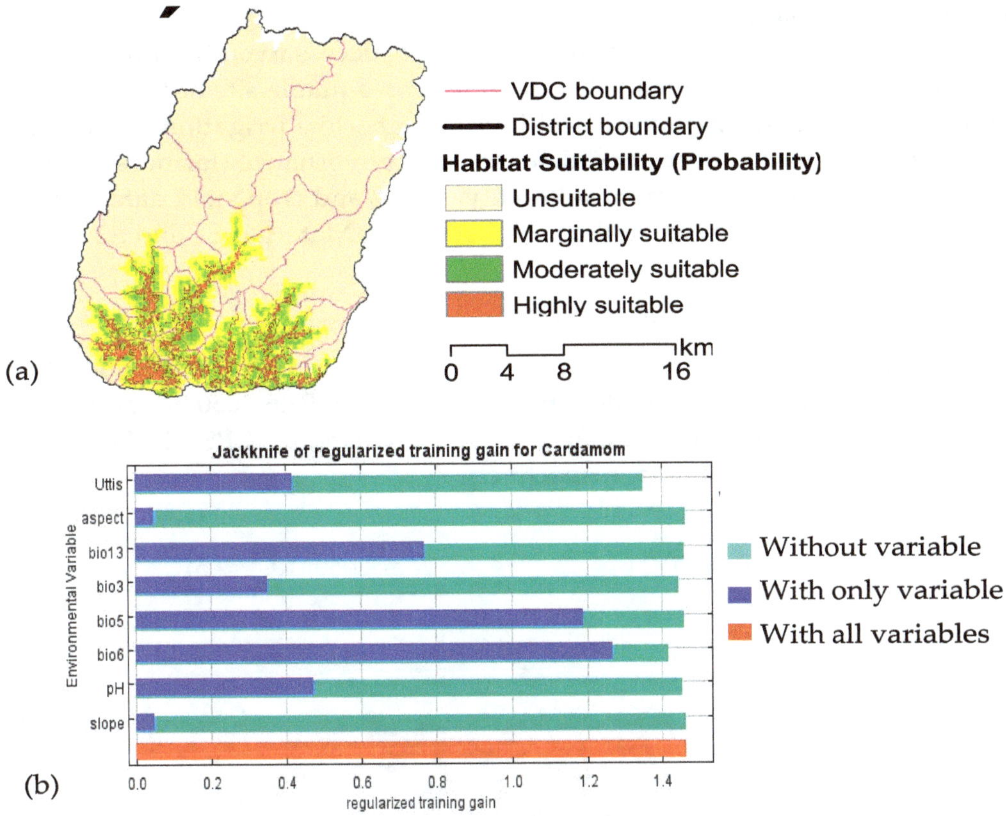

Figure 6. (**a**) Potential distribution of cardamom in the district, (**b**) Relative predictive power of different contributing variables based on the jackknife of regularized training gain in the Maxent model for large cardamom.

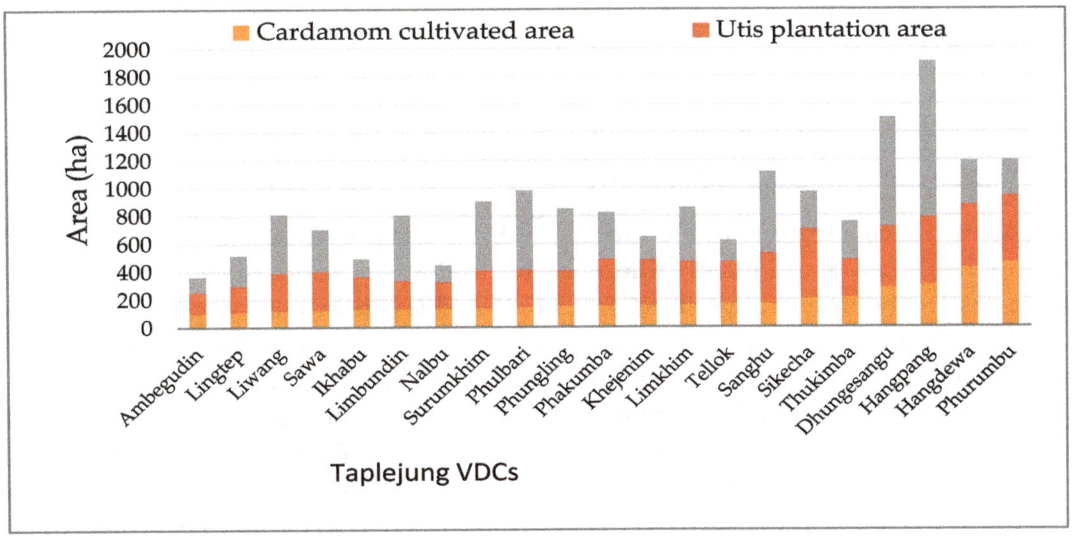

Figure 7. VDC-wise statistics to identify current utilization of the potential habitat.

The Uttis map, existing cultivated area, and current habitat suitability of large cardamom provided the means to look at the gaps and opportunities for cardamom cultivation at the VDC level. Figure 7 demonstrates that the current cultivation area of large cardamom is much less in most of the VDCs, though there are more Uttis areas and highly suitable areas. The cardamom cultivated area and Uttis area in Ambegudin, Hangdewa, Phurumbu, and Thukimba VDCs are much closer, whereas current highly suitable areas in Hangdewa, Ikhabu, Nalbu, Phurumbu, and Tellok VDCs are less than the existing cultivated area. The statistics describing gaps and opportunities for cardamom crop cultivation will support management plans across the administrative area.

Suitability maps help in selecting the areas for better success in cultivating any particular crop. However, productivity of a crop is not solely dependent on suitable site but also heavily depend on many other factors, such as infrastructure and investments (like irrigation, fertilization, and farm management), biological (insect, pests, and disease), and governance (farmers' training on growth management). A map on Large Cardamom suitability classes and crop yield data, of year 2011, across the VDCs of Taplejung District is given in the Supplementary Materials S1 and S2.

3.5. Suitable Habitat under Climate Change Scenerio

The predicted future habitat suitability for large cardamom in the district under RCP2.6-2050 and RCP8.5-2050 is shown in Figure 8. The model predicted under RCP2.6-2050 has accuracy with an AUC value of 0.94 and 0.914 for training and test data, respectively. Under RCP8.5-2050, the AUC values are 0.939 and 0.916 for training and test data, respectively.

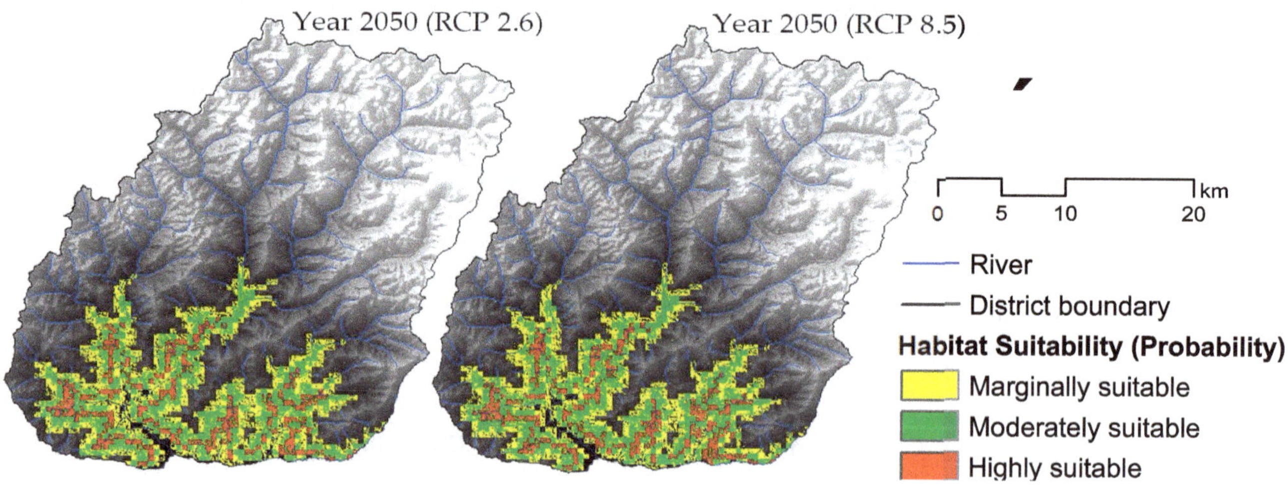

Figure 8. Distribution of large cardamom in Taplejung, Nepal, under future scenarios.

3.6. Projected Changes in the Suitable Habitat Area

Compared to the currently suitable areas, the total areas tend to decrease in both the future scenarios. However, there is no major difference in total areas of the highly suitable class (2%) in RCP 2.6 and a relatively high change in RCP 8.5 (12%). Elevation-wise, we can see there is loss of highly suitable area from the current scenario to both future scenarios, which for the lower is 80% and in higher elevation it is 94% (Table 2). The area tends to maintain a nearly steady-state at elevation range 1000–2000 m, which is ± 0.2%. In conclusion, total potentially suitable areas maintain a nearly steady-state under medium elevation range (1000–2000 m) for all scenarios, whereas the areas are decreasing under future scenarios for the low (500–1000 m) and high elevation (2000–3000 m) (Table 2).

This is the first study to explore current crop cultivation as well as the climate change impacts on the potential distribution of large cardamom crop. Models with AUC values more than 0.75 are considered robust, acceptable, and potentially useful for ecological niche model interpretation [56,57]. Our models obtained AUC values greater than 0.9 which shows our models are satisfactory. The spatial distribution trend under climate change may vary with species causing shifts, contraction, or expansions [58–61]. Our prediction showed suitable habitat for the crop is diminishing under future conditions compared to the current scenario.

Table 2. Areas (ha) of habitat suitability under different scenarios with elevation ranges.

		Current Cultivation (2016)	Current				RCP2.6 (2050)				RCP8.5 (2050)			
	Suitability		Marginal	Moderate	High	Total	Marginal	Moderate	High	Total	Marginal	Moderate	High	Total
Elevation Ranges (m)	500–1000	726	2673	1415	483	4571	2700	667	102	3469	2300	513	107	2920
	1000–2000	17,400	16,699	21,038	12,866	50,602	15,952	22,471	13,374	51,797	15,521	23,318	11,300	50,139
	2000–3000	4603	16,686	5325	331	22,342	11,626	1075	18	12,719	12,266	1051	20	13,336
	Total	22,729	36,058	27,778	13,679	77,515	30,278	24,213	13,494	67,985	30,087	24,882	11,426	66,395

Future changes in suitable habitat area are closely related to temperature and precipitation, which are steadily changing in Nepal over time. It is really important to understand the dynamics of these environmental variables. Stations data showed that the maximum temperature in Nepal increased at the rate of 0.056 °C/year for the period of 1975–2014 [62]. The rate was recorded higher at stations placed at higher altitude, whereas very wet and extremely wet days are diminishing significantly in the northern districts of Nepal [63]. Climate projections for temperature developed for entire Nepal by the Ministry of Forest and Environment (MoFE) [63] showed that the temperature trend is higher in high mountains than other regions in both medium- (2016–2045 for RCP 4.5 and RCP 8.5) and long-term (2036–2065 for RCP 4.5 and RCP 8.5) scenarios. The study found central and western parts of the country will be wetter than the eastern region despite the fact that average annual precipitation change tends to increase by 2.1% in the medium-term period and increase by 7.9% in the long-term period. As large cardamom needs cool and humid climatic conditions [39], most of the current habitat will be negatively affected in future considering the climate projection and data recorded in the past. Particularly, lower altitude would not be favorable for the crop as the temperature tends to reach higher degrees in these elevations. Although annual precipitation change increases in the country, warmer temperature in High Mountain would lead to it becoming drier as it would stimulate more evapotranspiration [58]. These projected drier conditions will negatively influence the habitat suitability of cardamom in the study area. Further, Gudade et al. [64] observed that declining soil nutrition and soil moisture are key drivers in reduction in crop productivity both in India and Nepal.

4. Conclusions

Mapping understory crops and respective habitat conditions is challenging particularly in a complex mountain terrain. The study demonstrated using a combination of expert knowledge-based mapping and species modelling based on bio-climatic factors to understand crop dynamics in a complex mountain terrain. The work applied object-based image analysis on freely available high-resolution satellite data of Sentinel-2A to identify Uttis tree cover in the study area. Use of textures and related indices made it possible to identify a particular tree species within the land cover map. The methodology developed in this study can be used for Uttis cover mapping in other areas of Nepal related to cardamom mapping.

This is the first study to demonstrate potential distribution of large cardamom in the region and the impact of climate change on the distribution. The study showed that crop habitat distribution patterns of understory crop could be modelled using Maxent and coupled with expert knowledge-based mapping to understand the environmental constraints and dynamics. The model under the current scenario showed potentially suitable areas for large cardamom, which has created scope to fill the gaps between farmers' practice in selection of farming land, and planners to develop sustainable land use planning and management. Current cultivation area is shown the highest in the medium suitability class. Farmers can change the farming area on high suitability area. The planning agencies can use the cardamom map and other information produced in this work for land use planning and crop management to optimize the crop sown area. Models under future scenarios showed that the overall suitable habitat is shrinking to mid-hill elevations (1000–2000 m). Therefore, in response to the projected loss of potential distribution, appropriate spatial planning for crop management and livelihood strategies needs to be developed for large cardamom farmers.

Author Contributions: S.M. performed the assessment and writing of the manuscript. F.M.Q. designed the study, performed field data collection, data analysis and writing of the manuscript. G.J. performed field data collection and data analysis. M.M. contributed in data analysis and manuscript writing. S.B. contributed to the writing discussion of results.

Acknowledgments: This research was supported by the Rural Livelihoods and Climate Change Adaptation in the Himalayas (Himalica) Programme, which is implemented by the International Center for Integrated Mountain Development (ICIMOD), and is funded by the European Union. The opinions expressed in the paper are those of the authors and do not necessarily reflect the views of their respective institutions.

References

1. Sharma, E.; Sharma, R.; Singh, K.K.; Sharma, G. A Boon for Mountain Populations: Large Cardamom Farming in the Sikkim Himalaya. *Mt. Res. Dev.* **2000**, *20*, 108–111. [CrossRef]
2. Sharma, G.; Joshi, S.R.; Gurung, M.B.; Chilwal, H.C. Package of Practices for Promoting Climate Resilient Cardamom Value Chains in Nepal. In *Training Manual*; International Centre for Integrated Mountain Development (ICIMOD): Kathmandu, Nepal, 2017.
3. Vijayan, A.K. Climate change and its impact of an productivity of large cardamom (*Amomum subutatum* Roxburgh). In Proceedings of the Stakeholders Consultation Workshop on Large Cardamom Development in Nepal, Pakhribas, Nepal, 20 April 2015; pp. 16–27.
4. Lenoir, J.; Gegout, J.C.; Marquet, P.A.; de Ruffray, P.; Brisse, H. A Significant Upward Shift in Plant Species Optimum Elevation during the 20th Century. *Science* **2008**, *320*, 1768–1771. [CrossRef]
5. Bertrand, R.; Lenoir, J.; Piedallu, C.; Riofrío-Dillon, G.; de Ruffray, P.; Vidal, C.; Pierrat, J.C.; Gégout, J.C. Changes in plant community composition lag behind climate warming in lowland forests. *Nature* **2011**, *479*, 517–520. [CrossRef]
6. Rijal, S.P. Impact of climate change on large cardamom-based livelihoods in Panchthar District, Nepal. *Third Pole J. Geogr. Educ.* **2013**, *13*, 33–38. [CrossRef]
7. Justice, C.O.; Becker-Reshef, I. *Developing a Strategy for Global Agricultural Monitoring in the Framework of Group on Earth Observations (GEO) Workshop Report*; FAO: Rome, Italy, 2007.
8. Soares, J.; Williams, M.; Jarvis, I.; Bingfang, W.; Leo, O.; Fabre, P.; Huynh, F.; Kosuth, P.; Lepoutre, D.; Parilar, J.S.; et al. *Strengthening Global Agricultural Monitoring—Improving Sustainable Data for Worldwide Food Security & Commodity Market Transparency*. 2011, pp. 1–16. Available online: http://www.earthobservations.org/documents/cop/ag_gams/201106_g20_global_agricultural_monitoring_initiative.pdf (accessed on 23 July 2019).
9. Asner, G.P.; Knapp, D.E.; Kennedy-Bowdoin, T.; Jones, M.O.; Martin, R.E.; Boardman, J.; Hughes, R.F. Invasive species detection in Hawaiian rainforests using airborne imaging spectroscopy and LiDAR. *Remote Sens. Environ.* **2008**, *112*, 1942–1955. [CrossRef]
10. Tuanmu, M.N.; Viña, A.; Bearer, S.; Xu, W.; Ouyang, Z.; Zhang, H.; Liu, J. Mapping understory vegetation using phenological characteristics derived from remotely sensed data. *Remote Sens. Environ.* **2010**, *114*, 1833–1844. [CrossRef]
11. Hall, R.J.; Peddle, D.R.; Klita, D.L. Mapping conifer understory within boreal mixedwoods from Landsat TM satellite imagery and forest inventory information. *For. Chron.* **2000**, *76*, 887–902. [CrossRef]
12. Johnston, S.E.; Henry, M.C.; Gorchov, D.L. Using Advanced Land Imager (ALI) and Landsat Thematic Mapper (TM) for the Detection of the Invasive Shrub *Lonicera maackii* in Southwestern Ohio Forests. *GIScience Remote Sens.* **2012**, *49*, 450–462. [CrossRef]
13. Resasco, J.; Hale, A.N.; Henry, M.C.; Gorchov, D.L. Detecting an invasive shrub in a deciduous forest understory using latefall Landsat sensor imagery. *Int. J. Remote Sens.* **2007**, *28*, 3739–3745. [CrossRef]
14. Becker, R.H.; Zmijewski, K.A.; Crail, T. Seeing the forest for the invasives: Mapping buckthorn in the Oak Openings. *Biol. Invasions* **2013**, *15*, 315–326. [CrossRef]
15. Leduc, M.B.; Knudby, A.J. Mapping wild leek through the forest canopy using a UAV. *Remote Sens.* **2018**, *10*, 70. [CrossRef]
16. Korpela, I.S. Mapping of understory lichens with airborne discrete-return LiDAR data. *Remote Sens. Environ.* **2008**, *112*, 3891–3897. [CrossRef]
17. Peckham, S.D.; Ahl, D.E.; Gower, S.T. Bryophyte cover estimation in a boreal black spruce forest using airborne lidar and multispectral sensors. *Remote Sens. Environ.* **2009**, *113*, 1127–1132. [CrossRef]
18. Singh, K.K.; Davis, A.J.; Meentemeyer, R.K. Detecting understory plant invasion in urban forests using LiDAR. *Int. J. Appl. Earth Obs. Geoinf.* **2015**, *38*, 267–279. [CrossRef]
19. Vierling, K.T.; Vierling, L.A.; Gould, W.A.; Martinuzzi, S.; Clawges, R.M. Lidar: Shedding new light on habitat characterization and modeling. *Front. Ecol. Environ.* **2008**, *6*, 90–98. [CrossRef]

20. Wang, T.J.; Skidmore, A.K.; Toxopeus, A.G. Improved understorey bamboo cover mapping using a novel hybrid neural network and expert system. *Int. J. Remote Sens.* **2009**, *30*, 965–981. [CrossRef]
21. Phillips, S.B.; Aneja, V.P.; Kang, D.; Arya, S.P. Modelling and analysis of the atmospheric nitrogen deposition in North Carolina. *Int. J. Glob. Environ. Issues* **2006**, *6*, 231–252. [CrossRef]
22. Graham, C.H.; Ferrier, S.; Huettman, F.; Moritz, C.; Peterson, A.T. New developments in museum-based informatics and applications in biodiversity analysis. *Trends Ecol. Evol.* **2004**, *19*, 497–503. [CrossRef]
23. Barik, S.K.; Adhikari, D. Predicting the geographical distribution of an invasive species (*Chromolaena odorata* L. (king) & H.E. Robins) in the Indian subcontinent under climate change scenarios. In *Invasive Alien Plants: An Ecological Appraisal for the Indian Subcontinent*; CABI: Wallingford, UK, 2011.
24. Niphadkar, M.; Nagendra, H.; Tarantino, C.; Adamo, M.; Blonda, P. Comparing Pixel and Object-Based Approaches to Map an Understorey Invasive Shrub in Tropical Mixed Forests. *Front. Plant Sci.* **2017**, *8*, 1–18. [CrossRef]
25. Blaschke, T. Object based image analysis for remote sensing. *ISPRS J. Photogramm. Remote Sens.* **2010**, *65*, 2–16. [CrossRef]
26. Mialhe, F.; Gunnell, Y.; Ignacio, J.A.F.; Delbart, N.; Ogania, J.L.; Henry, S. Monitoring land-use change by combining participatory land-use maps with standard remote sensing techniques: Showcase from a remote forest catchment on Mindanao, Philippines. *Int. J. Appl. Earth Obs. Geoinf.* **2015**, *36*, 69–82. [CrossRef]
27. Robiglio, V.; Mala, W.A. Integrating local and expert knowledge using participatory mapping and GIS to implement intergrated forest management options in Akok, Cameroon. *For. Chron.* **2005**, *81*, 392–397. [CrossRef]
28. Lauer, M.; Aswani, S. Integrating indigenous ecological knowledge and multi-spectral image classification for marine habitat mapping in Oceania. *Ocean Coast. Manag.* **2008**, *51*, 495–504. [CrossRef]
29. Elith, J.; Graham, C.H.; Anderson, R.P.; Dudı'k, M.; Ferrier, S.; Guisan, A.; Hijmans, R.J.; Huettmann, F.; Leathwick, J.R.; Lehmann, A.; et al. Novel methods improve prediction of species' distributions from occurrence data. *Ecography* **2006**, *29*, 129–151. [CrossRef]
30. Pearson, R.G.; Raxworthy, C.J.; Nakamura, M.; Townsend Peterson, A. Predicting species distributions from small numbers of occurrence records: A test case using cryptic geckos in Madagascar. *J. Biogeogr.* **2007**, *34*, 102–117. [CrossRef]
31. Kumar, S.; Stohlgren, T.J. Maxent modeling for predicting suitable habitat for threatened and endangered tree *Canacomyrica monticola* in New Caledonia. *J. Ecol. Nat. Environ.* **2009**, *1*, 094–098.
32. Shrestha, J.; Prasai, H.K.; Timilsina, K.P.; Shrestha, K.P.; Pokhrel, D.; Paudel, K.; Yadav, M. *Large Cardamom in Nepal: Production Practice and Economics, Processing and Marketing*; National Commercial Agriculture Research Program, Nepal Agriculture Research Council: Pakhribas, Dhankuta, Nepal, 2018.
33. McCall, M.K.; Minang, P.A. Assessing Participatory GIS for Community-Based Natural Resource Management: Claiming Community Forests in Cameroon. *Geogr. J.* **2005**, *171*, 340–356. [CrossRef]
34. Drusch, M.; Del Bello, U.; Carlier, S.; Colin, O.; Fernandez, V.; Gascon, F.; Hoersch, B.; Isola, C.; Laberinti, P.; Martimort, P.; et al. Sentinel-2: ESA's Optical High-Resolution Mission for GMES Operational Services. *Remote Sens. Environ.* **2012**, *120*, 25–36. [CrossRef]
35. Blaschke, T.; Lang, S.; Lorup, E.; Strobl, J.; Zeil, P. Object-Oriented Image Processing in an Integrated GIS/Remote Sensing Environment and Perspectives for Environmental Applications. *Umweltinf. Plan. Polit. Öffentl.keit Environ. Inf. Plan. Polit. Public* **2000**, *2*, 555–570.
36. Hay, G.J.; Blaschke, T.; Marceau, D.J.; Bouchard, A. A comparison of three image-object methods for the multiscale analysis of landscape structure. *ISPRS J. Photogramm. Remote Sens.* **2003**, *57*, 327–345. [CrossRef]
37. Belgiu, M.; Drăguţ, L. Comparing supervised and unsupervised multiresolution segmentation approaches for extracting buildings from very high resolution imagery. *ISPRS J. Photogramm. Remote Sens.* **2014**, *96*, 67–75. [CrossRef]
38. National Spice Crop Development Programme. *Annual Report of National Spice Crop Development Programme*; National Spice Crop Development Programme, Ministry of Agriculture Development, Government of Nepal: Khumaltar, Kathmandu, Nepal, 2009.
39. Uddin, K.; Shrestha, H.L.; Murthy, M.S.R.; Bajracharya, B.; Shrestha, B.; Gilani, H.; Pradhan, S.; Dangol, B. Development of 2010 national land cover database for the Nepal. *J. Environ. Manag.* **2015**, *148*, 82–90. [CrossRef]
40. Tadono, T.; Ishida, H.; Oda, F.; Naito, S.; Minakawa, K.; Iwamoto, H. Precise Global DEM Generation by ALOS PRISM. *ISPRS Ann. Photogramm. Remote Sens. Spat. Inf. Sci.* **2014**, *II–4*, 71–76. [CrossRef]

41. Micheal, H. Graham Confronting Multicollinearity in Ecological Multiple Regression. *Ecology* **2003**, *84*, 2809–2815.
42. Yang, X.Q.; Kushwaha, S.P.S.; Saran, S.; Xu, J.; Roy, P.S. Maxent modeling for predicting the potential distribution of medicinal plant, *Justicia adhatoda* L. in Lesser Himalayan foothills. *Ecol. Eng.* **2013**, *51*, 83–87. [CrossRef]
43. Naimi, B.; Hamm, N.A.S.; Groen, T.A.; Skidmore, A.K.; Toxopeus, A.G. Where is Positional Uncertainty a Problem for Species Distribution Modelling? Available online: https://onlinelibrary.wiley.com/doi/abs/10.1111/j.1600-0587.2013.00205.x (accessed on 11 July 2019).
44. Maharjan, S. Estimation and Mapping Above Ground Woody Carbon Stocks Using Lidar Data And Digital Camera Imagery in The Hilly Forest of Gorkha, Nepal. Master's Thesis, University of Twente, Enschede, The Netherlands, 2012.
45. Karna, Y.K.; Hussin, Y.A.; Gilani, H.; Bronsveld, M.C.; Murthy, M.S.R.; Qamer, F.M.; Karky, B.S.; Bhattarai, T.; Aigong, X.; Baniya, C.B. Integration of WorldView-2 and airborne LiDAR data for tree species level carbon stock mapping in Kayar Khola watershed, Nepal. *Int. J. Appl. Earth Obs. Geoinf.* **2015**, *38*, 280–291. [CrossRef]
46. Baniya, N.; Böehme, M.; Baniya, S. Physical land suitability assessment for the large cardamom amomum subulatum roxb. Cultivation in hills of Kathmandu valley. *Chin. J. Popul. Resour. Environ.* **2009**, *7*, 59–63. [CrossRef]
47. ECDF. *Alaichi Kheti and Prashodhan Hate Pustika*; Environment Conservation and Development Forum: Taplejung, Mechi Zone, Nepal, 2008.
48. Coro, G.; Magliozzi, C.; Ellenbroek, A.; Pagano, P. Improving data quality to build a robust distribution model for *Architeuthis dux*. *Ecol. Model.* **2015**, *305*, 29–39. [CrossRef]
49. Aubry, K.B.; Raley, C.M.; McKelvey, K.S. The importance of data quality for generating reliable distribution models for rare, elusive, and cryptic species. *PLoS ONE* **2017**, *12*, 1–17. [CrossRef]
50. Kumar, S.; Graham, J.; West, A.M.; Evangelista, P.H. Using district-level occurrences in MaxEnt for predicting the invasion potential of an exotic insect pest in India. *Comput. Electron. Agric.* **2014**, *103*, 55–62. [CrossRef]
51. Stockwell, D.R.B.; Peterson, A.T. Effects of sample size on accuracy of species distribution models. *Ecol. Model.* **2002**, *148*, 1–13. [CrossRef]
52. Dormann, C.F.; Elith, J.; Bacher, S.; Buchmann, C.; Carl, G.; Carré, G.; Marquéz, J.R.G.; Gruber, B.; Lafourcade, B.; Leitão, P.J.; et al. Collinearity: A review of methods to deal with it and a simulation study evaluating their performance. *Ecography* **2013**, *36*, 027–046. [CrossRef]
53. Evangelista, P.H.; Kumar, S.; Stohlgren, T.J.; Jarnevich, C.S.; Crall, A.W.; Norman, J.B.; Barnett, D.T. Modelling invasion for a habitat generalist and a specialist plant species. *Divers. Distrib.* **2008**, *14*, 808–817. [CrossRef]
54. Elith, J.; Graham, C.H.; Anderson, R.P.; Dudík, M.; Guisan, A.; Hijmans, R.J.; Huettmann, F.; Leathwick, J.R.; Lehmann, A.; Li, J.; et al. Novel Methods Improve Prediction of Species' Distributions from OccurrenceData. *Ecography.* **2006**, *29*, 129–151. [CrossRef]
55. Ortega-Huerta, M.A.; Peterson, A.T. Modeling ecological niches and predicting geographic distributions: A test of six presence-only methods. *Rev. Mex. De Biodivers.* **2008**, *79*, 205–216.
56. Pearce, J.; Ferrier, S. An evaluation of alternative algorithms for fitting species distribution models using logistic regression. *Ecol. Model.* **2000**, *128*, 127–147. [CrossRef]
57. Elith, J. *Quantitative Methods for Modeling Species Habitat: Comparative Performance and An Application to Australian Plants*; Springer: New York, NY, USA, 2000.
58. Li, R.; Xu, M.; Wong, M.H.G.; Qiu, S.; Sheng, Q.; Li, X.; Song, Z. Climate change-induced decline in bamboo habitats and species diversity: Implications for giant panda conservation. *Divers. Distrib.* **2014**, *21*, 379–391. [CrossRef]
59. Lamsal, P.; Kumar, L.; Aryal, A.; Atreya, K. Invasive alien plant species dynamics in the Himalayan region under climate change. *Ambio* **2018**, *47*, 697–710. [CrossRef]
60. Maclean, I.M.D.; Wilson, R.J. Recent ecological responses to climate change support predictions of high extinction risk. *Proc. Natl. Acad. Sci. USA* **2011**, *108*, 12337–12342. [CrossRef]
61. Ponce-Reyes, R.; Nicholson, E.; Baxter, P.W.J.; Fuller, R.A.; Possingham, H. Extinction risk in cloud forest fragments under climate change and habitat loss. *Divers. Distrib.* **2013**, *19*, 518–529. [CrossRef]

62. *DHM Observed Climate Trend Analysis in the Districts and Physiographic Zones of Nepal (1971–2014)*; Government of Nepal Ministry of Population and Environment Department of Hydrology and Meteorology: Kathmandu, Nepal, 2017.
63. *MoFE Climate Change Scenarios for Nepal for National Adaptation Plan (NAP)*; Ministry of Forests and Environment: Kathmandu, Nepal, 2019.
64. Gudade, B.A.; Harsha, K.N.; Vijayan, A.K.; Chhetri, P.; Deka, T.N.; Babu, S.; Singh, R. Effect of soil application of Zn, Mn and Mg on growth and nutrient content of large cardamom, Amomum subulatum Roxb at Sikkim. *Int. J. Farm Sci.* **2015**, *5*, 51–55.

8

Evaluation of Rapeseed Winter Crop Damage using UAV-Based Multispectral Imagery

Łukasz Jełowicki [1], Konrad Sosnowicz [2], Wojciech Ostrowski [3], Katarzyna Osińska-Skotak [3,*] and Krzysztof Bakuła [3]

1 OPEGIEKA Sp. z o.o., 82-300 Elbląg, Poland; lukasz.jelowicki@opegieka.pl
2 Skysnap Sp. z o.o., 02-001 Warsaw, Poland; konrad.sosnowicz@skysnap.pl
3 Department of Photogrammetry, Remote Sensing and Spatial Information Systems, Faculty of Geodesy and Cartography, Warsaw University of Technology, 00-661 Warsaw, Poland; wojciech.ostrowski@pw.edu.pl (W.O.); krzysztof.bakula@pw.edu.pl (K.B.)
* Correspondence: katarzyna.osinska-skotak@pw.edu.pl

Abstract: This research is related to the exploitation of multispectral imagery from an unmanned aerial vehicle (UAV) in the assessment of damage to rapeseed after winter. Such damage is one of a few cases for which reimbursement may be claimed in agricultural insurance. Since direct measurements are difficult in such a case, mainly because of large, unreachable areas, it is therefore important to be able to use remote sensing in the assessment of the plant surface affected by frost damage. In this experiment, UAV images were taken using a Sequoia multispectral camera that collected data in four spectral bands: green, red, red-edge, and near-infrared. Data were acquired from three altitudes above the ground, which resulted in different ground sampling distances. Within several tests, various vegetation indices, calculated based on four spectral bands, were used in the experiment (normalized difference vegetation index (NDVI), normalized difference vegetation index—red edge (NDVI_RE), optimized soil adjusted vegetation index (OSAVI), optimized soil adjusted vegetation index—red edge (OSAVI_RE), soil adjusted vegetation index (SAVI), soil adjusted vegetation index—red edge (SAVI_RE)). As a result, selected vegetation indices were provided to classify the areas which qualified for reimbursement due to frost damage. The negative influence of visible technical roads was proved and eliminated using OBIA (object-based image analysis) to select and remove roads from classified images selected for classification. Detection of damaged areas was performed using three different approaches, one object-based and two pixel-based. Different ground sampling distances and different vegetation indices were tested within the experiment, which demonstrated the possibility of using the modern low-altitude photogrammetry of a UAV platform with a multispectral sensor in applications related to agriculture. Within the tests performed, it was shown that detection using UAV-based multispectral data can be a successful alternative for direct measurements in a field to estimate the area of winterkill damage. The best results were achieved in the study of damage detection using OSAVI and NDVI and images with ground sampling distance (GSD) = 10 cm, with an overall classification accuracy of 95% and a F1-score value of 0.87. Other results of approaches with different flight settings and vegetation indices were also promising.

Keywords: damage detection; winter crop; UAV; multispectral imagery; vegetation indices; agricultural insurance; technical roads

---

1. Introduction

The European Union is the global leader in rapeseed production, and Poland is one of the largest rapeseed producers and processors in Europe. In 2019, it was second only to France [1]. In 2010–2019, the area of rapeseed sown in Poland ranged from 720,000 to 950,000 hectares, representing about 8–10% of the total crops, and about 95–96% of the oilseed acreage [2].

Due to the climate changes observed in recent years (i.e., snow-free winters), yield losses caused by poor wintering of plants are becoming more frequent. Low temperatures are the most dangerous for crops, combined with no snow or a thin layer of snow cover. Rapid warming causing thaws, followed by frosts, are also not conducive to wintering plants. The result of such winter weather is freezing winter crops, i.e., damage to the crops. If there is no snow cover, winter rapeseed may freeze at −10 °C [3]. Annual losses in rapeseed yields in 2015–2019 were estimated to range from 10% to 20% [2]. To assess damages for agricultural insurance, an assessment of its occurrence in the field is performed. These are imprecise measurements made based on a sample selected in the field whereby an expert determines the number of plants per square meter and takes this as a representation of one hectare of the affected field. The results of these measurements depend on the expert's experience. In the case of rapeseed crops, compensation is awarded when there are less than 15 healthy plants per square meter and the damage covers at least 10% of the field area. The damage threshold values (min. 10% and max. 50% of the parcel area) are particularly important because, according to the scope of insurance, they affect the final decision on the qualification of loss in crops (partial or total compensation). Since the methods of assessing damages used by insurance companies are not very accurate and are time consuming, it appears that the use of remote sensing techniques may be an excellent alternative.

Currently, research conducted in the area of remote sensing applications with the use of unmanned aerial vehicles (UAV) is often related to aspects of vegetation monitoring, particularly relating to forest and agricultural areas. UAV-based remote sensing in agriculture is particularly related to pest and disease detection, development of crops during their growth cycle, assessment of biomass, and water stress in plants [4]. To date, these phenomena have been assessed through satellite and airborne photogrammetry [5–8]. The breakthrough solutions in terms of image acquisition techniques are those using UAVs [9–11]. There is a need to receive continuous spatial information with a high level of accuracy and actuality, while retaining low operating costs. Drones can provide an alternative, providing high resolution images with a short revisit time, even every couple of hours, contrary to optical satellite systems or aerial images which have a much lower time resolution.

To date, analyses connected with object geometry, such as an assessment of parcel areas, landslide monitoring, and cropland and forest inventory, have been widely applied in many countries [12–14]). However, applications of sensors with more than three spectral bands (e.g., multispectral or hyperspectral cameras) mounted on unmanned aerial vehicles are not as commonly used. Research on implementation of these cameras and integration of various remote sensing techniques is a key area of interest for many groups of scientists [15–18].

Remote sensing techniques for vegetation mapping play an important role in precise agriculture [9,19] and phenotyping research of crops [20–22]. Remote monitoring of arable lands using spectral libraries (obtained from field measurements, satellite imagery, or aerial and close-range images) could be helpful in the analysis of plant growth [19,22], assessment of the size of harvest [23] and crop fertilizing needs [24,25], pest control [24], and the extraction of dead plants. For many years, the Remote Sensing Centre at the Institute of Geodesy and Cartography in Poland has carried out monitoring of farmland with the use of satellite imagery to forecast harvests [26]. The accuracy of the prepared models for the prediction of crop volumes using NOAA AVHRR data is estimated at 90–95%. A high correlation between a harvest's size and its spectral characteristics based on field measurements [27] is a clear indicator of the potential of remote sensing data for such research. This translates into numerous areas of scientific research, involving spectral vegetation indices in the assessment of the physical condition of flora [5,19,27,28].

The use of spectral indices aims to extract essential information about vegetation condition, and the indices should indicate a proper correlation between their values and biophysical parameters of plant cover, such as biomass or leaf area index (LAI) [5,19,27,29]. The most popular spectral indices in this area are those that use big reflectance differences between the near infrared region and the red band. This kind of index includes the normalized difference vegetation index (NDVI) and simple ratio (SR). Based on the research carried out to date, it is observed that there is a link between the above-mentioned parameters and the size of the biomass and LAI [5]. The outcome of scientific studies has confirmed that a high accuracy of the performed estimations (85%) [18,19] was found, and that there are strong and statistically significant relationships between NDVI (obtained from UAV) and crop biophysical variables (for LAI, R^2 was 0.95; for ground truth percent canopy cover, R^2 was 0.93). The obtained strong relationships suggest that UAV multispectral data can be used for estimation of LAI and percent canopy cover. Wei et al. (2017) [30] concluded in their research that derivative spectral indices formulated using optimized narrow wavebands were most effective in quantifying the changes in pigment and water content of leaves subjected to freezing injury.

Frost is one of the environmental stresses of plants. The plant response to frost can be a decrease in chlorophyll, inhibition of photosynthesis, altered leaf angle, and plant freezing [31]. The influence of frost on biochemical and biophysical changes, as well as spectral properties of various plant species, have been investigated [30,32–36]. It was found that freezing causes changes in the spatial differentiation of chlorophyll content on the lamina surface [32] and the structural changes of mesophyll cells [30]. In turn, these changes affect the spectral reflection of various plant parts, especially in blue-green, green, red, and near infrared radiation [30,33–36]. However, as the analyzed studies show, these changes depend on the species studied and the degree of freezing.

In order to analyze the effects of crop freezing in winter, it is essential that the influence of the spectral reflectance of soil should be included. For practical implementation, it is common to use the soil-adjusted vegetation index (SAVI). Because of the difficulty in choosing the optimal value of the coefficient describing soil brightness and the low sensitivity to a small amount of chlorophyll, the modified chlorophyll absorption ratio index (MCARI) was introduced. This parameter highlights an amount of chlorophyll absorption in the range of about 670 nm in relation to spectral characteristics centered at 550 and 700 nm [6]. With the use of band normalization in the spectral range of red edge and red (R700, R670), it is possible to minimize the influence of soil and extract information about vegetation without green pigments. Another method of implementation is to use the transformed chlorophyll absorption reflectance index (TCARI)/optimized soil adjusted vegetation index (OSAVI). The use of normalization enables the detection of biomass variability even at low LAI values [6]. Furthermore, Hunt's research (2008) [28] confirms the effectiveness of the green normalized difference vegetation index (GNDVI, [37]), which differentiates biomass content within the area of winter wheat crops. Moreover, studies carried out by many scientists provides evidence that GNDVI correlates significantly with the nutrient ingredients (mainly nitrogen compounds) [28,38]. Among the methods used, it is worth mentioning the fusion of chosen spectral indices. As an example, the use of NDVI, the modified soil-adjusted vegetation index (MSAVI)/the transformed soil-adjusted vegetation index (TSAVI) (in the case where spectral bands in the shortwave infrared (SWIR) range are available) and GNDVI enables the assessment of the spatial variability of cropland, including its impact on agricultural yields. The opportunities from using remote sensing data outlined above appear to be a reasonable solution to crop insurance, including estimating the winterkill losses. Multispectral sensors have been significantly miniaturized in recent years, and their current price allows for the commercialization of the presented method without significant financial outlay. In contrast to satellite data, the quality of which is determined by cloud cover range, the use of UAVs in time-series analysis is not as strongly related to weather conditions. This fact is essential for regions of moderate climate zones because winter crop inventories are conducted in the late autumn which is usually very foggy. Furthermore, an application of the appropriate methodology for data processing and analysis enables farmers and insurance providers to accurately estimate winterkill losses and the amounts of compensation paid by insurers. One of the most important steps in this process is damage detection at the beginning of the vegetation period.

The purpose of this study was to: (1) evaluate the possibility of using UAVs for estimating losses in rapeseed crops caused by poor wintering, (2) determine the impact of data acquisition parameters (mainly the altitude influencing the ground sampling distance (GSD)), and (3) present the methodology of data processing (proper vegetation index) to obtain the best results compared to the traditional method of field inventory.

2. Materials and Methods

The experiment was carried out to analyze the possibility of using UAV-based multispectral imagery in the detection of damaged rapeseed after winter and the estimation of the damage area. The experiment was carried out in cooperation with an insurance company. Figure 1 shows the scheme of the experiment performed. Data were registered during three test flights on different flight heights in order to evaluate the influence of image spatial resolution. Each time, remote sensing data were acquired in four spectral bands, namely green, red, red edge, and near infrared, which were used to calculate tested vegetation indices. RGB images were acquired and combined with high-resolution spectral data, which were later used for manual preparation of reference data. Training samples were obtained during the field campaign.

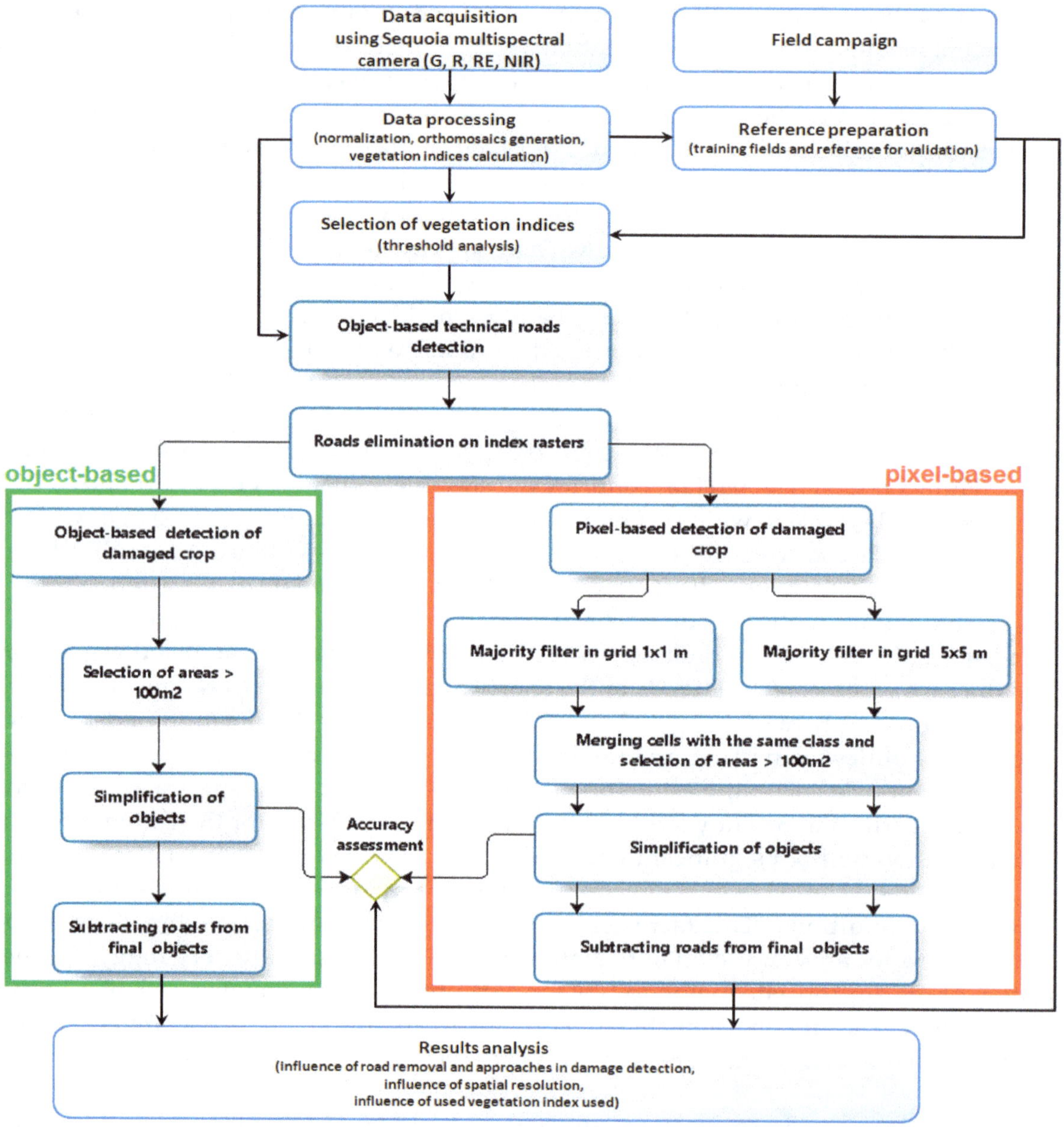

Figure 1. The scheme of the experiment performed.

In the methodology, there were a few crucial elements which are described in this section. Firstly, selected indices were tested to evaluate their usefulness in further analysis. The important part of the experiment was to delineate technical roads and replace those areas with a mean value from neighborhoods to exclude the effect of machinery influence (technical roads) on estimating damage. Another issue in the experiment was to define the influence of spatial resolution of the multispectral data on the obtained results.

2.1. Test Area and Data Used

For the planned tests, a 34 ha parcel with a rapeseed field was selected. It is situated in Olecko County in Northern Poland (Figure 2). This region is well known for its low temperatures and harsh climate throughout the year, hence, it was chosen for this experiment on the detection of winter crop damage. The winter of 2016/2017 in this region was characterized by high variability of weather conditions, with alternate periods of frost and thaw. During the period of severe frosts (from −12 to −27 °C) there was a thin layer of snow of several centimeters. This caused winter damage, which was confirmed during a field inspection.

Figure 2. Test area, rapeseed field in Olecko, Poland.

The testing data were acquired with the use of an eBee UAV equipped with a Sequoia multispectral camera. This allowed for the acquisition of data in four different spectral bands: green (530–570 nm), red (640–680 nm), red edge (730–740 nm), and near infrared (770–810 nm). Three flights were performed on 20 April 2017, at three different altitudes of 50, 100, and 200 m above the ground. This resulted in images with three different spatial resolutions—ground sampling distances (GSD) of 5, 10, and 20 cm, respectively. This allowed us to assess the impact of spatial resolution on the imagery in the detection of a crop's health when undamaged by frost. The overlapping of successive images along a flight strip (endlap) was 80%, and the overlap between flight strips (sidelap) was 70%. Flights were carried out in midday hours (12:00 p.m.–2:00 p.m.), with a cloudless or slightly cloudy sky. Wind speed was 1–2 m/s, air temperature was 5–6 °C, and humidity was 32–37%. The visibility was 35.5–43.8 km.

The orientation of images was performed in Pix4DMapper Pro software (Pix4D SA, Switzerland). This software allows the user to select the appropriate calculation template. For multispectral data calculation, the AG (agriculture) multispectral template is preferred. The program automatically recognizes the camera, retrieves its parameters from the database, groups the images into the appropriate spectral bands, and enables them to be calibrated. A useful feature of the Sequoia camera is that it can measure the amount of irradiance coming from the sun for each image and for each band with the sunshine sensor. This enables the user to take irradiance into account and normalize the value of the images amongst themselves. In the orientation process, self-calibration with 15 ground control points (control and check points) measured with a GNSS receiver was also performed to give a final root mean square error (RMS) of aerial triangulation of 10 cm. The result of the images processed in Pix4DMapper Pro were reflectance orthomosaics.

Based on the orthomosaics generated from the collected images in four spectral bands, different vegetation indices were calculated, based on which further experiments were conducted. To perform the correct classification of healthy and damaged crops, 23 training fields were created using data from field inspection and verified on RGB images. The training fields contained information about the number of healthy plants over a surface of 1 square meter (Figure 3). This helped to describe the relationship between the vegetation indices values and the number of plants in a surface unit, which resulted in thresholds for image classification.

Figure 3. Examples of the training fields (1 m × 1 m).

2.2. *Selection of Vegetation Indices*

The most important process in determining post-winter damage is the correct detection of areas where the crop has not grown properly, or where the plants have died. Remote sensing as a multidisciplinary technique is a tool with many indices, which can help test the health of plants and solve the problem of the value of compensation for farmers for damaged crops. The Sequoia camera was the equipment used on the UAV in this study and allowed recording in four spectral bands: green, red, red edge, and near infrared. As a result of a literature review, a set of vegetation indicators was identified for further analysis. Consequently, after preliminary tests, the final selection was based on three basic spectral indices of vegetation: NDVI, SAVI, and OSAVI. In addition, an interesting solution in the calculation of these indices was to replace the red channel with the red edge, which was supported by the multispectral camera used in the experiment. Such modification was carried out and three further indices were examined: NDVI_RE, SAVI_RE, OSAVI_RE.

The NDVI (normalized difference vegetation index) is one of the most basic and commonly used indices [39]. It is a measure of photosynthetic activity and determines the condition of the plant and its development stage. Plants absorb most of the red light that hits it while reflecting much of the near infrared light. When vegetation is dead or stressed it reflects more red light and less near infrared light. Based on these properties, NDVI is calculated as the normalized difference between the red and near infrared bands from an image. The basic formula for calculating the index value is as follows:

$$NDVI = \frac{NIR - RED}{NIR + RED} \quad (1)$$

where NIR is the near infrared band value and RED is the red band value.

The SAVI (soil adjusted vegetation index) is a modification of the NDVI index [40]. Due to changes in the numerator and denominator of the typical NDVI equation (soil brightness correction factor), the SAVI index further improves the final result for the influence of the soil brightness. For this reason, it is used in situations where a large part of the crop is not covered with lush vegetation. Therefore, the beginning of the growing season appears to be an excellent time for examining the condition of plants using this index. The formula for calculating SAVI is as follows:

$$\text{SAVI} = \frac{(1 + \text{L}) * (\text{NIR} - \text{RED})}{\text{NIR} + \text{RED} + \text{L}} \tag{2}$$

where L denotes the coverage of the vegetation area. In most cases, and particularly for the case of intermediate vegetation canopy levels, optimal results are achieved using L = 0.5. For this experiment, a value 0.5 was used during the tests.

The OSAVI (optimized soil adjusted vegetation index) was developed by Rondeaux et al. (1996) [41] and was first presented in the work entitled "Optimized Soil Adjusted Vegetation Index". This index is a modification of the SAVI that has been optimized for agricultural monitoring. It is more sensitive to changes in plant condition than the original SAVI and does not need a priori knowledge of the soil type. A value of 0.16 as the soil adjustment coefficient was selected as the optimal value to minimize variation with soil background. The formula for calculating OSAVI is as follows:

$$\text{OSAVI} = \frac{\text{NIR} - \text{RED}}{\text{NIR} + \text{RED} + 0.16} \tag{3}$$

In addition to the above three spectral indices, their modifications were tested in which the red channel was replaced by red edge. By analyzing spectral reflectance properties for plants in red edge wavelength, we can observe quick changes in reflection and can thus expect to see more discreet differences between healthy and damaged plants. To test this premise, NDVI_RE, SAVI_RE, and OSAVI_RE indices were calculated. The application of such indices with a modified channel has been discussed previously [42,43].

The correctness of damage recognition by the vegetation index was examined using training fields measured during a field inventory with the participation of insurance specialists. Twenty-tree sites were surveyed with an area of approximately 1 square meter and the condition of the plants was determined in their range. These fields were then characterized by experts as to whether they would qualify for compensation or if the plants in the field area were healthy. A training field was also specified that could be treated as two classes on the border: damaged and healthy crops. The training fields were ranked according to the condition of the plants. After calculating the mean values of the tested indices for each training field, the values obtained were compared with the reference from the field inventory. The threshold value was established based on the training fields which, according to experts, were found not to meet the conditions for granting compensation. Based on the healthy training field, threshold values for each index were used to classify (or not) training fields for compensation. All training fields were verified with an expert in the field. Their correctness was ensured so the outlier fields were eliminated in defining the thresholds. The threshold value was calculated as the mean value from the lowest mean value of the index for an accepted healthy crop training field and the highest mean value of the index for an accepted damaged crop training field. A comparison of results for training fields with defined thresholds is shown in Figure 4. As can be seen, the training fields were characterized by differentiation of the values of the vegetation indices. The highest standard deviation was noted in the fields with healthy plants, and the lowest in damaged areas. The greatest variability of the index values was related to the occurrence of plants of different sizes and condition. Training field no. 9 recorded higher NDVI and SAVI values due to the presence of weeds; furthermore, in training field no. 11, there were 10 healthy, well-developed plants, which also resulted in higher NDVI values.

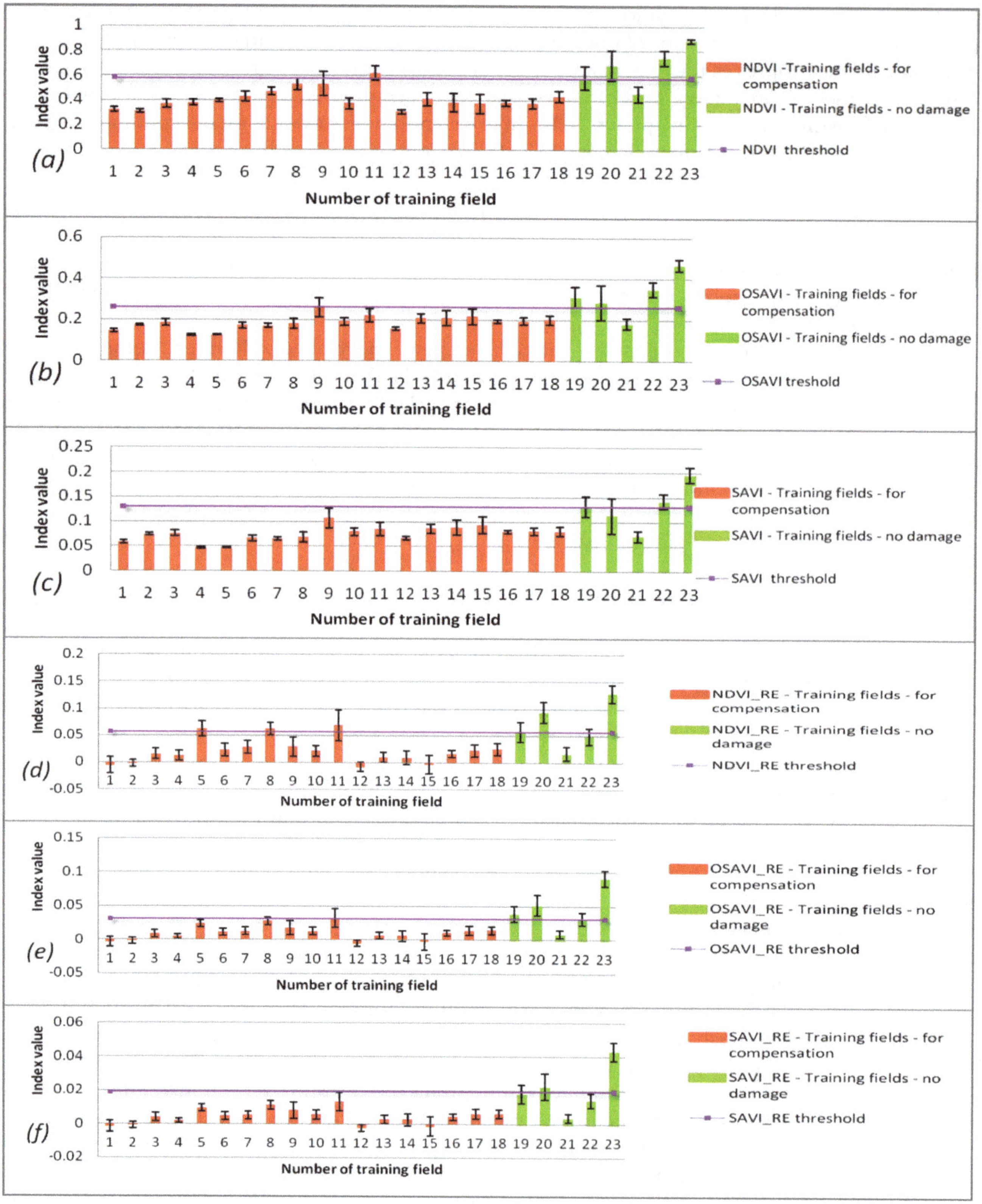

Figure 4. Interpretation of training fields with selected thresholds according to indices: Normalized Difference Vegetation Index (NDVI) (**a**), Optimized Soil Adjusted Vegetation Index (OSAVI) (**b**), Soil Adjusted Vegetation Index (SAVI) (**c**), Normalized Difference Vegetation Index—Red Edge (NDVI_RE) (**d**), Optimized Soil Adjusted Vegetation Index—Red Edge (OSAVI_RE) (**e**), and Soil Adjusted Vegetation Index—Red Edge (SAVI_RE) (**f**). The training fields are ranked according to the condition of plants. The chart shows the average value of index for each of the training fields with the standard deviation.

Based on the selected thresholds, referring to each index, Figure 5 illustrates the errors in training fields, verified by inspection of the field. The red areas represent the test fields that, according to specialists, should be classified as damaged areas. The green areas represent fields without visible frozen influence. The threshold shown by a horizontal line in Figure 4 determined the correctness of interpretation for each training field: red areas should be below the threshold and green above. If the situation is reversed, it means that tested index does not work properly. After summarizing the errors, the top 3 results were finally selected, i.e., NDVI, OSAVI, and OSAVI_RE, to carry out further experiments.

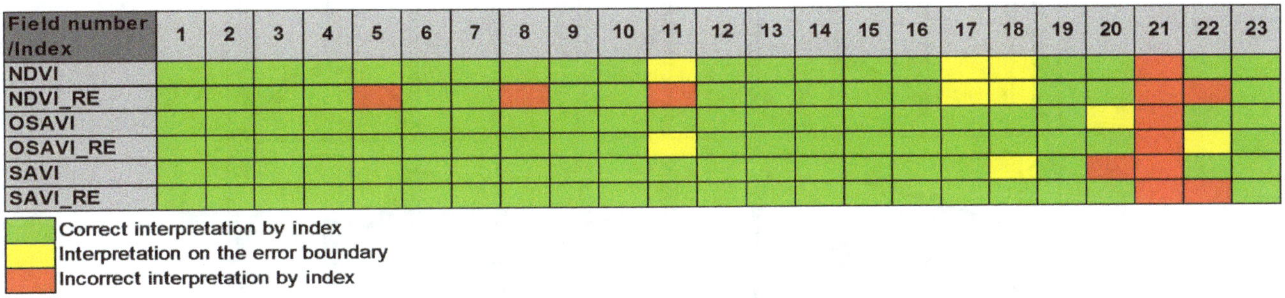

Figure 5. Errors of the classification for training fields intended for compensation process.

In addition, considering the training fields, it is clear that the worst results were obtained for field no. 21. This was a unique area with healthy but very small plants (significant influence of the soil background). This suggests that the accepted threshold method may be sensitive to the stage in which the plants are grown, and results in slight over detection of damaged areas.

2.3. *Elimination of Road Influence*

Regardless of the selected vegetation index, the influence of technical roads is clearly visible on the classification results. The classification of technical roads, which are visible as traces of tires on a part of the field in poor condition, could have a big impact on the final results of detection, because this area should not be included in the calculation of the area of rapeseed damaged by frost.

The main insurer's condition applied to an area for reimbursement is that the area of a detected polygon must cover some minimum area (i.e., larger than 100 m^2). Therefore, many smaller objects are removed from the analysis. However, small isolated damaged areas are sometimes related to technical roads, which can result in detection of one bigger area, and are thus mistakenly detected as an area eligible for reimbursement. To prevent such situations, it was decided to work on rasters without the influence of roads on the reimbursement area.

For this purpose, road detection was conducted, based on compositions consisting of NDVI and green, red, and red edge channels. Object-based classification was performed in eCognition software for each dataset (data with spatial resolution equal to 5, 10, and 20 cm). In such a classification, information about spectral values, object shape, and extent were used. This resulted in three different road images, one for each spatial resolution of 5, 10, and 20 cm. Detected road images differed significantly from one another. The most precise and complete images came from images of 5 cm, and the worst from the lowest resolution (see Figure 6).

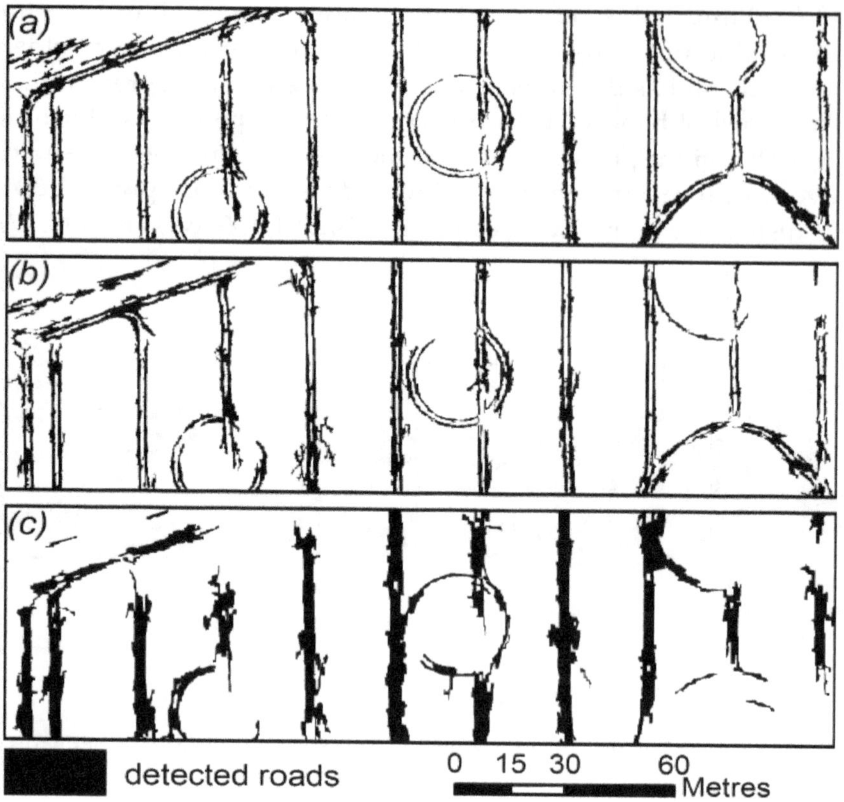

Figure 6. Detected roads on orthomosaics with ground sampling distance (GSD) of 5 cm (**a**), 10 cm (**b**), and 20 cm (**c**).

The biggest problems with areas of roads are presented in Figure 7, where there is no visible border between road and bare soil, which is why some tires traces were not detected. However, these places do not need to be detected because they do not cause the described problem of connecting polygons. Subsequently, a road mask was subtracted from the index rasters and the resulting holes were replaced with a mean value from corresponding cells (see Figure 7). Thus, the negative influence of technical roads on the final results of classification and detection of regions for reimbursement was reduced.

2.4. *Approaches for Classification*

The detection of damaged areas was performed using three different vegetation indices (NDVI, OSAVI, OSAVI_RE) in three different classification approaches, which were used to avoid the influence of technical roads on the result of winter damage crop estimation. The first method was a simple object-based classification to delineate the areas eligible for reimbursement. The main assumption behind this method was that it should appropriately suit borders of reference polygons which represent real situations and were vectorized without the generalization of polygons.

The second and third approach exploited pixel-based classification with thresholds. Index images were intersected with grids of cell sizes of 1×1 m and 5×5 m. For cells, a mean value of the index was assigned, and each cell was classified and merged with corresponding cells with the same class, creating homogenous regions. Thus, the resulting polygon geometry was simplified. By using those two approaches, we could also simulate a decrease in the resolution of input data.

In every approach, objects larger than 100 m^2 were finally recognized as detected damages. However, smaller regions lying within 5 m of other suspected damaged objects were also used, if the sum of their area was bigger than 100 m^2.

The possibility of proper detection of crops damaged by frost was examined using many variants. The influence of road elimination was examined, in addition to using a vegetation index for detection and spatial resolution of collected data. This resulted in 42 variants.

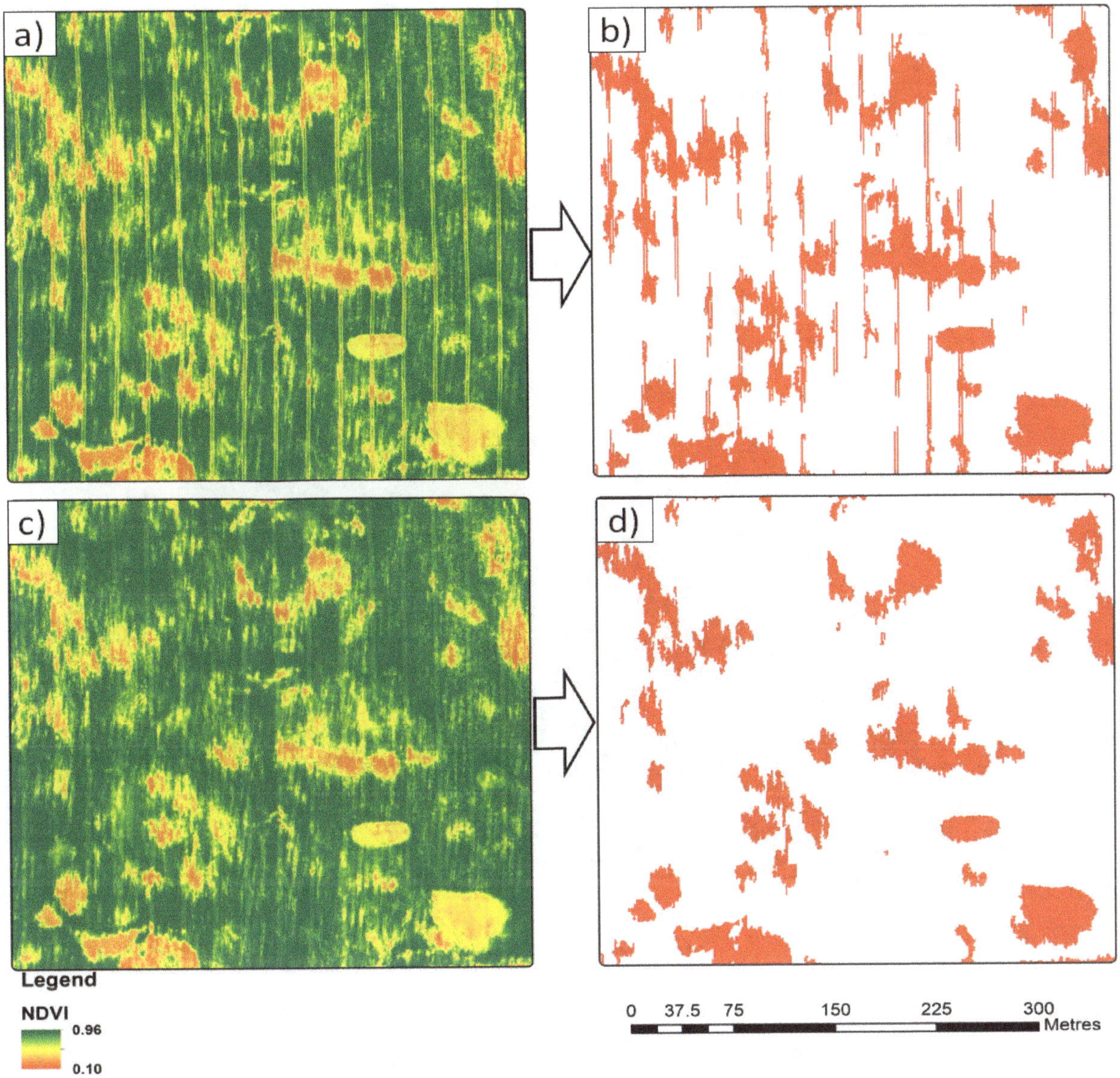

Figure 7. Presentation of impact of tires traces shown on NDVI images and the results of classification of winter damage for crops: (**a**) NDVI image, (**b**) NDVI image after elimination of influence of roads, (**c**,**d**) the results of classification of winter damages for crop based on the NDVI image, before and after elimination of influence of roads.

2.5. *Accuracy Assessment*

To estimate the accuracy of the final indications of which areas in each variant were eligible for reimbursement, a reference image was created by vectorization of damaged crops on an orthomosaic (see Figure 8).

The results were compared with reference areas by creating an error matrix which allowed for calculating such parameters as producer's accuracy (PA, completeness), user's accuracy (UA, correctness), F1-score, and overall accuracy (OA) [44,45]. The F1-score is the harmonic mean of precision and sensitivity and is usually used as an accuracy measure of a dichotomous model [45], so it is suitable for one-class delineation. In compensation assessment, the accurate area size of detected damage is much more important for the insurance company than its position, so another parameter was also used. This was the basic ratio of detected damaged areas to the reference area, and for the purpose of this study we named it "area index" (AI).

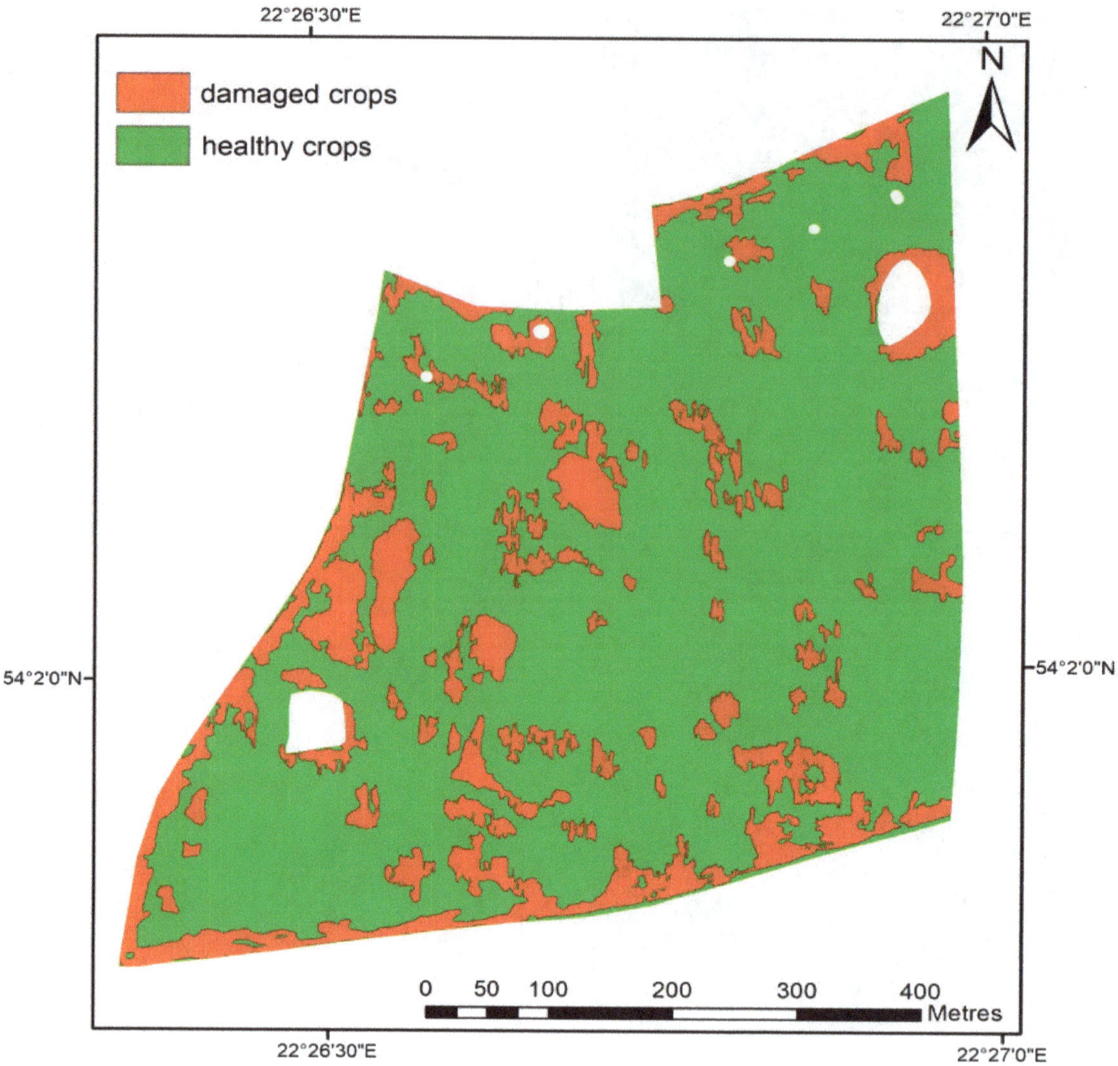

Figure 8. Reference data: crop field map manually vectorized.

3. Results

The results of all variants for damaged rapeseed detection are presented in Table A1 in Appendix A. Figure 9 illustrates the comparison of the overall accuracy, the F1-score, and AI for all variants of the experiment. The visual effect of the classification for selected variants is presented in Figure 10.

Most variants of classification allow detection of winterkill with OA above 90% (Figure 9a). UA and PA ranged between 70–80% and 90–100%, respectively (Table A1 in Appendix A). This means that the algorithm detects almost every polygon from our reference set but also mistakenly indicates other areas (overestimation). The poorer result in UA may indicate some mistakes in the reference set, which was difficult to vectorize objectively in a field, and because the algorithm may overestimate the detection of crop losses, which is confirmed by the area of damage often exceeding 110% to 130% (Figure 9c). However, it should be noted that the overestimation relates to transition areas, which are inherently difficult to clearly separate (see Figure 11). Generally, the best results were achieved using OSAVI and NDVI indices (the highest UA, OA, and F1-score values). The highest accuracy (measured by F1-score and OA) with the smallest overestimation of the area of losses in crops (AI) was obtained for a spatial resolution of 10 cm. The worst results (low OA and F1, and high AI) were obtained for OSAVI_RE, using images with GSD = 5 cm. The area for compensation was overestimated by 29–31%, which may be caused, for example, by inaccuracies in the georeference of an image with a spatial resolution of 5 cm.

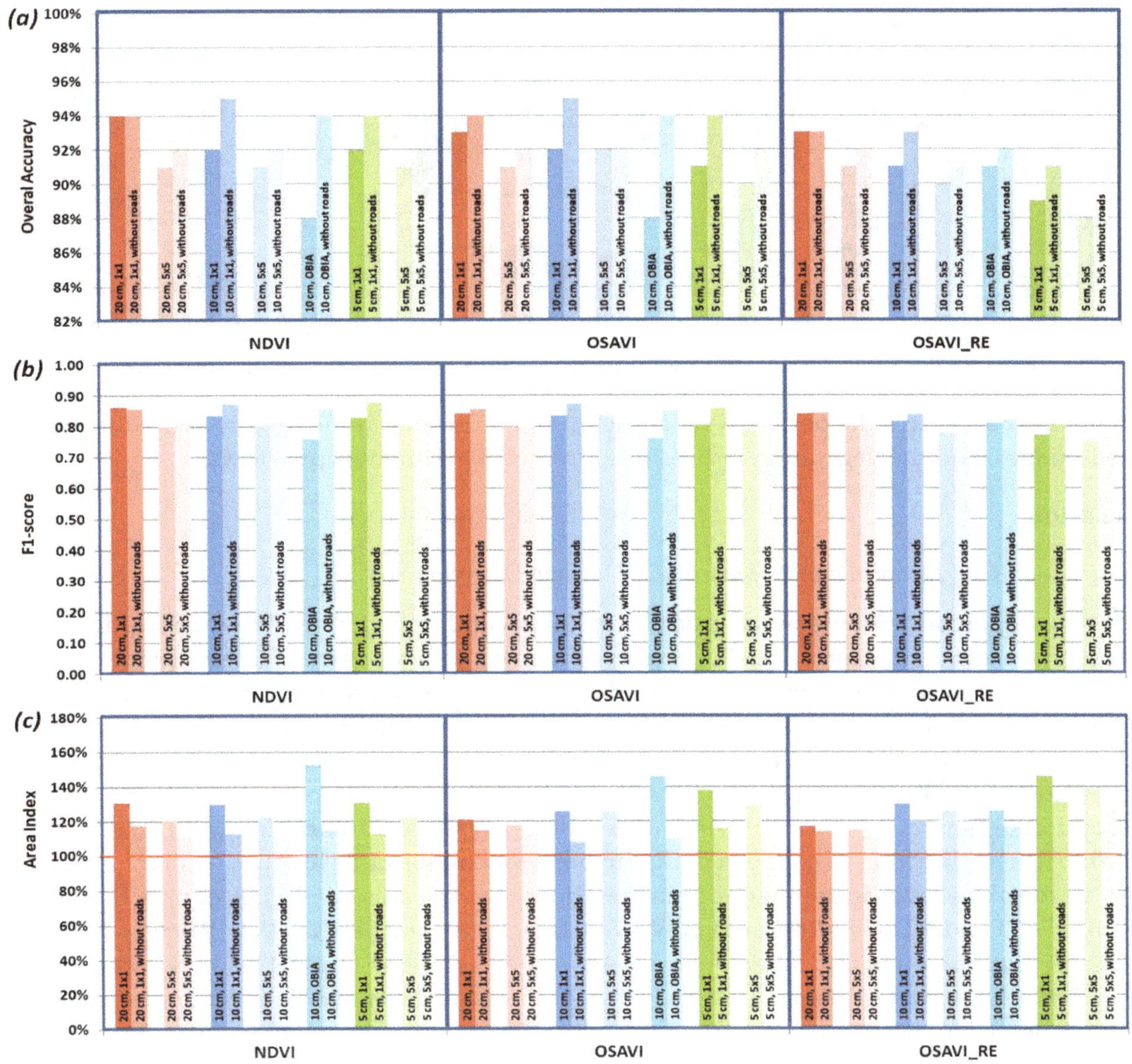

Figure 9. The comparison of the overall accuracy (OA) (**a**), the F1-score (**b**), and area index (AI) (**c**) of all variants of classifications.

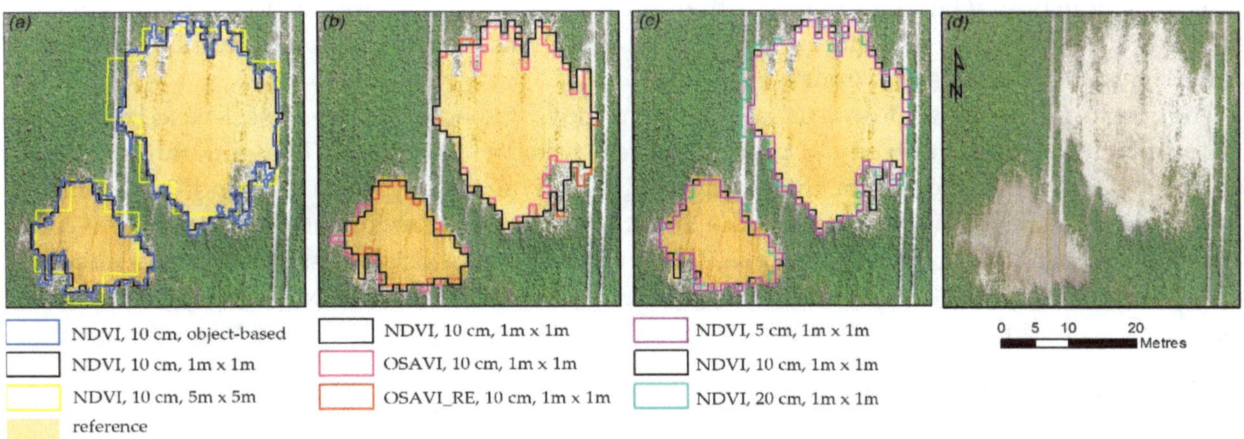

Figure 10. Small fragment of true color orthomosaic with reference mask and detected areas of damaged rapeseed without the influence of roads: (**a**) in three different approaches, (**b**) with three different indices, (**c**) with three different resolutions (**d**) true color orthomosaic.

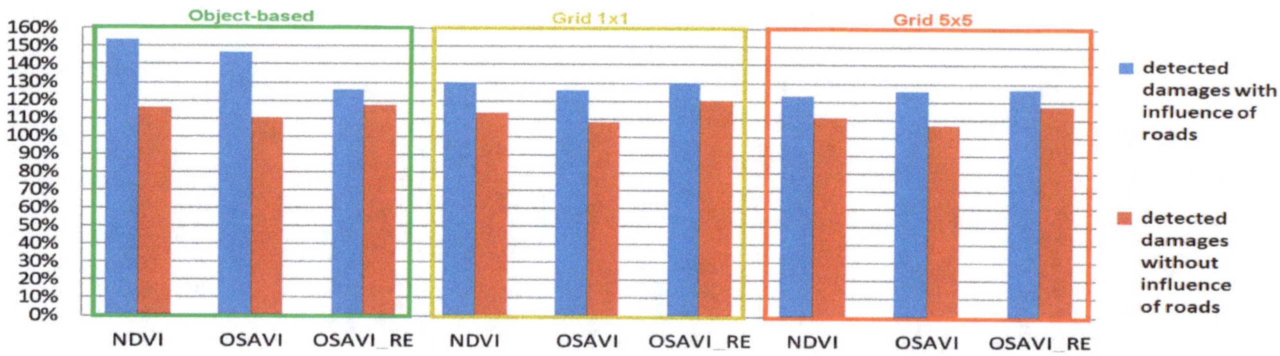

Figure 11. Area index (AI) in three approaches with 10 cm resolution of input data.

In the next section, a deeper analysis of the influence of road removal, the spatial resolution of collected images, and the vegetation indices used in this experiment is provided.

4. Discussion

4.1. Influence of Road Removal and Approaches in Damage Detection

As expected, road removal was a crucial step for obtaining better results. It allowed us to avoid many errors in the classification of smaller objects connected with roads (see Figure 7). It also prevented the incorrect enlargement of detected losses by incorporating road areas in detected polygons.

The negative influence of roads is highlighted in all three approaches on 10 cm input rasters. In both the second (grid 1 × 1 m) and the third (grid 5 × 5 m) approaches, each index increases UA by 3–5 percentage points, with a slight decrease of PA (around three percentage points) with only one exception of a higher outlier (10 percentage points for OSAVI in grid 5 × 5 m) (see Table A1 in Appendix A). The direction of changes in the first approach (object-based) are similar. The decrease of PA is on the same level as in other approaches, however, the increase in UA equals 17 percentage points on OSAVI and NDVI indices, and only two percentage points for OSAVI _RE, which is due to a much bigger UA value for this index before road removal (73–63% and 64% for NDVI and OSAVI, respectively). Although the differences in the range of increase of UA are high, the discrepancies in the final results between approaches 1 and 2 do not exceed three percentage points. This is also reflected by the area index (see Figure 11). The area index's values decrease in all three approaches after using the road removal technique. For OSAVI and NDVI indices, the difference is over two times higher in the object-based approach, but for OSAVI_RE, the decrease is on the same level as in the pixel-based approaches, which may be due to the fact that in the OSAVI_RE variant, technical roads are slightly less visible and are harder to distinguish in the images. This results in a lower number of roads that qualified as winter damage polygons.

The conducted analysis indicates that without removing road influence, object-based classification is a significantly worse method for proper winter damage detection. The generalization that is performed during the process of averaging the cell value in approaches 2 and 3 (resolution of grids 1 and 5 m, respectively), causes the rejection of some fraction of roads, which is not the case in the object-based approach because it provides a more precise classification. Because of this characteristic, a higher resistance to road influence in the OSAVI_RE index is most observable in the object-based scenario.

4.2. Influence of Spatial Resolution

The comparison of PA and UA for a 1 × 1 m grid with respect to vegetation indices is shown in Figure 12. Comparing NDVI results in this approach, we can assume that the smaller the GSD (ground sampling distance of obtained images), the better the results (an increase in the F1-score is observed). However, UA and PA values are almost equal for resolutions of 5 and 10 cm. The visible difference comes only at the 20 cm raster (a decrease of three percentage points in UA and 0.02 in F1-score value). In the case of the OSAVI index, we can confirm that results are slightly better on the 10 cm raster than 5 cm, but 20 cm still results in the worst outcome. The OSAVI_RE index trend is contrary to NDVI, with most of the indicator's values rising with decreasing pixel size. In the second approach, a 20 cm input OSAVI_RE raster results in a F1-score value of 0.84 and an OA equal to 93% (Table A1 in Appendix A), which is only slightly worse than the results obtained from other indices. The detected damage area is also more similar to the reference data than in comparable variants. This information allows us to suppose that, with an increasing size of GSD, OSAVI_RE could be the best index to use in the whole process. Generally, it can be assumed that any tested spatial resolution of UAV-based imagery is enough for winter damage assessment for insurance purposes and there are few differences in terms of accuracy in the tested approaches in the detection of these areas without plants.

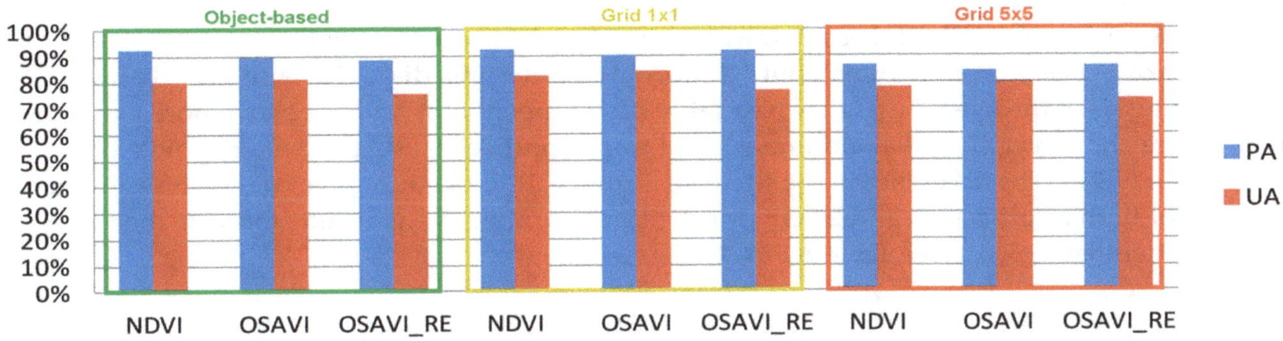

Figure 12. Producer's (PA) and user's (UA) accuracy of the detected damaged crops, without the influence of roads, with three different resolution of input data in the second approach (1 × 1 m).

For the results obtained in the described experiments with NDVI, decreasing the size of GSD provided better results, while the research of Candiago et al. (2015) [9] showed that the use of such a high resolution can be a problem when applying algorithms originally developed for aerial and satellite data. In this paper, three approaches were tested with OBIA, or resampling to grid sizes of 1 × 1 m and 5 × 5 m, which limited this problem.

The highest overall accuracy of detection wais achieved using the NDVI index on two different input data sets with 5 and 20 cm GSD (see Table A1 in Appendix A). For rasters with 10 cm resolution, the better choice was the OSAVI index, which was clearly confirmed in the majority of variants. However, the difference between the overall accuracy of damage detection using NDVI and OSAVI is not significant (1–2%). The usefulness of the NDVI vegetation index to determine the vegetation fraction (the percentage of green vegetation per unit of ground surface) and the number of plants per unit of ground surface is confirmed by the results of other authors (among others [19,22]). In our experiment, the results related to OSAVI_RE index are slightly better.

4.3. *Influence of Vegetation Index Used*

Figure 13 presents the results of damage detection for different vegetation indices in the case of the three approaches. Analysis of this figure allows us to draw a few crucial conclusions. Firstly, we can state that the producer's accuracy is stable in all three approaches regardless of chosen vegetation index as input data for classification. Its values vary from 0 to 3 percentage points in the same approach. However, approach 3 (grid 5 × 5 m) provides smaller absolute PA (around 85%) and UA (77%) values than the other approaches (mean values of 91% and 80%, respectively). This is mostly the result of much stronger generalization in this approach. Secondly, it is observed that there are significant differences in the user's accuracy between the chosen indices in all three approaches.

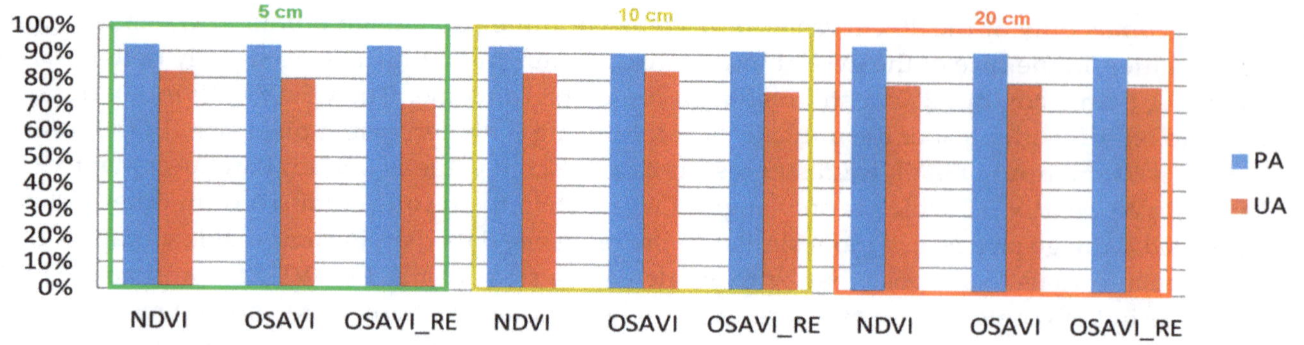

Figure 13. Producer's (PA) and user's (UA) accuracy of results without the influence of roads, with three different approaches performed on 10 cm input data.

In each variant, OSAVI_RE has the lowest value. OSAVI provides slightly better UA than NDVI (difference of 1–2 percentage points), but NDVI compensates this disparity with slightly higher PA values. Analysis of the overall accuracy (see Table A1 in Appendix A) indicates that those two indices are the same, however, the area of damage in all three approaches is 5 percentage points better for OSAVI than NDVI. The difference in the F1-score is only 0.01. Nonetheless, for data with a resolution of 5 cm (for both approaches) and 20 cm (for the third approach), the results indicate that NDVI is slightly better than OSAVI (see Table A1 in Appendix A).

5. Conclusions

The experiments performed demonstrate that low-altitude remote sensing can be used for the assessment of damaged areas in the case of insurance compensation. The case study used multispectral UAV images for the damage detection of rapeseed caused during winter. Data were acquired in April, when plant vegetation starts. This is the most important period in estimating damage since it directly exposes the effect of the winter period on vegetation. The presented methodology includes selecting possible vegetation indices, ground sampling resolutions, and approaches for technical road influence elimination, which can be used as an effective alternative for direct measurements in a field that currently are often used during the refurbishment process.

The analysis of GSD images (in the range of 5–20 cm) demonstrates a relatively small (a few percentage points) influence on the obtained results of winterkill detection. The best approximation of the area of the winter damage to crops was obtained using images with GSD of 10 cm. Results for most variants indicate that there is no need to acquire data with a higher resolution than 10 cm. This is a crucial conclusion for UAV application when the size of the area and the GSD of the images are considered within mission planning

The conclusion of the presented experiment is of great practical importance because cameras with four spectral channels (R, G, B, NIR) are becoming increasingly popular and relatively inexpensive. This provides the possibility to choose various vegetation indices appropriately for the analyzed crop and the purpose of the study related to the monitoring of the crop condition. The best results were obtained using NDVI and OSAVI vegetation indices; however, results from other indices were not significantly worse.

Referring to the utilization of the red edge spectral band, an optimized Soil-Adjusted Vegetation Index with the use of the red edge channel instead of the red channel is the most resilient index to the negative influence of technical roads, which is crucial when road detection cannot be conducted. The optimized soil-adjusted vegetation index with red edge channel (OSAVI_RE) may be the best for crop condition monitoring using lower spatial resolution data.

During tests, it was also necessary to develop and implement elimination of technical roads in calculations of rapeseed area damaged with frost. Including the detected area in the classification helps in the reliable calculation of the area for compensation. Without removing road influence, object-based classification is a significantly worse method, but elimination of road influence can be processed using both object- and pixel-based approaches. In tests with NDVI and OSAVI, overestimation of 10–30% was noticed for calculations without the elimination of the impact of roads, and all of the remote sensing-based approaches in all experiments were characterized by overestimation of at least a few percent considering the Area Index.

The implementation of the presented solutions for delimiting areas for refurbishment in the comprehensive damage estimation method, which also includes periodic monitoring and time analysis, can significantly improve the compensation process, and provide it with scientific certainty for obtained results.

Author Contributions: Conceptualization, K.S., W.O., and K.B.; methodology, W.O.; software, K.S.; validation, W.O.; formal analysis, K.O.-S.; investigation, Ł.J.; resources, K.S. and W.O.; writing—original draft preparation, Ł.J., W.O., K.O.-S., and K.B.; writing—review and editing, K.B. and K.O.-S.; visualization, K.O.-S. and Ł.J.; supervision, project administration and funding acquisition, K.O.-S. and K.B. All authors have read and agreed to the published version of the manuscript.

Acknowledgments: We would like to thank Skysnap Sp. z o.o. for providing the data from the Sequoia camera and participation in the calculations and analyzes, and Paulina Bartkowiak for the help in reference data preparation.

Appendix A

Table A1. The results of all variants within the experiment, assessed by producer's accuracy (PA) and user's accuracy (UA) for class damaged crops, and area index (AI), F1-score, and overall accuracy (OA) of classification for damaged crops and healthy crops.

	Resolution			5 cm					5 cm							
	Approach			Grid 1 × 1					Grid 5 × 5							
	Index	**UA**	**PA**	**AI**	**F1**	**OA**	**UA**	**PA**	**AI**	**F1**	**OA**					
without influence of roads	**NDVI**	83%	93%	113%	0.88	94%	78%	87%	111%	0.82	92%					
	OSAVI	80%	93%	116%	0.86	94%	76%	87%	114%	0.81	92%					
	OSAVI_RE	71%	93%	131%	0.81	91%	68%	88%	129%	0.77	89%					
with influence of roads	**NDVI**	73%	96%	131%	0.83	92%	73%	90%	123%	0.81	91%					
	OSAVI	69%	96%	138%	0.80	91%	70%	90%	129%	0.79	90%					
	OSAVI_RE	65%	95%	146%	0.77	89%	65%	90%	139%	0.75	88%					
	Resolution			10 cm					10 cm					10 cm		
	Approach			Grid 1 × 1					Grid 5 × 5					Object-Based		
	Index	**UA**	**PA**	**AI**	**F1**	**OA**	**UA**	**PA**	**AI**	**F1**	**OA**	**UA**	**PA**	**AI**	**F1**	**OA**
without influence of roads	**NDVI**	82%	93%	113%	0.87	95%	78%	86%	111%	0.82	92%	80%	92%	115%	0.86	94%
	OSAVI	84%	91%	108%	0.87	95%	79%	84%	106%	0.81	92%	81%	90%	110%	0.85	94%
	OSAVI_RE	77%	92%	120%	0.84	93%	73%	86%	117%	0.79	91%	76%	89%	117%	0.82	92%
with influence of roads	**NDVI**	74%	96%	130%	0.84	92%	73%	90%	123%	0.81	91%	63%	96%	153%	0.76	88%
	OSAVI	75%	94%	126%	0.83	92%	75%	94%	126%	0.83	92%	64%	94%	146%	0.76	88%
	OSAVI_RE	72%	94%	130%	0.82	91%	70%	88%	126%	0.78	90%	73%	91%	126%	0.81	91%
	Resolution			20 cm					20 cm							
	Approach			Grid 1 × 1					Grid 5 × 5							
	Index	**UA**	**PA**	**AI**	**F1**	**OA**	**UA**	**PA**	**AI**	**F1**	**OA**					
without influence of roads	**NDVI**	79%	94%	118%	0.86	94%	78%	86%	111%	0.82	92%					
	OSAVI	80%	92%	115%	0.86	94%	76%	86%	113%	0.81	92%					
	OSAVI_RE	79%	90%	114%	0.84	93%	76%	85%	112%	0.80	92%					
with influence of roads	**NDVI**	76%	100%	131%	0.86	94%	73%	89%	121%	0.80	91%					
	OSAVI	77%	93%	121%	0.84	93%	74%	87%	118%	0.80	91%					
	OSAVI_RE	78%	91%	117%	0.84	93%	75%	86%	115%	0.80	91%					

References

1. EUROSTAT DATA. 2019. Available online: https://ec.europa.eu/eurostat/databrowser/view/tag00100/default/table?lang=en (accessed on 17 July 2020).
2. GUS. Statistical Yearbook of Agriculture. Warsaw. 2020. Available online: https://stat.gov.pl (accessed on 17 July 2020).
3. Szulc, K. Wymarzanie Rzepaku Ozimego, Farmer.pl. 2018. Available online: https://www.farmer.pl/produkcja-roslinna/rosliny-oleiste/wymarzanie-rzepaku-ozimego,76896.html (accessed on 17 July 2020).
4. Wójtowicz, M.; Wójtowicz, A.; Piekarczyk, J. Application of remote sensing methods in agriculture. *Commun. Biometry Crop Sci.* **2016**, *11*, 31–50.
5. Brown, R.J.; Staenz, K.; McNairn, H.; Hopp, B.; Van Acker, R. Application of high-resolution optical imagery to precision agriculture. In Proceedings of the International Symposium Geomatics in the Era of RADARSAT (GER'97), Ottawa, ON, Canada, 25–30 May 1997; p. 9.
6. Haboudane, D.; Miller, J.R.; Tremblay, N.; Zarco-Tejada, P.J.; Dextraze, L. Integrated narrow-band vegetation indices for prediction of crop chlorophyll content for application to precision agriculture. *Remote Sens. Environ.* **2002**, *81*, 416–426. [CrossRef]
7. Doraiswamy, P.C.; Moulin, S.; Cook, P.W.; Stern, A. Crop yield assessment from remote sensing. *Photogramm. Eng. Remote Sens.* **2003**, *69*, 665–674. [CrossRef]
8. Rembold, F.; Atzberger, C.; Savin, I.; Rojas, O. Using low resolution satellite imagery for yield prediction and yield anomaly detection. *Remote Sens.* **2013**, *5*, 1704–1733. [CrossRef]
9. Candiago, S.; Remondino, F.; De Giglio, M.; Dubbini, M.; Gattelli, M. Evaluating multispectral images and vegetation indices for precision farming applications from UAV images. *Remote Sens.* **2015**, *7*, 4026–4047. [CrossRef]
10. Gago, J.; Douthe, C.; Coopman, R.; Gallego, P.; Ribas-Carbo, M.; Flexas, J.; Medrano, H. UAVs challenge to assess water stress for sustainable agriculture. *Agric. Water Manag.* **2015**, *153*, 9–19. [CrossRef]
11. Holman, F.H.; Riche, A.B.; Michalski, A.; Castle, M.; Wooster, M.J.; Hawkesford, M.J. High throughput field phenotyping of wheat plant height and growth rate in field plot trials using UAV based remote sensing. *Remote Sens.* **2016**, *8*, 1031. [CrossRef]
12. Puliti, S.; Ørka, H.O.; Gobakken, T.; Næsset, E. Inventory of Small Forest Areas Using an Unmanned Aerial System. *Remote Sens.* **2015**, *7*, 9632–9654. [CrossRef]
13. Lucieer, A.; Jong, S.M.D.; Turner, D. Mapping landslide displacements using Structure from Motion (SfM) and image correlation of multi-temporal UAV photography. *Prog. Phys. Geogr.* **2014**, *38*, 97–116. [CrossRef]
14. Wei, Z.; Han, Y.; Li, M.; Yang, K.; Yang, Y.; Luo, Y.; Ong, S.-H. A Small UAV Based Multi-Temporal Image Registration for Dynamic Agricultural Terrace Monitoring. *Remote Sens.* **2017**, *9*, 904. [CrossRef]
15. Khot, L.R.; Sankaran, S.; Carter, A.H.; Johnson, D.A.; Cummings, T.F. UAS imaging-based decision tools for arid winter wheat and irrigated potato production management. *Int. J. Remote Sens.* **2016**, *37*, 125–137. [CrossRef]
16. Kipp, S.; Mistele, B.; Baresel, P.; Schmidhalter, U. High-throughput phenotyping early plant vigour of winter wheat. *Eur. J. Agron.* **2014**, *52*, 271–278. [CrossRef]
17. Zhou, J.; Pavek, M.J.; Shelton, S.C.; Holden, Z.J.; Sankaran, S. Aerial multispectral imaging for crop hail damage assessment in potato. *Comput. Electron. Agric.* **2016**, *127*, 406–412. [CrossRef]
18. Salamí, E.; Barrado, C.; Pastor, E. UAV flight experiments applied to the remote sensing of vegetated areas. *Remote Sens.* **2014**, *6*, 11051–11081. [CrossRef]
19. Shi, Y.; Thomasson, J.A.; Murray, S.C.; Pugh, N.A.; Rooney, W.L.; Shafian, S.; Rajan, N.; Rouze, G.; Morgan, C.L.S.; Neely, H.L.; et al. Unmanned Aerial Vehicles for High-Throughput Phenotyping and Agronomic Research. *PLoS ONE* **2016**, *11*, e0159781. [CrossRef] [PubMed]
20. Zaman-Allah, M.; Vergara, O.; Araus, J.L.; Tarekegne, A.; Magorokosho, C.; Zarco-Tejada, P.J.; Hornero, A.; Hernández Albà, A.; Das, B.; Craufurd, P.; et al. Unmanned aerial platform-based multi-spectral imaging for field phenotyping of maize. *Plant Methods* **2015**, *11*, 1–10. [CrossRef] [PubMed]
21. Yang, G.; Liu, J.; Zhao, C.; Li, Z.; Huang, Y.; Yu, H.; Xu, B.; Yang, X.; Zhu, D.; Zhang, X.; et al. Unmanned Aerial Vehicle Remote Sensing for Field-Based Crop Phenotyping: Current Status and Perspectives. *Front. Plant Sci.* **2017**, *8*, 1–26. [CrossRef]

22. Ni, J.; Yao, L.; Zhang, J.; Cao, W.; Zhu, Y.; Tai, X. Development of an Unmanned Aerial Vehicle-Borne Crop-Growth Monitoring System. *Sensors* **2017**, *17*, 502. [CrossRef]
23. Shanahan, J.F.; Schepers, J.S.; Francis, D.D.; Varvel, G.E.; Wilhelm, W.W.; Tringe, J.M.; Schlemmer, M.R.; Major, D.J. Use of Remote-Sensing Imagery to Estimate Corn Grain Yield. *Agron. J.* **2001**, *93*, 583–589. [CrossRef]
24. Zhang, C.; Walters, D.; Kovacs, J.M. Applications of Low Altitude Remote Sensing in Agriculture upon Farmers' Requests—A Case Study in Northeastern Ontario, Canada. *PLoS ONE* **2014**, *9*, e112894. [CrossRef]
25. Hunt, E.R.; Hively, W.D.; Daughtry, C.S.; McCarty, G.W.; Fujikawa, S.J.; Ng, T.L.; Yoel, D.W. Remote sensing of crop leaf area index using unmanned airborne vehicles. In Proceedings of the Pecora 17 Symposium, Denver, CO, USA, 18–20 November 2008.
26. Bochenek, Z.; Dąbrowska-Zielińska, K.; Ciołkosz, A.; Drupka, S.; Boken, V.K. *Monitoring Agricultural Drought in Poland. Monitoring and Predicting Agricultural Drought*; Oxford University Press: Oxford, UK, 2005; pp. 171–180.
27. Piekarczyk, J.; Sulewska, H.; Szymańska, G. Winter oilseed-rape yield estimates from hyperspectral radiometer measurements. *Quaest. Geogr.* **2011**, *30*, 77–84. [CrossRef]
28. Hunt, E.R.; Hively, W.D.; Fujikawa, S.J.; Linden, D.S.; Daughtry, C.S.T.; McCarty, G.W. Acquisition of NIR-Green-Blue Digital Photographs from Unmanned Aircraft for Crop Monitoring. *Remote Sens.* **2010**, *2*, 290–305. [CrossRef]
29. Mathews, A.J.; Jensen, J.L.R. Visualizing and quantifying vineyard canopy LAI using an unmanned aerial vehicle (UAV) collected high density structure from motion point cloud. *Remote Sens.* **2013**, *5*, 2164–2183. [CrossRef]
30. Wei, C.; Huang, J.; Wang, X.; Blackburn, G.A.; Zhang, Y.; Wang, S.; Mansaray, L.R. Hyperspectral characterization of freezing injury and its biochemical impacts in oilseed rape leaves. *Remote Sens. Environ.* **2017**, *195*, 56–66. [CrossRef]
31. Jones, H.G.; Vaughan, R.A. *Remote Sensing of Vegetation—Principles, Techniques, and Applications*; Oxford University Press: New York, NY, USA, 2010.
32. Nicotra, A.B.; Hofmann, M.; Siebke, K.; Ball, M.C. Spatial patterning of pigmentation in evergreen leaves in response to freezing stress. *Plant Cell Environ.* **2003**, *26*, 1893–1904. [CrossRef]
33. Wu, Q.; Zhu, D.; Wang, C.; Ma, Z.; Wang, J. Diagnosis of freezing stress in wheat seedlings using hyperspectral imaging. *Biosyst. Eng.* **2012**, *112*, 253–260. [CrossRef]
34. Flower, K.; Boru, B.; Nansen, C.; Jones, H.; Thompson, S.; Lacoste, C.; Murphy, M. *Proof of Concept: Remote Sensing Frost-Induced Stress in Wheat Paddocks*; Grains Research and Development Corporation: Canberra, Australia, 2014.
35. Murphy, M.E.; Boruff, B.; Callow, J.N.; Flower, K.C. Detecting Frost Stress in Wheat: A Controlled Environment Hyperspectral Study on Wheat Plant Components and Implications for Multispectral Field Sensing. *Remote Sens.* **2020**, *12*, 477. [CrossRef]
36. Xie, Y.; Wang, C.; Yang, W.; Feng, M.; Qiao, X.; Song, J. Canopy hyperspectral characteristics and yield estimation of winter wheat (Triticum aestivum) under low temperature injury. *Sci. Rep.* **2020**, *10*, 244. [CrossRef]
37. Gitelson, A.A.; Kaufman, Y.J.; Merzlyak, M.N. Use of a green channel in remote sensing of global vegetation from EOS-MODIS. *Remote Sens. Environ.* **1996**, *58*, 289–298. [CrossRef]
38. Daughtry, C.S.T.; Walthall, C.L.; Kim, M.S.; Brown de Colstoun, E.; McMurtrey, J.E., III. Estimating corn leaf chlorophyll concentration from leaf and canopy reflectance. *Remote Sens. Environ.* **2000**, *74*, 229–239. [CrossRef]
39. Rouse, J.W. *Monitoring the Vernal Advancement and Retrogradation (Green Wave Effect) of Natural Vegetation*; Technical Report; NASA: Washington, DC, USA, 1972.
40. Huete, A.R. A soil-adjusted vegetation index (SAVI). *Remote Sens. Environ.* **1988**, *25*, 295–309. [CrossRef]
41. Rondeaux, C.; Steven, M.D.; Baret, F. Optimized of Soil-adjusted vegetation indices. *Remote Sens Environ.* **1996**, *55*, 95–107. [CrossRef]
42. Schuster, C.; Förster, M.; Kleinschmit, B. Testing the red edge channel for improving land-use classifications based on high-resolution multi-spectral satellite data. *Int. J. Remote Sens.* **2012**, *33*, 5583–5599. [CrossRef]

43. Li, F.; Miao, Y.; Feng, G.; Yuan, F.; Yue, S.; Gao, X.; Liu, Y.; Liu, B.; Ustin, S.L.; Chen, X. Improving estimation of summer maize nitrogen status with red edge-based spectral vegetation indices. *Field Crop. Res.* **2014**, *157*, 111–123. [CrossRef]
44. Congalton, R. A Review of Assessing the Accuracy of Classifications of Remotely Sensed Data. *Remote Sens. Environ.* **1991**, *37*, 35–46. [CrossRef]
45. Koga, Y.; Miyazaki, H.; Shibasaki, R. A CNN-based Method of vehicle detection from aerial images using hard example mining. *Remote Sens.* **2018**, *10*, 124. [CrossRef]

9

Mapping of Olive Trees using Pansharpened QuickBird Images: An Evaluation of Pixel- and Object-Based Analyses

Isabel Luisa Castillejo-González

Department of Graphic Engineering and Geomatics, University of Cordoba, Campus de Rabanales, 14071 Córdoba, Spain; ilcasti@uco.es

Abstract: This study sought to verify whether remote sensing offers the ability to efficiently delineate olive tree canopies using QuickBird (QB) satellite imagery. This paper compares four classification algorithms performed in pixel- and object-based analyses. To increase the spectral and spatial resolution of the standard QB image, three different pansharpened images were obtained based on variations in the weight of the red and near infrared bands. The results showed slight differences between classifiers. Maximum Likelihood algorithm yielded the highest results in pixel-based classifications with an average overall accuracy (OA) of 94.2%. In object-based analyses, Maximum Likelihood and Decision Tree classifiers offered the highest precisions with average OA of 95.3% and 96.6%, respectively. Between pixel- and object-based analyses no clear difference was observed, showing an increase of average OA values of approximately 1% for all classifiers except Decision Tree, which improved up to 4.5%. The alteration of the weight of different bands in the pansharpen process exhibited satisfactory results with a general performance improvement of up to 9% and 11% in pixel- and object-based analyses, respectively. Thus, object-based analyses with the DT algorithm and the pansharpened imagery with the near-infrared band altered would be highly recommended to obtain accurate maps for site-specific management.

Keywords: Á Trous algorithm; conservation agriculture; crop inventory; remote sensing; spectral-weight variations in fused images

---

1. Introduction

Nowadays, one of the most important objectives in agriculture is to perform precision agriculture (PA) in most possible scenarios to control efficiently the input data and, consequently, reduce the production cost and the environmental pollution produced by this activity. To facilitate and assist this change in farm management, government institutions tend to regulate and encourage different techniques based on PA. The European Commission is greatly concerned about the new challenges in agriculture and promote the changes by different legal instruments and key texts [1]. Diverse action areas such as farming, protection of natural or agricultural environment, food safety, security and traceability, or even climate change mitigation are regulated. Some of the recommended or mandatory practices are supported in an accurate control of the spatial distribution of crops. To obtain economical funds from the common agricultural policy [2], PA promotes among many other actions, the application of site-specific management or integrated management systems in crops production to reduce the use of fertilisers, herbicides, or pesticides and the establishment of certain conservation agro-environmental measures such as cover crops in olive orchards [3]. To control the correct application of PA techniques, different monitoring systems were developed. The expensive, time-consuming, and imprecise system based on sample and ground visit to small percentage of fields has forced a search for new techniques that reduce costs and increase the controlled area, maintaining high accuracy in the analyses. Remote sensing data can significantly improve the deficiencies of ground visits, allowing accurate maps.

Several studies have focused on addressing diverse PA topics by remote sensing, such as obtaining accurate maps of crops [4–6]; detecting the location of weeds [7–9], pests [10–12], and diseases [13–15] to apply site-specific management, or determining the level of water stress to design optimal irrigation systems [16–18]. Nevertheless, most of the precision agriculture studies with remote sensing analysed the characteristics of herbaceous crops, as these types of crops usually cover all the field and are easier to study with digital imagery. Woody crops present very different spectral responses between tree canopies, soils, and other covers presented in the field. Thus, very high spatial resolution images are needed. Most of the studies which aim is to characterize the architecture of the trees used airborne or Unmanned Aerial Vehicle (UAV) images to obtain the canopy information [19–21]. Nevertheless, few studies used satellite imagery [22–24] and most of them were aid with LiDAR information [25–27].

High resolution satellite imagery can be useful to accurately map tree canopies. Companies that distribute Earth Observation Satellite images usually offer the user community two separate products: a high-spatial resolution panchromatic image and a low-spatial resolution multispectral one. While the multispectral image facilitates the discrimination of land covers types, the panchromatic image allows to delimit accurately each land cover [28]. To use simultaneously the advantages of both resolutions in one image, fusion techniques have been developed. The pansharpen fusion method allows the injection of spatial detail information from the panchromatic image into each band of the multispectral image [29]. These new characteristics of the pansharpened image can help to accurately delineate the tree canopies to apply PA techniques.

Supervised classification methods are extensively used in land use classification studies [30]. These procedures extrapolate the spectral characteristics obtained from the image training sites defined for classification by the user to other areas of the image. Nowadays, there are classification routines based on spectral or angular distances, probability analysis, and more advanced data mining techniques. There is no one ideal classification routine. The most appropriate method is determined by the needs and requirements of each study [31]. Many of the remote sensing classifications are based on pixels as the minimum spatial information unit. These analyses provide very good results in homogeneous land uses. Nevertheless, the increase of spatial resolution causes an increase in the intraclass spectral variability and a reduction in classification performance and accuracy when pixel-based analyses are used [32]. This is particularly true when the pixel size is significantly smaller than the average size of the objects of interest [33]. To overcome this problem, different segmentation techniques, in which adjacent pixels are grouped into spectrally and spatially homogeneous objects, have been developed. The main segmentation algorithms can be classified into two general classes: edge-based and object-merging algorithms [34]. Most of the segmentation procedure developed are object merging algorithms, which take some pixels as seeds and grow the regions around them based on certain homogeneity criteria [35]. Since Kettig and Landgrebe [36], the object-based approach has hardly been used in favour of easier pixel-based analyses. Some researchers have reported that the segmentation techniques used in classifications reduce the local variation caused by textures, shadows and shape in forestry trees [37,38] and agricultural trees [39] classifications. However, object-based classifications in typical agricultural dryland Mediterranean areas to map accurately olive tree canopies using only high spatial resolution satellite imagery are lacking.

Therefore, the main objective of this paper was to evaluate the potential of four supervised classification routines, applied to pixel- and object-based classifications, to delineate olive tree canopies using a pansharpened QuickBird image. An additional goal was to check the effect of the variation of the spectral weight in the pansharpen process, to emphasise the spectral information of different wavelengths over another.

2. Materials and Methods

2.1. Study Area and Satellite Image Acquisition

This study was focused on dryland olive (*Olea europaea* L.) orchards representative of the typical continental Mediterranean climate [40]. The analysis was conducted in five olive orchards fields named A, B, C, D and E (Figure 1) located near Montilla, province of Córdoba (Andalusia, southern Spain, centre UTM *X*-*Y* coordinates 355,746–4,164,520 m, datum WGS84). This agricultural region has a typical continental Mediterranean climate characterized with short mild winters and long dry summers [41]. The study sites were located in a farmer-managed area where the farmers made decisions individually. Thus, different characteristics of the studied fields such as size and morphology of olive crowns, presence or not of vegetation cover or soil tillage, were found. The total areas of the A, B, C, D and E fields were 2.16 ha, 22.80 ha, 15.66 ha, 24.55 ha and 28.44 ha, respectively.

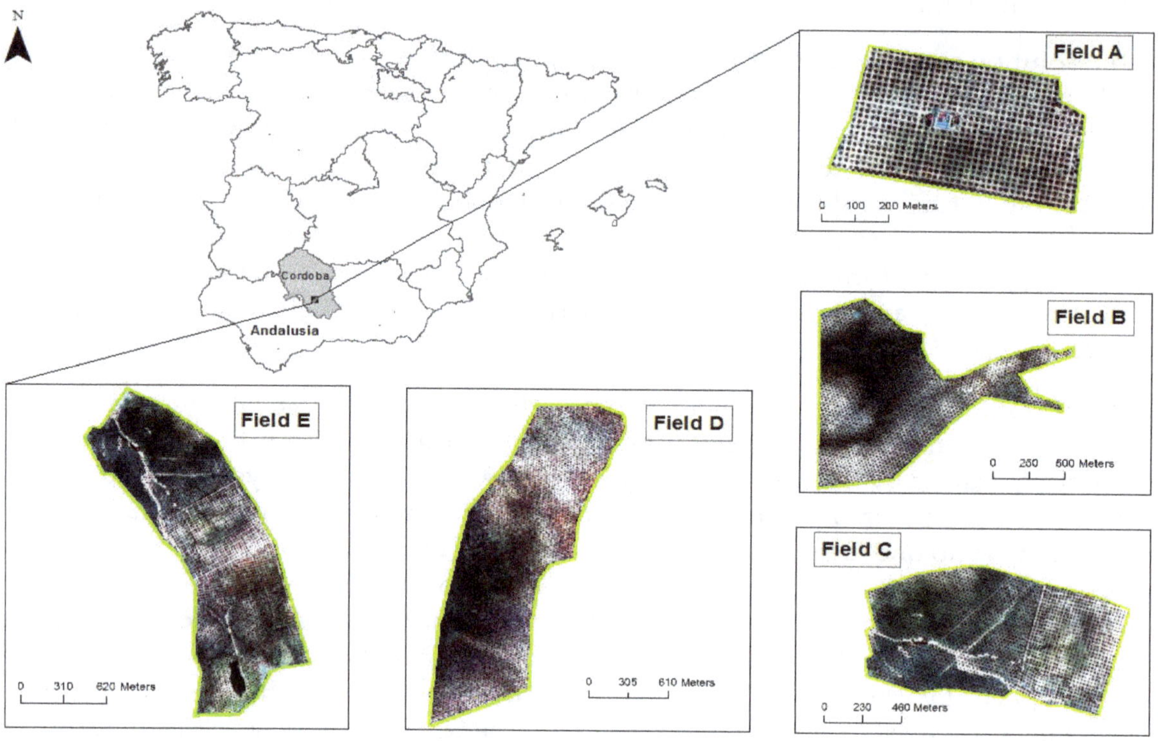

Figure 1. Location of the study area in Andalusia, southern Spain. Detailed olive orchards fields are depicted by QuickBird pansharpened images.

On 10 July 2004, a QuickBird (QB) satellite digital image was acquired for the study area. The QB satellite provided four multispectral bands (blue, B: 450–520 nm; green, G: 520–600 nm; red, R: 630–690 nm; and near-infrared, NIR: 760–900 nm) with a spatial resolution of 2.8 m, and a panchromatic band (PAN: 450–900 nm) with a spatial resolution of 0.7 m. Radiometric resolution of the QB image was 11 bit. A QB Standard image product was ordered, which included radiometric, sensor and geometric corrections previously carried out by the distributor [42]. No atmospheric corrections were needed as long as each classification was carried out in a single date image on the same relative scale [43].

2.2. Data Fusion: Pansharpening of Multispectral Images

To obtain an image of high spectral and spatial resolution, a pansharpening process was carried out with the QB bands. The pansharpen techniques allow to obtain new bands with the spectral resolution of the multispectral bands and the spatial resolution of the panchromatic band. In this study, a weighting variant of the fusion algorithm based on the wavelet transform calculated using the Á Trous algorithm was used to fuse the multispectral and panchromatic bands [44]. This fusion method consists basically of successive convolutions between the analyzed image and a low-pass filter called the scaling function, which commonly is the b3-spline function. The filters to be applied in the subsequent decomposition levels are obtained from the filter applied in the previous level, intercalating it with zeros in the rows and columns. The wavelet coefficients are obtained from the difference between two consecutive decomposition levels.

Knowing that the red and especially the near infrared bands are very important in the discrimination of vegetation, these two bands were weighted differently. Thus, to evaluate the effect of different weighted pansharpened bands in the classifications, three weight combinations were proposed: (a) 1-1-1-1, (b) 1-1-5-5 and (c) 1-1-1-10 as weight factor for B-G-R-NIR bands, respectively. As a result of this fusion or pansharpening process, three different images with four multispectral bands (B, G, R and NIR) and with a spatial resolution of 0.7 m were obtained.

Usually, the global quality of the resulting pansharpened image is estimated with the ERGAS index (Erreur Relative Globale Adimensionnelle de Synthèse) [45]. This relative dimensionless global error in synthesis offers a global picture of the spectral quality of the fused product. Nevertheless, in pansharpening techniques, high spectral quality implies low spatial quality and vice versa, which suggests the necessity to control not only the spectral quality of the process, but also the spatial. Thus, a spatial index based on ERGAS concepts and translated to the spatial domain [46] was used.

2.3. *Segmentation*

A segmentation procedure was performed to partition the QB pansharpened images into homogeneous objects using the Fractal Net Evolution Approach (FNEA) segmentation algorithm [47] (Figure 2). This algorithm allows the multiresolution bottom-up region-merging segmentation, a process in which individual pixels merge to objects in successive fusing iterations. The merging process continues until a threshold derived from the user-defined parameters is reached. The result is an image in which the pixels are aggregated in highly homogeneous objects at an arbitrary resolution.

(a) (b) (c)

Figure 2. Multiresolution segmentation of pansharpened QuickBird (QB) imagery in field A. Pansharpen weight (B-G-R-NIR): **(a)** 1-1-1-1; **(b)** 1-1-5-5; and **(c)** 1-1-1-10.

The segmentation process can be controlled by the weighting of the input data and the definition of three parameters. The scale parameter controls the size of the objects, while the colour and shape parameters define the importance of the spectral and morphological information involved in the object generation, respectively. The setting of the segmentation parameters were determined by testing different segmentation input scenarios to evaluate their ability to delineate olive crowns. For each field, the first parameter to adjust was the scale parameter to control the size of the objects depending on the characteristics of the field. Then, with the scale parameter fixed, the spectral and morphological weight of the information was defined. The morphological information is divided into two characteristics, the compactness and the smoothness of the objects. For this study, as the trees crowns present a compact structure, these two characteristics were fixed in all scenarios tested to 0.8 and 0.2 for compactness and smoothness, respectively.

The segmentation procedure generates not only the mean spectral information of the objects, which is derived from the spectral information of the pixels that form each object, but also a large amount of data divided mainly in three categories: spectral, morphological and textural. In this study, only some spectral and morphological variables derived from the segmentation process were used to characterize the olive orchards. Therefore, to perform this analysis eight object-based features, three spectral and five morphological variables, were included (Table 1). The spectral feature *Mean* is calculated for each multispectral band independently, obtaining 4 final features: Mean (Blue), Mean (Green), Mean (Red) and Mean (Near-Infrared).

Table 1. Object-based features derived from segmentation.

Categories	Features	Brief Description
Spectral	Mean	Mean of the intensity values of all pixels forming an image object
	NDVI	Normalized Difference Vegetation Index [48]
	RDVI	Renormalized Difference Vegetation Index [49]
Shape	Area	Number of pixels forming an image object
	Asymmetry	Relative length of an image object compared to a regular ellipse polygon
	Border index	Ratio between the border lengths of the image object and the smallest enclosing rectangle
	Length	Multiplication between the number of pixels and the length-to-width ratio of an image object
	Width	Ratio between the number of pixels and the length-to-width ratio of an image object

2.4. Classification and Accuracy Assessments

Both pixel- and object-based analyses with four different supervised classifiers were conducted on the five olive orchards fields. The classification algorithms were Minimum Distance (MD), Spectral Angle Mapper (SAM), Maximum Likelihood (ML) and Decision Tree (DT). MD classifies each pixel in the category that presents the minimum spectral distance between the spectral signal of the pixel and the spectral average of the class. The spectral distance is determined by Euclidean distance in N-bands spectral space [50]. Similarly, SAM measures spectral similarity but assigns the category of each pixel to the class that presents the minimum spectral angle, instead the spectral distance. The spectral angle between two spectra is calculated by taking the arccosine of the dot product of the two spectral vector [51]. ML creates classification rules based on probabilistic algorithms considering the spectral average of each class and the variance. This algorithm assigns a pixel to the most probable class and thus minimizes the probability of error using Bayesian theory [52]. Finally, DT classifier creates models of decision based on conditional control statements. In this study, the DT classification was performed with the data mining C4.5 algorithm, a top-down inductor of decision trees that expands nodes in depth-first order for each step using the divide-and-conquer strategy [53]. Ground-truth land use was randomly defined to substantiate and validate the classification procedures. For each field, a sampling with distant and independent locations was digitized directly on the image. Approximately 25% of the sampled surface were used to collect the spectral signature in the training process, and the remaining 75% were used to assess the accuracy of the classifications. To avoid any subjective estimation, the training and verification procedure did not change in any of the classifications.

To determine the accuracy obtained with every classifier in each olive orchard field, the confusion matrix of the classification and the Kappa test were analysed. The confusion matrix compares the percentage of classified pixels of each class with the verified ground truth class, indicating the correct assessment and the errors among the studied classes [38]. In addition to detailed accuracies obtained in every classification category, the confusion matrix obtain the overall accuracy (OA), which indicates the overall percentage of correctly classified pixels in the classification. The Kappa test yields the Kappa coefficient (K), which determines if the results obtained in the classification are significantly better than the results obtained in a random classification. The combination of both accuracy values is more conservative than a simple percent agreement value [54,55].

In pixel-based classification it is frequent to observe isolated-misclassified pixels dispersed within the area of another class. To reduce this commonly named salt and pepper noise and increase the accuracy of the classifications, a majority filter of 3 × 3 was applied to improve the classified maps. In object-based classification this noise is practically eliminated when pixels group in the segmentation process.

The pansharpened QB bands were generated with the IJFUSION software (Polytechnic University of Madrid, Spain). To obtain the segmented bands used in object-based classifications, the eCognition Developer 8 software (Definiens AG) was used. The Weka 3.8 software (University of Waikato, New Zealand) determined the decision tree sequences. Finally, the software ENVI 5.1 (Harris Geospatial Solutions) was used to carry out all the pixel- and object-based classifications.

3. Results and Discussion

3.1. Data Fusion: Pansharpening of Multispectral Images

Three pansharpen procedures were performed with the entire QB image. To control the quality of the process, two ERGAS indexes were calculated. Table 2 shows the spectral and spatial ERGAS indexes obtained in the study for each pansharpen combination. The spectral ERGAS index showed values of 0.72, 1.84 and 1.83 whereas the spatial ERGAS index exhibited slightly higher values of 1.3, 1.73, and 1.86, for pansharpen B-G-R-NIR combinations 1-1-1-1; 1-1-5-5 and 1-1-1-10, respectively. Being a value of 0 of each ERGAS index the maximum quality, the lower the ERGAS value, the higher the spectral quality of fused images. ERGAS error lower than 3 are considered as good quality for fused product [56]. In this study, the ERGAS errors obtained were considerably low (lower than 2 in all combinations), which implied a high spectral and spatial quality in the images obtained [57]. This premise is important in this type of study, as an accurate isolation of olive crowns need a very high spatial resolution but without losing the spectral information, especially in complex areas where a mixture of olive tree, natural cover, and soil spectral data can be observed. The increase of the ERGAS indexes values in the pansharpened image, where the red and near-infrared bands were over-weighed, was predictable. Gonzalo and Lillo-Saavedra [44] conceived the pansharpen algorithm performed in this study to apply the exact weighted factor to every band to obtain same spatial and spectral quality of the image, it means, to obtain the "best fused image". In this study, and controlling that the quality of the fused images does not exceed the ERGAS index limit, the weight of the bands was based on the necessity of the study but not in the control of the pansharpen quality.

Table 2. Spectral and spatial indexes to control the quality of the pansharpened images.

Pansharpen Weight (B-G-R-NIR)	Spectral ERGAS	Spatial ERGAS
1-1-1-1	0.72	1.13
1-1-5-5	1.84	1.73
1-1-1-10	1.83	1.86

3.2. Segmentation

For the fifteen pansharpened fields analysed, three pansharpen combinations × five olive orchard fields, a considerable number of input parameters were tested to obtain the criteria that provided the most satisfactory inputs scenarios (Table 3). Each olive orchard field presented different characteristics, which involved different segmentation parameters. As the aim of the segmentation is to obtain objects with a similar or smaller area than an olive tree canopy, the values of the scale parameters were low, ranging from 12 in the fields A and E to 25 in the fields B and C. The pansharpen images with the weight 1-1-1-1 always required a smaller scale parameter than the other two pansharpen images. In the segmentation process, the spectral information (colour) presented more weight than the morphology of the objects (shape) in most of the scenarios evaluated, although colour never exceeded the 70% of the weight. Only in the field D, the percentage of both types of information was divided equally (50%).

As mentioned in Section 2.3, the compactness and smoothness characteristics of the morphology of the objects were fixed to 0.8 and 0.2, respectively.

Table 3. Most satisfactory segmentation parameters obtained for each pansharpened field.

Field	Pansharpen Weight (B-G-R-NIR)	Scale Parameter	Colour	Shape	Compactness	Smoothness
A	1-1-1-1	12	0.6	0.4	0.8	0.2
	1-1-5-5	20	0.7	0.3	0.8	0.2
	1-1-1-10	20	0.7	0.3	0.8	0.2
B	1-1-1-1	15	0.7	0.3	0.8	0.2
	1-1-5-5	25	0.7	0.3	0.8	0.2
	1-1-1-10	25	0.5	0.5	0.8	0.2
C	1-1-1-1	15	0.6	0.4	0.8	0.2
	1-1-5-5	25	0.7	0.3	0.8	0.2
	1-1-1-10	17	0.6	0.4	0.8	0.2
D	1-1-1-1	14	0.6	0.4	0.8	0.2
	1-1-5-5	14	0.5	0.5	0.8	0.2
	1-1-1-10	14	0.5	0.5	0.8	0.2
E	1-1-1-1	12	0.6	0.4	0.8	0.2
	1-1-5-5	19	0.7	0.3	0.8	0.2
	1-1-1-10	22	0.7	0.3	0.8	0.2

3.3. *Olive Orchard Fields Classification*

The accuracy assessments, OA and K, of the different pixel- and object-based classifications carried out in the different weighted pansharpened images of every field are displayed in Table 4. Figure 3 shows an example of the least and the most accurate olive crowns classifications in an individual field. Table 4 reveals slight differences between classifiers, yielding most of the scenarios evaluated very high classification accuracies and showing that olive trees canopies could be discriminated very accurately in most of the test carried out. All the classifiers achieved high comparable classification results, although some classifiers stood out above others. In pixel-based classifications, MP exhibited the highest average of OA and K with a value of 94.2% and 0.89, respectively. In object-based classification, MP obtained the highest accuracies values in eight of the 15 images analysed and DT obtained the highest precision in the remaining seven classifications. Nevertheless, the average results were slightly higher with the DT classifier offering OA and K values of 96.6% and 0.94%, while ML obtained values of 95.3% and 0.91%. ML and DT classification algorithms exhibited high reliability in all classifications performed, while SAM and MD yielded more erratic results, showing always the lowest accuracies of each analysis. As an example, in pixel-based classification of field E, MD obtained the highest OA values for the pansharpen combinations 1-1-5-5 and 1-1-1-10 with 93.4% and 95.2%, respectively, whereas this classifier showed very low OA value for the pansharpen combination 1-1-1-1, with a value of 77.3%. Nevertheless, all the olive orchard fields could be classified very accurately with at least one classifier, showing OA values greater than 90%. The results observed in this study satisfied the commonly accepted requirements to consider an accurate classification when the OA value is at least an 85% [58] and the Kappa coefficient exceeds the 0.75 [59]. Such high accuracies are essential if the olive orchard map obtained is going to be used in precision agriculture to design site specific management. To emphasise one classifier from all of them, ML classifier could be selected considering that yielded one of the highest precisions in all classifications performed and that the computational and expertise requirements involved in this classification method is lower than the other most accurate classifier, DT, a data mining algorithms which demands deeper knowledge. Additionally, ML algorithm is implemented in most of the image processing software.

Table 4. Classification accuracies of olive orchards at the five fields analysed in the three pansharpened images using different classification algorithms.

		Analyses															
		Pixel-Based								Object-Based							
Field	Image [1]	MD [2]		SAM		ML		DT		MD		SAM		ML		DT	
		OA [3]	K	OA	K	OA	K	OA	K	OA	K	OA	K	OA	K	OA	K
A	1-1-1-1	94.4	0.91	92.6	0.88	95.0	0.92	91.7	0.86	96.2	0.93	96.6	0.94	97.8	0.96	94.9	0.91
	1-1-5-5	91.7	0.86	90.8	0.84	94.4	0.91	98.7	0.97	96.4	0.94	94.5	0.91	97.9	0.96	98.9	0.98
	1-1-1-10	89.4	0.82	88.0	0.79	92.8	0.87	98.6	0.98	91.7	0.86	95.7	0.93	98.8	0.98	98.7	0.98
B	1-1-1-1	91.3	0.83	94.2	0.89	98.7	0.97	95.7	0.91	88.8	0.78	91.3	0.82	99.1	0.98	96.9	0.95
	1-1-5-5	95.2	0.91	95.4	0.91	98.7	0.98	92.1	0.89	94.6	0.89	97.3	0.94	99.3	0.99	99.3	0.98
	1-1-1-10	95.7	0.86	86.5	0.73	97.5	0.95	94.3	0.90	94.1	0.88	77.9	0.59	98.7	0.97	98.5	0.97
C	1-1-1-1	90.4	0.81	93.4	0.87	97.7	0.95	96.5	0.94	88.8	0.78	88.4	0.77	97.5	0.95	97.0	0.94
	1-1-5-5	95.8	0.92	93.1	0.86	97.7	0.95	92.4	0.89	95.6	0.91	96.5	0.93	97.9	0.96	98.7	0.98
	1-1-1-10	95.5	0.91	85.3	0.71	96.6	0.93	94.1	0.90	94.9	0.90	64.9	0.31	98.5	0.97	98.6	0.97
D	1-1-1-1	69.7	0.39	75.4	0.51	88.1	0.76	84.8	0.80	78.9	0.58	86.1	0.72	79.7	0.59	87.2	0.82
	1-1-5-5	87.5	0.75	87.9	0.76	94.2	0.88	88.5	0.84	87.1	0.74	85.6	0.71	88.4	0.77	98.3	0.96
	1-1-1-10	89.5	0.79	85.3	0.71	92.3	0.85	87.3	0.83	85.7	0.72	88.0	0.76	91.6	0.83	98.7	0.97
E	1-1-1-1	77.3	0.55	76.7	0.54	86.7	0.73	86.8	0.83	82.0	0.64	79.2	0.59	89.4	0.79	89.0	0.84
	1-1-5-5	93.4	0.87	91.5	0.83	93.0	0.86	90.7	0.89	97.1	0.94	96.8	0.94	97.5	0.95	97.9	0.96
	1-1-1-10	95.2	0.91	88.1	0.76	89.9	0.80	88.8	0.85	96.1	0.92	82.1	0.64	97.3	0.95	97.1	0.95

[1] Pansharpen weight (B-G-R-NIR); [2] Method of classification: MD, Minimum Distance; SAM, Spectral Angel Mapper; ML, Maximum Likelihood; DT, Decision Tree; [3] Accuracy values: OA, overall accuracy (%); K, Kappa coefficient.

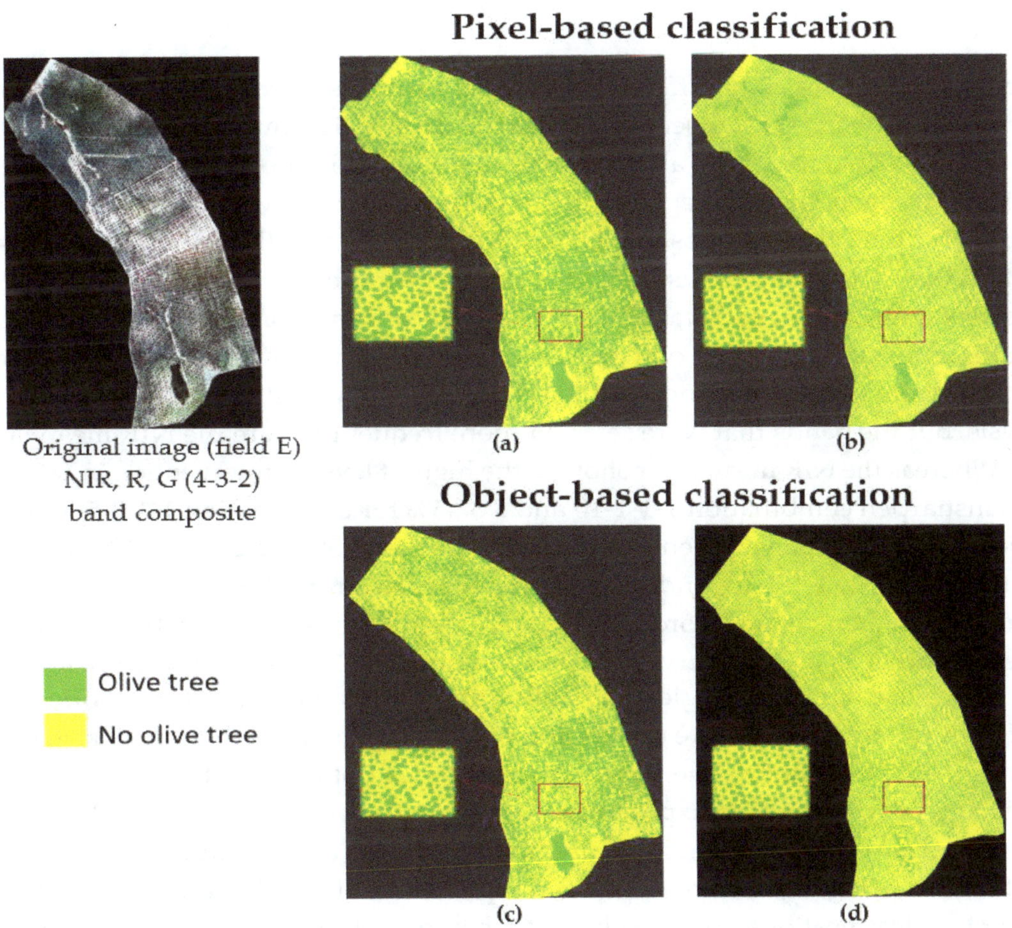

Figure 3. Result of the least (**a**,**c**) and most accurate (**b**,**d**) olive orchard classifications of field E.

Between pixel- and object-based analyses, no clear difference was observed. Despite that object-based classifications analysed a greater number of segmented variables than pixel-based, the precision of the classifications yielded were slightly higher. In object-based analyses, the increase of average OA values was approximately 1% for all classifiers except DT, which improved its general performance with an average increase of 4.5% and showed the most accurate average OA (96.6%). Twelve of the fifteen pansharpen combinations classified improved the accuracies when the object-based analysis were performed, but these increases usually were minimal, not exceeding in many cases the 1% of improvement. The most significant variations between object- and pixel-based analyses were observed in fields D and E. A 6.4% of improvement was observed in the pansharpen combination 1-1-1-10 of the field D, when the most accurate pixel-based analysis was performed with the MP classifier and yielded an OA value of 92.3%, whereas the most accurate object-based classification was obtained with DT algorithm and obtained an OA of 98.7%. Similarly, accuracy increases superior to 4% could be observed with the pansharpen combination 1-1-5-5 of both, field D and E, obtaining the greatest OA values of 98.3% and 97.9% with DT algorithm. Little advantage when applying object-based techniques can be observed in other studies focused on precision agriculture when pixel-based analyses offer reasonable performance. Pérez-Ortiz et al. [60] tested different scenarios of pixel- and object-based classifications to detect weeds in sunflower crops and obtained similar results, with improvements of up to 6%. Similarly, Castillejo González et al. [61] evaluated pixel- and object based classifications to distinguish late-season wild oat weed patches in wheat fields and suggested that the small sizes of the objects and the excellent behaviour of the classification algorithms in the pixel-based classifications did not produce a significant improvement over the precision obtained in the object-based classifications.

In object-based analyses eight different segmented variables were classified, which implies more information that can enhance or worsen the capacity to distinguish among categories. MD, SAM, and ML algorithms give equal weight to all the variables involved in the classification process, and use all these variables in the process, independently of the level of improvement or deterioration that the classification can suffer. Nevertheless, the DT algorithm analyses all the variables and selects only the information that really help to distinguish among categories, increasing its efficiency. Figure 4 shows the percentage of use for each variable in DT classifications. From the eight segmented variables managed in this study, DT algorithm only used six different variables in the total set of analyses, and only two or three were necessary in most of the DT classifications. All spectral variables were used in the DT analysis, but the bands that were selected more frequently were the NIR mean layer and the NDVI index. Whereas the NIR mean layer showed the highest level of intervention in DT analysis with a 44.4% for pansharpen combination 1-1-1-10 and a 37.5% for combination 1-1-5-5, the NDVI index band was used in the three pansharpen combinations with a 27.3%, 22.2%, and 12.5% of intervention for combinations 1-1-1-1, 1-1-1-10, and 1-1-5-5, respectively. From the five geometrical segmented variables, only Width, Length, and Border index were used in those classifications. With feature was the most useful, with a 45.5% of intervention for combination 1-1-1-1 and a 12.5% for combination 1-1-1-5. The scarce use of morphological variables can be explained because the olive is a tree that presents different canopy architecture depending on characteristics such as age, farm management, variety, pruning, etc. (Figure 5). The very high accuracies obtained with limited variables in DT classifications agrees with the idea exposed in [62], when they concluded that DT tends to be very efficient and robust when a large volume of predicted variables are introduced in a model, generally performing fast and being insensitive to noise in input data. This behaviour explain the rise of the accuracies that DT classifications exhibited in object-based analyses compared to ML algorithm.

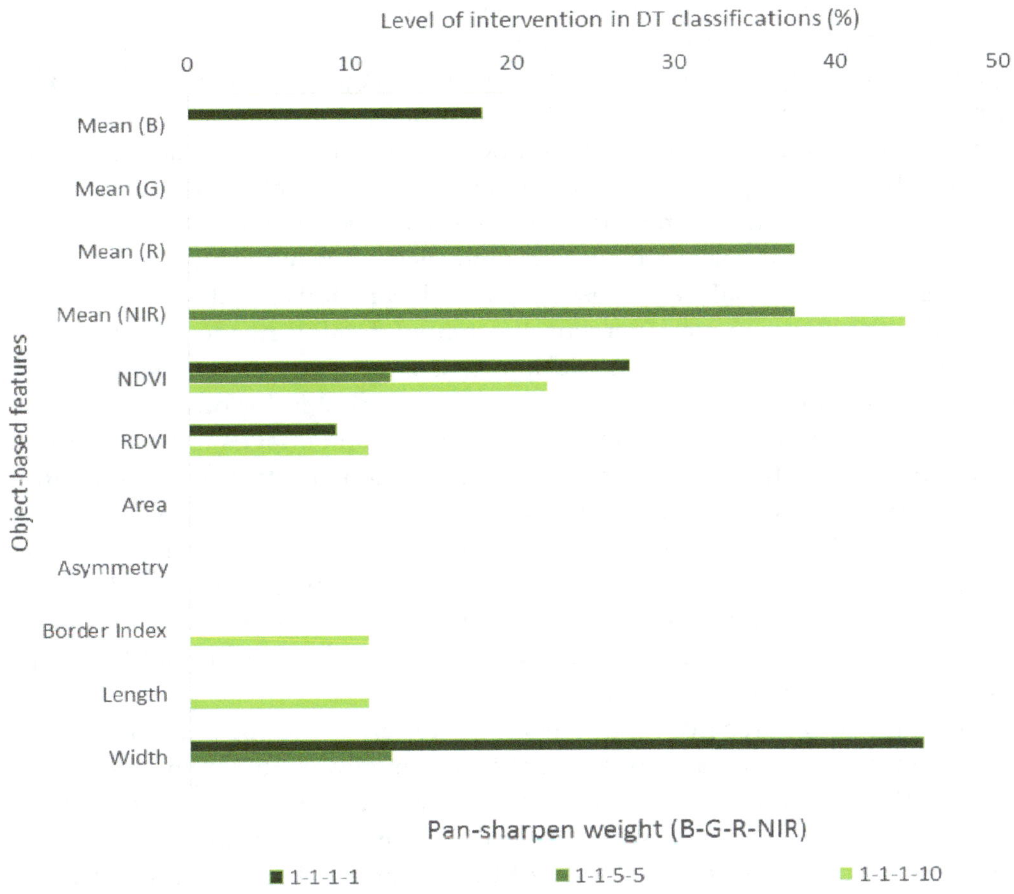

Figure 4. Relative contribution of object-based variables for DT classifications.

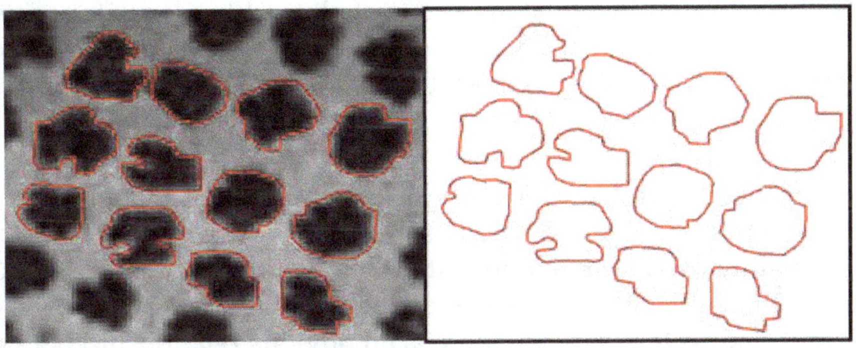

Figure 5. Example of differences in the morphology of the olive crowns.

The pansharpen process was necessary to obtain enough spatial resolution in the images to accurately distinguish the olive trees canopies. The original idea of alter the spectral information of the pansharpened images to emphasize the most useful spectral bands to distinguish vegetation from other land uses exhibited satisfactory results, offering increases of accuracy in three of the five fields studied. This improvement is especially significant in fields D and E, spectral complex fields which showed more difficult to isolate the olive trees. Field D showed the greatest improvement with the spectrally altered images. In pixel-based classifications, increases of approximately 6% and 4% were observed among the combination 1-1-1-1 (OA of 88.1%), and the combinations 1-1-5-5 (94.2%) and 1-1-1-10 (92.3%). More prominent were the increases observed in object-based classifications, where the combination 1-1-1-1 obtained an OA of 87.2% whereas the pansharpen combinations 1-1-5-5 and 1-1-1-10 showed OA values of 98.3% and 98.7%, respectively, which implied accuracy increases higher than 11%. Similarly, field E showed important accuracy increases of 6% and 9% in pixel-based analyses and 8% and 7% in object-based classifications, for 1-1-5-5 and 1-1-1-10 pansharpened combinations, respectively. Although studies based on pansharpened images performed with different spectral weights were not found, the use of equal weight of the multispectral bands in the pansharpen process were evaluated in agronomic scenarios. When homogeneous land uses such as herbaceous crops were analysed with pansharpen imagery, the improvements in classification accuracy due to the use of more spectral and spatially detailed imagery could not really be considered as remarkable regarding the multispectral imagery [63]. Nevertheless, in land uses with high intraclass variability such as woody crops, where the individual trees must be isolated from soil and other covers presented in the field to design a site-specific crop management, the spatial detail is needed. García-Torres et al. [64] analyzed the capacity of different spatial resolutions to isolate olive trees and suggested that imagery with spatial resolutions from 0.25 to 1.5 m were generally suitable for olive grove characterization with olive trees of over 3–4 m^2. Scarce studies used pansharpen images to isolate trees, being more frequent to obtain the information of the tree based on the fusion of different types of information such as multispectral, hyperspectral or LiDAR data or with the analysis of very high spatial resolution UAV imagery. Johnson et al. [65] tested different pansharpened processes to map residential area trees and damaged oak trees. Since a hybrid approach including two pansharpening methods produced the best results in this study, they recommend users planning to process the pansharpen themselves rather than purchase pansharpened imagery directly from the image vendor, so that they can incorporate variations in the methodology for their analysis.

An accurate map of olive orchards fields enable the design of site-specific management treatments and can contribute to the follow up and assessment of agri-environmental regulations. Further studies will focus on establishing a hierarchical classification system that aims at discriminating all olive orchard fields present in the entire QB scene while evaluating the potential of image sharpening in classifying different land uses. In the first level of the hierarchical analysis, the olive orchard fields will be identified and isolated from the other land uses included in the whole studied region to, subsequently, characterize the olive trees of each field. Additionally, with each tree individually discriminated, different agri-environmental indicators of olive orchards related with the number and area of the trees (e.g., potential productivity of each tree and potential production of each plot) and related with bare soil and other vegetation covers (e.g., risk of erosion and run-off) can be predicted.

4. Conclusions

The olive production sector is characterized by a large number of small operators which directly affect production. The results of the present study show that pansharpened multi-spectral QuickBird imagery can be successfully used to map delineation of olive tree canopies. The knowledge of the accurate location and delimitation of the olive trees can be used as a basis for the precision management of fertilizers, pesticides and watering, since there is an obvious relationship between tree size and potential productivity based on the requirements of nutrients, watering doses and plant protection products such as fungicides. Although all classification algorithms tested offered accurate results, ML and DT were the two most precise classifiers. Knowing that the red and especially the near infrared bands are very important in the discrimination of vegetation, the alteration of the weight of these spectral bands in the pansharpen process enhanced significantly the accuracy of the olive trees delineation, with an average increase improvement of up to 9% and 11% in pixel- and object-based analyses, respectively.

With regard to recommending one methodology to define the olive trees crowns, two considerations should be made: the improvement in accuracy obtained and the computational or expertise requirements involved in the process. The decision of whether or not to carry out more complex analyses will depend on the importance of achieving maximum accuracy and the ratio of cost/efficiency wished to obtain in the objectives. If one aimed to create a crop inventory, then the performance of pixel-based classification with ML classifier and standard pansharpened images would be the best choice. However, if it desires to produce a map that is ready to be used for precision agriculture decision-making procedures, object-based analyses with the DT algorithm and the pansharpened imagery with the near-infrared band altered would be highly recommended.

Acknowledgments: I extend thanks to Javier Carrasco for his assistance with data processing.

References

1. Precision Agriculture in Europe. Legal, Social and Ethical Considerations. European Parliament (European Parliamentary Research Service). Available online: http://www.europarl.europa.eu/RegData/etudes/STUD/2017/603207/EPRS_STU(2017)603207_EN.pdf (accessed on 24 October 2018).
2. The European Parliament and the Council of the European Union. Regulation 1306/2013 of the European Parliament and of the Council of 17 December 2013 on the financing, management and monitoring of the common agricultural policy and repealing Council Regulations (EEC) No 352/78, (EC) No 165/94, (EC) No 2799/98, (EC) No 814/2000, (EC) No 1290/2005 and (EC) No 485/2008 OJ L 347, 20.12.2013. *Off. J. Eur. Union* **2013**, *347*, 549–607.
3. The European Parliament and the Council of the European Union. Regulation 1107/2009 of the European Parliament and of the Council of 21 October 2009 concerning the placing of plant protection products on the market and repealing Council Directives 79/117/EEC and 91/414/EEC. *Off. J. Eur. Union* **2009**, *309*, 1–50.
4. de Castro, A.I.; Jiménez-Brenes, F.M.; Torres-Sánchez, J.; Peña, J.M.; Borra-Serrano, I.; López-Granados, F. 3-D characterization of vineyards using a novel UAV imagery-based OBIA procedure for precision viticulture applications. *Remote Sens.* **2018**, *10*, 584. [CrossRef]
5. Fitzgerald, G.J.; Maas, S.J.; Detar, W.R. Spider mite detection and canopy component mapping in cotton using hyperspectral imagery and spectral mixture analysis. *Precis. Agric.* **2004**, *5*, 275–289. [CrossRef]
6. Yang, C.; Everitt, J.H.; Bradford, J.M. Comparison of QuickBird satellite imagery and airborne imagery for mapping grain sorghum yield patterns. *Precis. Agric.* **2006**, *7*, 33–44. [CrossRef]
7. López-Granados, F.; Torres-Sánchez, J.; Serrano-Pérez, A.; de Castro, A.I.; Mesas-Carrascosa, F.; Peña, J. Early season weed mapping in sunflower using UAV technology: Variability of herbicide treatment maps against weed thresholds. *Precis. Agric.* **2016**, *17*, 183–199. [CrossRef]
8. Huang, Y.; Reddy, K.N.; Fletcher, R.S.; Pennington, D. UAV low-altitude remote sensing for precision weed management. *Weed Technol.* **2018**, *32*, 2–6. [CrossRef]
9. López-Granados, F.; Torres-Sánchez, J.; De Castro, A.; Serrano-Pérez, A.; Mesas-Carrascosa, F.; Peña, J. Object-based early monitoring of a grass weed in a grass crop using high resolution UAV imagery. *Agron. Sustain. Dev.* **2016**, *36*, 67. [CrossRef]
10. Du, Q.; Chang, N.; Yang, C.; Srilakshmi, K.R. Combination of multispectral remote sensing, variable rate technology and environmental modeling for citrus pest management. *J. Environ. Manag.* **2008**, *86*, 14–26. [CrossRef]
11. Percival, D.C.; Gallant, D.; Harrington, T.; Brown, G. Potential for commercial unmanned aerial vehicle use in wild blueberry production. *Acta Hortic.* **2017**, *1180*, 233–240. [CrossRef]
12. Rai, M.; Ingle, A. Role of nanotechnology in agriculture with special reference to management of insect pests. *Appl. Microbiol. Biotechnol.* **2012**, *94*, 287–293. [CrossRef] [PubMed]
13. Herrmann, I.; Vosberg, S.K.; Ravindran, P.; Singh, A.; Chang, H.; Chilvers, M.I.; Conley, S.P.; Townsend, P.A. Leaf and canopy level detection of fusarium virguliforme (sudden death syndrome) in soybean. *Remote Sens.* **2018**, *10*, 426. [CrossRef]
14. Hillnhütter, C.; Mahlein, A.; Sikora, R.A.; Oerke, E. Remote sensing to detect plant stress induced by heterodera schachtii and rhizoctonia solani in sugar beet fields. *Field Crops Res.* **2011**, *122*, 70–77. [CrossRef]
15. Santoso, H.; Gunawan, T.; Jatmiko, R.H.; Darmosarkoro, W.; Minasny, B. Mapping and identifying basal stem rot disease in oil palms in north sumatra with QuickBird imagery. *Precis. Agric.* **2011**, *12*, 233–248. [CrossRef]
16. Fisher, D.K.; Hinton, J.; Masters, M.H.; Aasheim, C.; Butler, E.S.; Reichgelt, H. Improving irrigation efficiency through remote sensing technology and precision agriculture in se Georgia. In Proceedings of the ASAE Annual International Meeting 2004, Otawa, ON, Canada, 1–4 August 2004; pp. 3461–3470.

17. Gago, J.; Douthe, C.; Coopman, R.E.; Gallego, P.P.; Ribas-Carbo, M.; Flexas, J.; Escalona, J.; Medrano, H. UAVs challenge to assess water stress for sustainable agriculture. *Agric. Water Manag.* **2015**, *153*, 9–19. [CrossRef]
18. Rossini, M.; Fava, F.; Cogliati, S.; Meroni, M.; Marchesi, A.; Panigada, C.; Giardino, C.; Busetto, L.; Migliavacca, M.; Amaducci, S.; et al. Assessing canopy PRI from airborne imagery to map water stress in maize. *ISPRS J. Photogramm. Remote Sens.* **2013**, *86*, 168–177. [CrossRef]
19. Gonzalez-Dugo, V.; Zarco-Tejada, P.; Nicolás, E.; Nortes, P.A.; Alarcón, J.J.; Intrigliolo, D.S.; Fereres, E. Using high resolution UAV thermal imagery to assess the variability in the water status of five fruit tree species within a commercial orchard. *Precis. Agric.* **2013**, *14*, 660–678. [CrossRef]
20. Launeau, P.; Kassouk, Z.; Debaine, F.; Roy, R.; Mestayer, P.G.; Boulet, C.; Rouaud, J.; Giraud, M. Airborne hyperspectral mapping of trees in an urban area. *Int. J. Remote Sens.* **2017**, *38*, 1277–1311. [CrossRef]
21. Torres-Sánchez, J.; López-Granados, F.; Serrano, N.; Arquero, O.; Peña, JM. High-throughput 3-D monitoring of agricultural-tree plantations with unmanned aerial vehicle (UAV) technology. *PLoS ONE* **2015**, *10*, e0130479. [CrossRef]
22. Kim, C.; Hong, S. Identification of tree species from high resolution satellite imagery by using crown parameters. Presented at the SPIE—The International Society for Optical Engineering 2008, Cardiff, Wales, UK, 15–18 September 2008. [CrossRef]
23. Molinier, M.; Astola, H. Feature selection for tree species identification in very high resolution satellite images. In Proceedings of the 2011 IEEE International Geoscience and Remote Sensing Symposium (IGARSS), Vancouver, BC, Canada, 24–29 July 2011; pp. 4461–4464.
24. Arockiaraj, S.; Kumar, A.; Hoda, N.; Jeyaseelan, A.T. Identification and quantification of tree species in open mixed forests using high resolution QuickBird satellite imagery. *J. Trop. For. Environ.* **2015**, *5*. [CrossRef]
25. Caughlin, T.T.; Rifai, S.W.; Graves, S.J.; Asner, G.P.; Bohlman, S.A. Integrating LiDAR-derived tree height and Landsat satellite reflectance to estimate forest regrowth in a tropical agricultural landscape. *Remote Sens Ecol. Conserv.* **2016**, *2*, 190–203. [CrossRef]
26. Hawryło, P.; Wezyk, P. Predicting growing stock volume of scots pine stands using Sentinel-2 satellite imagery and airborne image-derived point clouds. *Forests* **2018**, *9*, 274. [CrossRef]
27. Vierling, L.A.; Vierling, K.T.; Adam, P.; Hudak, A.T. Using satellite and airborne LiDAR to model woodpecker habitat occupancy at the landscape scale. *PLoS ONE* **2013**, *8*, e80988. [CrossRef] [PubMed]
28. González-Audícana, M.; Saleta, J.L.; Catalán, R.G.; García, R. Fusion of multispectral and panchromatic images using improved IHS and PCA mergers based on wavelet decomposition. *IEEE Trans. Geosci. Remote Sens.* **2004**, *42*, 1291–1299. [CrossRef]
29. Otazu, X.; González-Audícana, M.; Fors, O.; Núñez, J. Introduction of sensor spectral response into image fusion methods. Application to wavelet-based methods. *IEEE Trans. Geosci. Remote Sens.* **2005**, *43*, 2376–2385. [CrossRef]
30. Ma, L.; Li, M.; Ma, X.; Cheng, L.; Du, P.; Liu, Y. A review of supervised object-based land-cover image classification. *ISPRS J. Photogramm. Remote Sens.* **2007**, *130*, 277–293. [CrossRef]
31. Smits, P.C.; Dellepiane, S.G.; Schowengerdt, R.A. Quality assessment of image classification algorithms for land-cover mapping: A review and a proposal for a cost-based approach. *Int. J. Remote Sens.* **1999**, *20*, 1461–1486. [CrossRef]
32. Immitzer, M.; Atzberger, C.; Koukal, T. Tree species classification with random forest using very high spatial resolution 8-band WorldView-2 satellite data. *Remote Sens.* **2012**, *4*, 2661–2693. [CrossRef]
33. Whiteside, T.G.; Boggs, G.S.; Maier, S.W. Comparing object-based and pixel-based classifications for mapping savannas. *Int. J. Appl. Earth Obs. Geoinf.* **2011**, *13*, 884–893. [CrossRef]
34. Funck, J.W.; Zhong, Y.; Butler, D.A.; Brunner, C.C.; Forrer, J.B. Image segmentation algorithms applied to wood defect detection. *Comput. Electron. Agric.* **2003**, *41*, 157–179. [CrossRef]
35. Yu, Q.; Gong, P.; Clinton, N.; Biging, G.; Kelly, M.; Schirokauer, D. Object-based detailed vegetation classification with airborne high spatial resolution remote sensing imagery. *Photogramm. Eng. Remote Sens.* **2006**, *72*, 799–811. [CrossRef]
36. Kettig, R.L.; Landgrebe, D.A. Classification of multispectral image data by extraction and classification of homogeneous objects. *IEEE Trans. Geosci. Electron.* **1976**, *14*, 19–26. [CrossRef]
37. Hulet, A.; Roundy, B.A.; Petersen, S.L.; Jensen, R.R.; Bunting, S.C. An object-based image analysis of pinyon and juniper woodlands treated to reduce fuels. *Environ. Manag.* **2014**, *53*, 660–671. [CrossRef] [PubMed]

38. MacLean, M.G.; Congalton, R.G. Using object-oriented classification to map forest community types. In Proceedings of the American Society for Photogrammetry and Remote Sensing Annual Conference, Milwaukee, WI, USA, 1–4 May 2011.
39. Jiménez-Brenes, F.M.; López-Granados, F.; Castro, A.I.; Torres-Sánchez, J.; Serrano, N.; Peña, J.M. Quantifying pruning impacts on olive tree architecture and annual canopy growth by using UAV-based 3D modelling. *Plant Methods* **2017**, *13*, 55. [CrossRef] [PubMed]
40. International Olive Council (IOC). Olive Growing and Nursery Production. Available online: http://www.internationaloliveoil.org/projects/paginas/Section-a.htm (accessed on 26 November 2018).
41. Ayerza, R.; Steven Sibbett, G. Thermal adaptability of olive (*Olea europaea* L.) to the arid Chaco of Argentina. *Agric. Ecosyst. Environ.* **2001**, *84*, 277–285. [CrossRef]
42. DigitalGlobe. Information Products: Standard Imagery. Available online: https://dg-cms-uploads-production.s3.amazonaws.com/uploads/document/file/21/Standard_Imagery_DS_10-7-16.pdf (accessed on 20 November 2018).
43. Song, C.; Woodcock, C.E.; Seto, K.C.; Lenney, M.P.; Macomber, S.A. Classification and change detection using Landsat TM data: When and how to correct atmospheric effects? *Remote Sens. Environ.* **2001**, *75*, 230–244. [CrossRef]
44. Gonzalo, C.; Lillo-Saavedra, M. A directed search algorithm for setting the spectral–spatial quality trade-off of fused images by the wavelet Á Trous method. *Can. J. Remote Sens.* **2008**, *34*, 367–375. [CrossRef]
45. Ranchin, T.; Wald, L. Fusion of high spatial and spectral resolution images: The ARSIS concept and its implementation. *Photogramm. Eng. Remote Sens.* **2000**, *66*, 49–61.
46. Lillo-Saavedra, M.; Gonzalo, C. Spectral or spatial quality for fused satellite imagery? A trade-off solution using the wavelet Á Trous algorithm. *Int. J. Remote Sens.* **2006**, *27*, 1453–1464. [CrossRef]
47. Baatz, M.; Schäpe, A. Multiresolution segmentation: An optimization approach for high quality multi-scale image segmentation. In Proceedings of the 12th Symposium for Applied Geographic Information Processing (Angewandte Geographische Informationsverarbeitung XII) AGIT 2000, Salzburg, Austria, 5–7 July 2000.
48. Rouse, J.W., Jr.; Haas, R.; Schell, J.; Deering, D. Monitoring vegetation systems in the Great Plains with ERTS. *NASA Spec. Publ.* **1974**, *351*, 309–317.
49. Roujean, J.L.; Breon, F.M. Estimating PAR absorbed by vegetation from bidirectional reflectance measurements. *Remote Sens. Environ.* **1995**, *51*, 375–384. [CrossRef]
50. Hodgson, M.E. Reducing the computational requirements of the minimum-distance classifier. *Remote Sens. Environ.* **1998**, *25*, 117–128. [CrossRef]
51. Hecker, C.; Van Der Meijde, M.; Van Der Werff, H.; Van Der Meer, F.D. Assessing the influence of reference spectra on synthetic SAM classification results. *IEEE Trans. Geosci. Remote Sens.* **2008**, *46*, 4162–4172. [CrossRef]
52. Strahler, A.H. The use of prior probabilities in maximum likelihood classification of remotely sensed data. *Remote Sens. Environ.* **1980**, *10*, 135–163. [CrossRef]
53. Quinlan, R. *C4-5: Programs for Machine Learning*; Morgan Kaufmann: San Mateo, CA, USA, 1993.
54. Congalton, R.G. A review of assessing the accuracy of classifications of remotely sensed data. *Remote Sens. Environ.* **1991**, *37*, 35–46. [CrossRef]
55. Rogan, J.; Franklin, J.; Roberts, D.A. A comparison of methods for monitoring multitemporal vegetation change using thematic mapper imagery. *Remote Sens. Environ.* **2002**, *80*, 143–156. [CrossRef]
56. Wald, L. *Data Fusion. Definition and Architectures- Fusion of Images of Different Spatial Resolutions*; Presses de l'Ecole, Ecole des Mines de Paris: Paris, France, 2002; p. 200, ISBN 2-911762-38-X.
57. Gonzalo-Martín, C.; Lillo-Saavedra, M. Quickbird image fusion by a multirresolution-multidirectional joint image representation. *IEEE Lat. Am. Trans.* **2007**, *5*, 32–37. [CrossRef]
58. Foody, G.M. Status of land cover classification accuracy assessment. *Remote Sens. Environ.* **2002**, *80*, 185–201. [CrossRef]
59. Monserud, R.A.; Leemans, R. Comparing global vegetation maps with the kappa statistic. *Ecol. Model.* **1992**, *62*, 275–293. [CrossRef]
60. Pérez-Ortiz, M.; Gutiérrez, P.A.; Peña, J.M.; Torres-Sánchez, J.; Hervás-Martínez, C.; López-Granados, F. An experimental comparison for the identification of weeds in sunflower crops via unmanned aerial vehicles and object-based analysis. *Lect. Notes Comput. Sci.* **2015**, *9094*, 252–262. [CrossRef]

61. Castillejo-González, I.L.; Peña-Barragán, J.M.; Jurado-Expósito, M.; Mesas-Carrascosa, F.J.; López-Granados, F. Evaluation of pixel- and object-based approaches for mapping wild oat (*Avena sterilis*) weed patches in wheat fields using QuickBird imagery for site-specific management. *Eur. J. Agron.* **2014**, *59*, 57–66. [CrossRef]
62. Brodley, C.E.; Friedl, M.A. Decision tree classification of land cover from remotely sensed data. *Remote Sens. Environ.* **1997**, *61*, 399–409. [CrossRef]
63. Castillejo-González, I.L.; López-Granados, F.; García-Ferrer, A.; Peña-Barragán, J.M.; Jurado-Expósito, M.; de la Orden, M.S.; González-Audicana, M. Object- and pixel-based analysis for mapping crops and their agro-environmental associated measures using QuickBird imagery. *Comput. Electron. Agric.* **2009**, *68*, 207–215. [CrossRef]
64. García Torres, L.; Peña-Barragán, J.M.; López-Granados, F.; Jurado-Expósito, M.; Fernández-Escobar, R. Automatic assessment of agro-environmental indicators from remotely sensed images of tree orchards and its evaluation using olive plantations. *Comput. Electron. Agric.* **2008**, *61*, 179–191. [CrossRef]
65. Johnson, B.A.; Tateishi, R.; Hoan, N.T. Satellite image pansharpening using a hybrid approach for object-based image analysis. *ISPRS Int. J. Geo-Inf.* **2012**, *1*, 228–241. [CrossRef]

10

Integrating Sentinel-2 Imagery with AquaCrop for Dynamic Assessment of Tomato Water Requirements

Anna Dalla Marta [1,*], Giovanni Battista Chirico [2], Salvatore Falanga Bolognesi [3], Marco Mancini [1], Guido D'Urso [2], Simone Orlandini [1], Carlo De Michele [3] and Filiberto Altobelli [4]

1 Department of Agriculture, Food, Environment and Forestry, University of Florence, Piazzale delle Cascine 18, 50144 Firenze, Italy
2 Department of Agricultural Sciences, University of Naples "Federico II", Via Università 100, 80055 Portici (NA), Italy
3 Ariespace s.r.l., Spin off Company, University of Napoli "Federico II", Centro Direzionale IS A3, 80143 Napoli, Italy
4 CREA Research Centre for Agricultural Policies and Bioeconomy, Via Po 14, 00198 Roma, Italy
* Correspondence: anna.dallamarta@unifi.it

Abstract: A research study was conducted in an open field tomato crop in order to: (i) Evaluate the capability of Sentinel-2 imagery to assess tomato canopy growth and its crop water requirements; and (ii) explore the possibility to predict crop water requirements by assimilating the canopy cover estimated by Sentinel-2 imagery into AquaCrop model. The pilot area was in Campania, a region in the south west of Italy, characterized by a typical Mediterranean climate, where field campaigns were conducted in seasons 2017 and 2018 on processing tomato. Crop water use and irrigation requirement were estimated by means of three different methods: (i) The AquaCrop model; (ii) an irrigation advisory service based on Sentinel-2 imagery known as IRRISAT and (iii) assimilating the canopy cover estimated by Sentinel-2 imagery into AquaCrop model Sentinel-2 imagery proved to be effective for monitoring canopy growth and for predicting irrigation water requirements during mid-season stage of the crop, when the canopy is fully developed. Conversely, the integration of the Sentinel-2 imagery with a crop growth model can contribute to improve the irrigation water requirement predictions in the early and development stage of the crop, when the soil evaporation is not negligible with respect to the total evapotranspiration.

Keywords: fractional cover; irrigation; satellite; crop simulation model; AquaCrop

---

1. Introduction

Worldwide significant progress has been made to utilize precision agriculture for irrigation as a mean to increase water use efficiency or decrease the water footprint in irrigated agriculture [1,2]. The progress is mainly restricted to advances at the plot scale and individual systems such as installations for drip irrigation or central pivots. As well known, actual crop evapotranspiration (ET) is a major term of water budget in agriculture and it is the main variable used to determine crops water requirement. Beside their massive progress during the recent years, accurate field measurements (soil moisture, plant-based sensors, etc.) are very scarce because sensors and measuring devices are expensive, their use requires specific expertise and complex maintenance and are typically limited to experimental stations. For this reason, many attempts have been made for developing indirect ET estimation methods based on crop data at a field scale easily available across large regions. Concordantly, the use of remotely sensed data has become more common to monitoring and controlling activities at different spatial and temporal scales including precision farming. Surface energy balance methods based on satellite observations in the thermal band [3] have been developed and applied in many areas with the aim of determining actual evapotranspiration and assessing the water balance of irrigated areas and the corresponding water accounting practices [4,5].

Sentinel-2 mission, launched by European Space Agency as part of the Copernicus program (http://www.copernicus.eu/) [6], has been a great step forward for continuous crop monitoring. Sentinel-2 carries a sensor, which captures data at 10, 20 and 60-m spatial resolution over 13 spectral bands and with a very high temporal resolution of five days at the equator. Thanks to the characteristics of these new technologies, commercial services started to develop for operational applications. In this context, Irrigation Advisory Services (IASs) for optimizing water management rapidly grow. Among them, IRRISAT is a fully operative satellite-based IAS provided in Campania Region, which combines Copernicus Sentinel-2A data with daily weather data for estimating irrigation water requirements. Since 2007 it has been active over the area, providing evidence of high efficiency for water saving (up to 30%) [7].

Moreover, the state of the art suggests that significant progresses in saving irrigation volumes at the farm level can be probably attained by assessing crop water requirements through an optimal combination of crop satellite images with a crop growth model since satellite images provide information concerning the current state of the crop canopy and a crop model is able to simulate the biophysical processes of the growing crop.

Ultimately, examples are implemented to derive crop water requirements from satellite estimates of biophysical parameters assimilated into agro-meteorological models [8] to monitor the nitrogen status and to apply fertilizer with variable rates or to derive agronomic variables [9].

Since the rapid development and availability of products with different spatial and temporal resolutions, the integration of remote sensing data into crop growth models has increased in recent years. Among the existing crop models, AquaCrop has been widely used for assessing water requirements and optimal irrigation scheduling for different crops and environments. AquaCrop, in fact, is a crop water productivity model developed by the Land and Water Division of FAO (Food and Agriculture Organization) in 2009 [10,11]. It simulates yield response to water of crops and it is mainly used to increase water efficiency practices in agricultural production

Linker and Ioslovich [12] have successfully shown the possibility of assimilating ground measured canopy cover data (using digital images above the canopy) during the growing season within the AquaCrop model for potato and cotton, based on the Extended Kalman Filter algorithm. However, this innovative procedure requires field data collection during the season, which is budget-and time-consuming but today, it can be facilitated and improved using Earth Observation data, especially from Sentinel-2 satellite sensors whose temporal and spatial resolution is adapted for parcel monitoring. Jin et al. [13] used spectral-based biomass values computed with on field spectral measurement data to calibrate the AquaCrop model with a particle swarm optimization (PSO) algorithm for winter wheat in China.

The same PSO algorithm was used for winter wheat by Silvestro et al. [14] and Jin et al. [15] to assimilate optical and radar satellite data into the AquaCrop model. Several coupling methods have been categorized and detailed by Jin et al. [16]. One of these methods is by substituting crop growth models variables with remotely sensed data [17]. A similar approach using remote sensing data was used to calibrate the AquaCrop model with the leaf area index from MODIS (Moderate Resolution Imaging Spectroradiometer) for winter wheat in Italy by Trombetta et al. [18]. It was only applied to large agricultural parcels, due to MODIS's limited spatial resolution. However, their use in real/operational scenarios is generally limited by data availability, i.e., crop initial conditions, planting date and application of inputs. Under this scenario, a possible alternative for assisting the models in reproducing the actual processes in the field is the use of algorithms relating to remote sensing data and key canopy biophysical parameters in the crop growth models.

This study aims to evaluate:

The capability of Sentinel-2 imagery to assess canopy growth of a tomato open field crop and the corresponding irrigation water requirement by means of Sentinel-2 imagery;

The possibility to predict crop water requirements by assimilating the fractional cover estimated by Sentinel-2 imagery into the AquaCrop model.

2. Materials and Methods

2.1. Test Site

The research was carried out in Frignano, located in Caserta province (Campania, Italy) (Figure 1). The area is characterized by a warm-temperate climate with an average annual temperature of 15.2 °C and 900 mm of annual cumulated precipitation, with the rainiest period occurring in winter and driest period in summer.

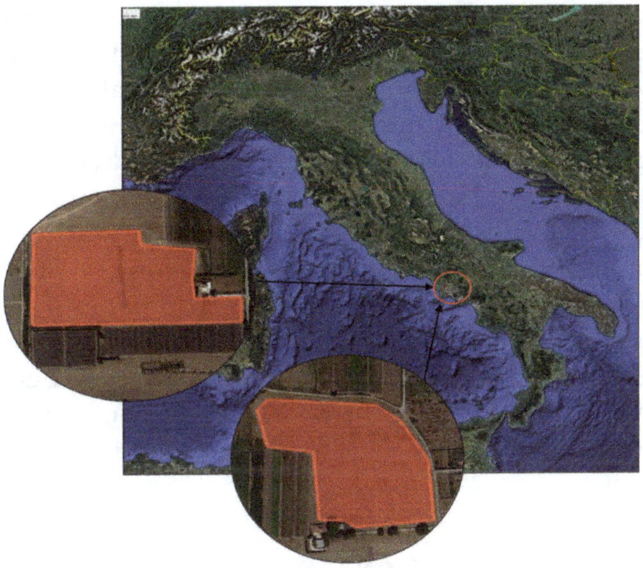

Figure 1. Localization of the pilot area and the two experimental fields.

Processing tomato (*Solanum lycopersicum* L.) was cultivated in years 2017 (41°00′26.33″ N 14°10′13.93″ E) and 2018 (41°01′39.60″ N 14°10′32.87″ E) in two parcels of 4 ha each. Soil samples (three samples were averaged for analysis) were examined for assessing the main physical–chemical properties, including gravel (%), soil texture, bulk density (t/m^3), pH and organic matter (%), at a depth of 15 cm. In 2018 soil characteristics were obtained by the soil map of Region Campania (pedological maps 1:50,000; http://agricoltura.regione.campania.it/pedologia/suoli.html). SPAW software (Soil–Plant–Air–Water; v6.02.75, United States Department of Agriculture-USDA, Washington, DC, USA) was used for assessing soil hydraulic properties, such as soil water retention, soil hydraulic conductivity, field capacity and plant available water.

Soil was ploughed at 40 cm depth, and tomato seedlings were transplanted on April 9 in continuous double rows with 33.5–40.0 cm space between plants, 50 cm between rows, 1.10–1.20 cm between double rows, with a final plant density of 32,000 and 33,500 plants/ha in year 2017 and 2018, respectively.

Fertilization was applied before transplanting (broadcast) with 120 kg/ha of diammonium phosphate (18-46-0) and 400 kg/ha of organic-mineral fertilizer (10-5-12); at transplanting (localized) with 160 kg/ha of diammonium phosphate and 20 kg/ha of seed sprint H5 (12-43-0).

Meteorological daily data of maximum and minimum temperature (°C), relative humidity (%), wind speed (m s^{-1}) and precipitation (mm) were collected at a complete weather station located in the study area for the two cropping seasons.

Plants were watered by light driplines, with 30 cm dripper spacing and 2 L/h flow rate at 1 bar.

Based on the farmer best knowledge, a total volume of 130 L/plant of water, corresponding to 4160 m^3/ha, and 120 L/plant corresponding to 4020 m^3/ha of irrigation was given throughout the growing period in 2017 and 2018, respectively. Such volumes were then considered as reference values against which the different estimation methods (AquaCrop, IRRISAT, integration of Sentinel 2A imagery into AquaCrop model) were tested.

2.2. Satellite-Based Assessment of Crop Water Requirement: IRRISAT

The remote sensing technique employed is the one employed by IRRISAT, a satellite-based irrigation advisory service developed in Italy and operational since 2007 in Southern Italy [7]. The service aims at providing farmers and water managers with real time information on crop water needs, which are estimated by combining high resolution data from Earth Observation satellites with weather data for the calculation of crop water requirements. More recently, IRRISAT has been combined with numerical weather predictions for forecasting crop water needs up to five days in advance [19,20]. Information is delivered in near-real time (24 h) to users (farmers, Water User Association and water agencies) by means of a dedicated WebGIS accessible from PC, tablets and smartphones (https://www.irrisat.com/en/). In 2016, the operational irrigation advisory service of the Campania region has reached about 2000 farmers with a total irrigation area larger than 80,000 hectares: The achieved water saving has been estimated to be larger than 30% [7]. In 2016 the Italian Ministry of Agriculture has listed IRRISAT among the applicable methodologies for the estimation of the irrigation volume, complying with the EU Water Framework Directive.

The IRRISAT methodology is summarized in Figure 2. Crop potential evapotranspiration (ETp) is computed with the Penman–Monteith equation, with crop parameters albedo (α) and leaf area index (LAI) derived from processing Sentinel-2A/B images in the visible and infrared regions [7] while assuming fixed values for the stomatal resistance (sr $\approx$ 100 sm^{-1}) and crop height (hc = 0.4 m) for herbaceous crop [8]. Following this approach, the calculation of ETp requires standard meteorological data (daily air temperature, relative humidity, solar radiation, wind speed and precipitation), LAI and surface α.

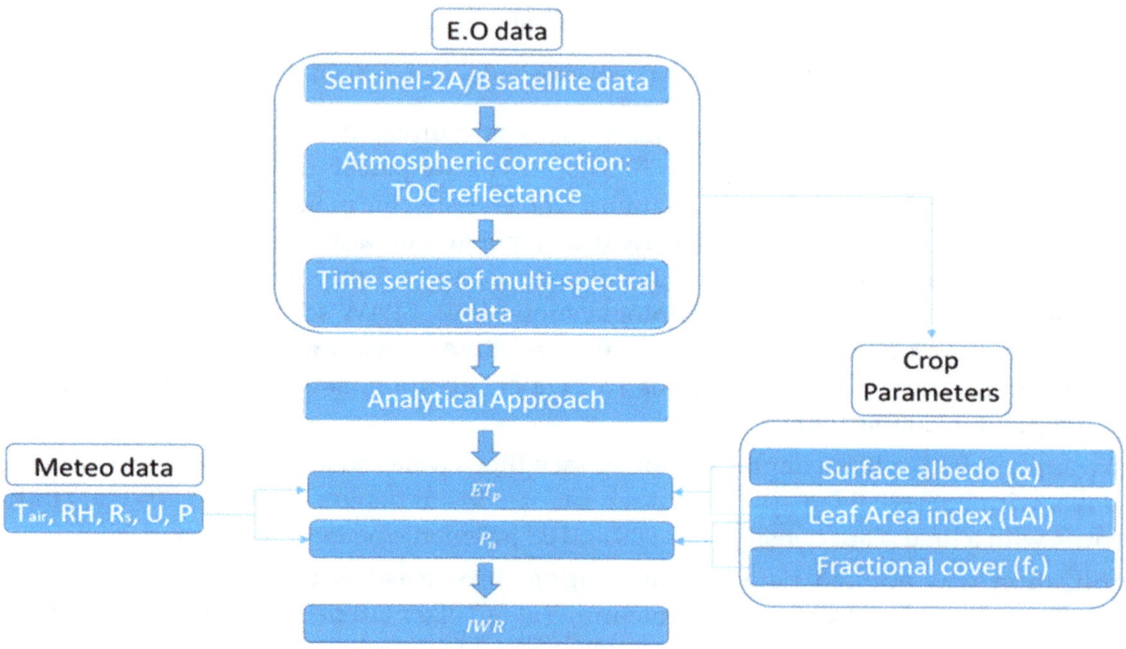

Figure 2. Flowchart shows dataset and processing procedure required to estimate crop potential evapotranspiration (ETp), and irrigation water requirements (IWR), Earth Observation-based direct FAO-56 Penman-Monteith equation method. Legend: Ta = air temperature, RH = relative humidity; Rs = solar radiation; U = wind speed; P = precipitation.

The irrigation water requirement (*IWR*) is then calculated as the difference between *ETp* and the crop effective rainfall (*Pn*), according to the following equation:

$$\mathrm{IWR} = \mathrm{ETp} - \mathrm{Pn} \tag{1}$$

It is assumed that capillary rise does not contribute to root zone soil moisture in the summer season, as usually occurs in the southern European regions. Runoff and deep drainage are assumed to be negligible considering the low amount of rainfall during the two growing seasons. *Pn* is obtained by reducing the precipitation above canopy (P) by a quantity that depends on canopy development, according to an empirical function of the LAI and the fractional vegetation cover fc, as reported in Vuolo et al. [7].

In this study, 10 and 21 multispectral high-resolution images from Sentinel-2A and 2B have been acquired during 2017 and 2018, respectively. The multi spectral instrument (MSI) on board of Sentinel-2A/2B captures data at 10, 20 and 60 m of spatial resolution over 13 spectral bands with a very high temporal resolution of five days at the equator. Individual Sentinel-2 granules Level-1C (processed at the top-of-atmosphere reflectance) were acquired from Copernicus Open Access Hub (https://scihub.copernicus.eu/), already ortho-rectified in UTM/WGS84 (image tiles of 100×100 km^2). The information gathered by Sentinel-2 system (orbit, attitude, date accuracy and viewing directions of all detectors) are exploited for geolocating all Sentinel-2 pixels with an accuracy of about 11 m for about 97% of the cases, which is about the size of one Sentinel-2 pixel. The standard need for multi-temporal registration errors is 0.3 pixels, and the current performances show that for more than 50% of the cases, the performance does not meet that requirement. The resolution is estimated to be three times the registration error, thus the resolution Sentinel-2 time series is around 30 m.

Level-1C products were processed into Level-2A-Bottom-of-Atmosphere (BoA) reflectance-data using the ESA's Sen2Cor v2.5.5 tool (http://step.esa.int/main/third-party-plugins-2/sen2cor/sen2cor_v2-5-5/). Sen2Cor tool performs the atmospheric, terrain and cirrus correction of Top-Of-Atmosphere Level 1C input data, and creates Bottom-Of-Atmosphere, optionally terrain and cirrus corrected reflectance images; additional, aerosol optical thickness, water vapor, scene classification maps and quality indicators for cloud and snow probabilities.

In order to obtain homogeneous and comparable products as time series, all value-added products (LAI, α and fc) are calculated based on atmospherically corrected Level-2A data. LAI and fc are calculated by S2ToolBox [21], an artificial neural network (ANN) algorithm, trained by using radiative transfer simulations from PROSPECT [22] and SAIL [23] models, and tailored for Sentinel-2 data. The algorithm requires eight Sentinel-2 spectral bands (B3–B7, B8a, B11 and B12) at 10 and 20 m (pixel size), which are all resampled to 10 m to derive LAI and fc. Experimental studies have shown the accuracy of this approach for LAI estimation in different environments and crops [24,25]. In this study, average and variance of LAI and fc at parcel scale were assessed by taking a minimum of 50 pixels falling within each parcel, after excluding pixels affected by boundary effects or cloudiness, according to the quality indicator provided by S2ToolBox.

The broadband surface albedo has been calculated, when the observed surface is considered as Lambertian, as the integration of at-surface reflectance across the shortwave spectrum [26], as shown in equation:

$$\alpha = \sum\nolimits_{bi} \left| \rho_{bi} \cdot \omega_{bi} \right| \tag{2}$$

where α is albedo, ρ_{bi} is surface reflectance for a given band *bi* at Level-2A Sentinel-2 surface reflectance, ω_{bi} is the weighting coefficient representing the solar radiation fraction derived from the solar irradiance spectrum [26] within the spectral range (spectral response curves) for bands *bi* and is calculated as equation:

$$\omega_{bi} = \frac{\int_{LO_{bi}}^{UP_{bi}} R_{s\lambda} \cdot d\lambda}{\int_{0.4}^{2.4} R_{s\lambda} \cdot d\lambda} \tag{3}$$

where $R_{s\lambda}$ is extra-terrestrial irradiance for wavelength λ (μm); and UP_{bi} and LO_{bi} are upper and lower wavelength bounds for Sentinel-2A/B band *bi*, respectively.

2.3. Model-Based Assessment of Crop Water Requirement: AquaCrop

AquaCrop simulates crop yield in four steps: Crop development, crop transpiration, biomass production and yield formation. It calculates the daily soil water balance and divides evapotranspiration into soil evaporation and crop transpiration. AquaCrop describes the foliage development of the crop by the canopy cover (CC), that is formally equivalent to the fractional cover (fc) estimated by Sentinel-2 imagery, i.e., it is the fraction of soil surface covered by the green canopy. Hereinafter, we use the two terms canopy cover (CC) and fractional cover (fc) just to distinguish the two variables, respectively derived with AquaCrop and Sentinel-2 imagery.

Transpiration is a function of CC, while evaporation is proportional to the area of soil not covered by vegetation. The CC is multiplied by reference evapotranspiration (ETo), determined by the FAO Penman–Monteith equation, and the crop coefficient (Kc) to calculate potential crop transpiration. Actual transpiration (Ta) is calculated starting from potential one by accounting for water stress. Then, Ta is used for the calculation of crop biomass though its multiplication with water productivity normalized for climate. By using a harvest index (HI), crop yield is obtained by the biomass. To describe the effect of water stress, the model considers different thresholds of water available to the root zone. The first affects leaf canopy expansion (slowing down); the second threshold affects canopy senescence (quickening); the third is referred to as stomata closure (increase) and so to transpiration. Stress coefficients (Ks) range between 1 (no stress) and 0 (complete stress) and are multiplicative factors of the target process.

Model parameters are grouped into two classes: Conservative and non-conservative. Conservative parameters are not dependent on local and management conditions: Canopy growth (CGC) and canopy decline (CDC) coefficients; full canopy crop transpiration coefficient (Kc); biomass WP and soil water depletion thresholds. Non-conservative parameters vary depending on crop and field management, soil type, and climate (sowing date and density, length of crop cycle and phenological stages, maximum canopy cover, etc.). Such parameters can be either retrieved from AquaCrop literature or calibrated by the user (e.g., by means of field experiments). Main AquaCrop outputs are crop production (biomass and yield) and crop water use.

AquaCrop Implementation

In this study, a limited number of AquaCrop parameters were partly calibrated with field observations, including management information: Transplant dates and densities, flowering date and duration, starting of senescence, maturity, and final yield were used for local calibration of the model. For simulating irrigation, the model was set in net irrigation requirement mode, which estimates the crop water requirement based on a selected threshold of allowed root zone (water) depletion (RZD). In order to reproduce the irrigation method adopted by the farmer, drip irrigation was simulated to ensure that RZD was always above 50% of the readily available water (RAW).

2.4. Assimilation-Based Assessment of Crop Water Requirement

The third method for assessing crop water requirements was based on the integration of Sentinel-2 crop derived data with AquaCrop. The fractional cover (fc) estimated by Sentinel-2 has been sequentially assimilated into AquaCrop, by direct insertion, in place of the canopy cover (CC) simulated by the model. The sequential direct insertion is applied under the assumption that a continuous update of one crop model state based on remote observations can reduce the biases induced by the model simplifications of the processes and environmental conditions influencing the crop growth dynamics.

Crop CC simulated by AquaCrop along the growing season and the fc values measured by satellite were compared and the differences were statistically analyzed by means of the Pearson correlation coefficient (r), root mean square error (RMSE%), Nash–Sutcliffe model efficiency coefficient (EF) and Willmott index of agreement (d).

3. Results and Discussion

3.1. Test Site Characteristics

The parcel cultivated in 2017 had a Vitric Phaeozems (Eutric) soil (WRB classification), with a loamy texture (USDA classification), while soil was Vitric Cambisols with a loamy-sand texture in 2018. Soils were both deep, well drained and with no fertility constraints (Table 1).

Table 1. Main soil characteristics of the two soils.

Parameter	Unit	2017	2018
Texture		Loam	Silty loam
Gravel	(vol%)	<5	<5
Saturation	(vol%)	52	46
Field Capacity	(vol%)	29	33
Wilting Point	(vol%)	10	13
Cation Exchange Capacity	(meq/100g)	32	24
Bulk Density	(t/m^3)	1.1	1.3
pH in H_2O	U.pH	6.9	7.1
Organic Matter	(%)	2.6	2.7

Considering the climatology of the study area, average temperature of 15.2 °C and precipitation of 900 mm (period 1982–2012), year 2017 was relatively dry, with a cumulated rainfall of 579 mm and 16.6 °C of average temperature, whereas 2018 was relatively wet and warm, with 1047 mm of rainfall and an average temperature of 17.25 °C (Figure 3). Observed processing tomato yield (expressed in dry matter) was 7.20 t/ha in 2017 and 7.35 t/ha in 2018.

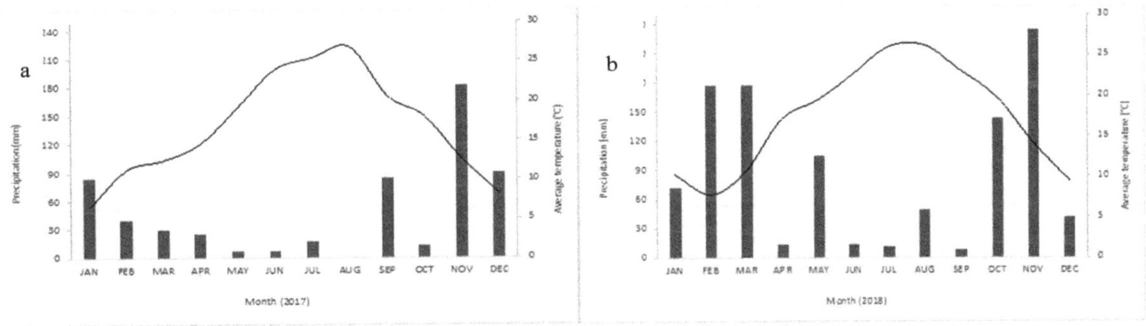

Figure 3. Temperature (line) and precipitation (bars) in the experimental area in year 2017 (**a**) and 2018 (**b**).

3.2. AquaCrop Calibration and Implementation

AquaCrop parameters were calibrated to obtain the best fit between field observations and simulations, both in 2017 and 2018 (Table 2). AquaCrop simulated yields were 7.23 and 7.60 t/ha (dry weight) in 2017 and 2018, with an error of 0.42% and 3.40%, respectively (Table 3). The growth of the green canopy simulated by the model was compared with the fc values observed by IRRISAT. In both seasons the simulated growing curve fitted well with the satellite observations, although an underestimation for the initial canopy cover (late April–early May) and an overestimation during the last part of the growing season (July) was observed (Figure 4). Nevertheless, the statistical analysis (RMSE, EF, d) showed a very good agreement between simulated CC and fc, which correlation was highly significant in both years ($\alpha = 0.001$; Table 4; Figure 5). In 2017 the spatial variability of the crop canopy within the study parcel was larger than in 2018, as testified by the larger variance of the fc values retrieved in 2017.

Table 2. Input crop file used for AquaCrop simulations (HI: Harvest index, GDD: Growing degree days).

Crop Parameter	Unit	Value
Max canopy cover	GDD	686
Flowering	GDD	612
Flowering duration	GDD	531
Length building up HI	GDD	448
Senescence	GDD	1013
Maturity	GDD	1358
Max rooting depth	GDD	612
Initial canopy cover	%	0.64
Maxi canopy cover	%	75
Reference HI	%	63
Base temperature	°C	5
Upper temperature	°C	30
Canopy expansion Upper		0.15
Canopy expansion Lower		0.55
Stomatal closure		0.5
Early canopy senescence		0.7

Table 3. Crop and water balance variables (Tr: Crop transpiration, E: Soil evaporation, ETp: Potential evapotranspiration, IWR: Irrigation water requirement, WP_{ET}: Evapotranspiration water productivity, WP_{IWR}: Irrigation water requirement water productivity).

		Yield (t/ha)	Tr (mm)	E (mm)	ETp (mm)	IWR (mm)	WP_{ET} (kg/m^3)	WP_{IWR} (kg/m^3)
2017	Observed	7.20				416		1.73
	IRRISAT				450	450		
	AquaCrop	7.23	345	192	537	461	1.35	1.57
	Assimilation	8.23	372	165	537	461	1.52	1.79
2018	Observed	7.35				402		1.82
	IRRISAT				349	298		
	AquaCrop	7.60	291	137	428	332	1.78	2.29
	Assimilation	7.34	273	139	412	317	1.78	2.31

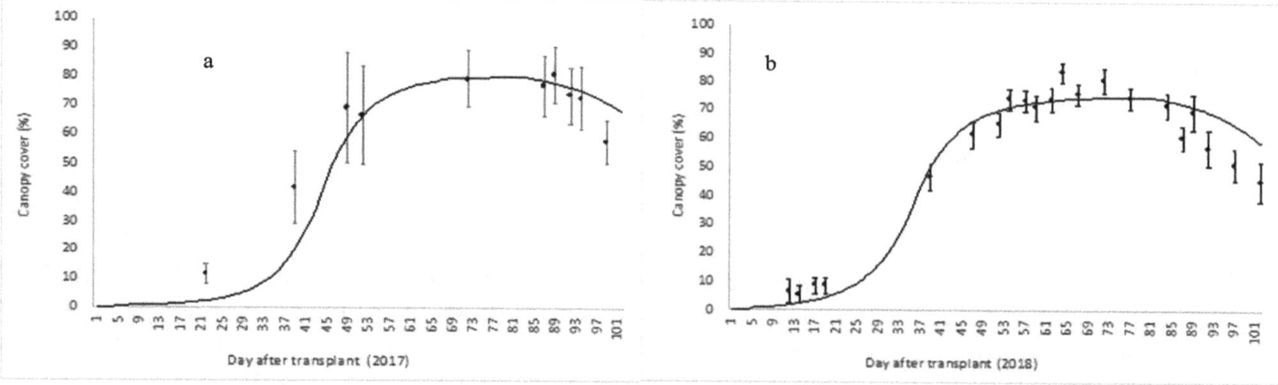

Figure 4. Canopy cover of tomato simulated by AquaCrop, after calibration with field data (line) and corresponding fractional cover values (dots) retrieved by Sentinel-2 imagery during 2017 (**a**) and 2018 (**b**) growing seasons, with corresponding standard deviations.

Table 4. Evaluation of canopy cover simulation results: Number of observations/simulations (n), Pearson correlation coefficient (r), root mean square error (RMSE%), Nash–Sutcliffe model efficiency coefficient (EF) and Willmott index of agreement (d).

	n	*r*	RMSE	EF	d
2017	10	0.95	9.10	0.8	0.96
2018	22	0.97	8.10	0.91	0.98

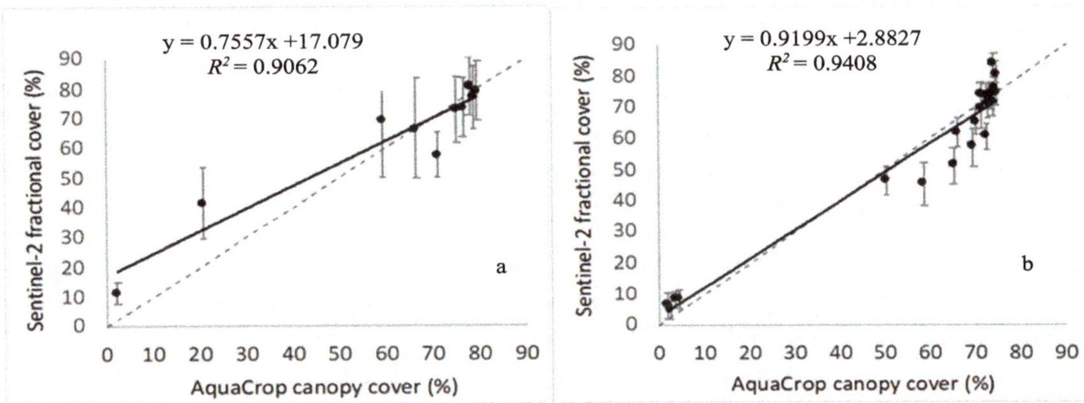

Figure 5. Linear correlation between tomato canopy cover simulated by AquaCrop and fractional cover retrieved by Sentinel-2 imagery in 2017 (**a**) and 2018 (**b**). Standard deviation (bars) and 1:1 line (dashed line) are reported.

3.3. *Estimation of Crop Water Irrigation Requirements*

During the 2017 crop growing season, spanning from June 9th to July 20th, the accumulated rainfall was 40.4 mm, the estimated reference evapotranspiration (ETo) was 553 mm, while crop transpiration (Tr) and soil evaporation (E) estimated by AquaCrop were 345 and 192 mm, respectively. In such conditions, AquaCrop modeled an irrigation water requirement (IWR) of 461 mm corresponding to 144 L/plants.

In year 2018, wetter conditions occurred, with an accumulated rainfall of 125 mm and an estimated ETo of 439 mm. Under such conditions, both crop Tr (291 mm) and E (137 mm) were lower. Accordingly, a lower IWR, 332 mm, was estimated by the model. Being the final yields very similar, the consequent evapotranspiration (WP_{ET}) was higher in 2018 than 2017 (Tab 3).

In Figure 6, three different daily ETp series were compared, respectively produced by calibrated AquaCrop, IRRISAT and AquaCrop after replacing CC with fc estimated by IRRISAT.

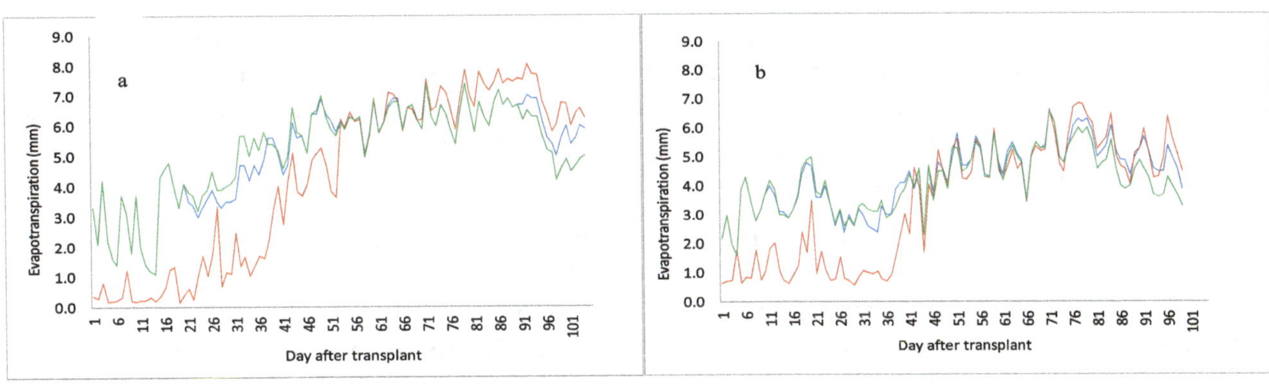

Figure 6. Crop evapotranspiration estimated by AquaCrop (blue), IRRISAT (red) and direct insertion (green) of fc retrieved by Sentinel-2 imagery into AquaCrop, in 2017 (**a**) and 2018 (**b**).

The agreement between AquaCrop and IRRISAT ETp estimates is excellent in the mid-season stage, between about the 50th and 85th day after transplant, when the crop canopy is fully developed. This result is remarkable considering that IRRISAT is fully based on remote observations weather data, and not relying on field or crop data. Moreover, IRRISAT predictions are not based on a continuous assessment of the crop dynamics, rather it provides ETp estimates based on crop parameters retrieved from the last available image, thus with a delay of five or more days, depending on the sky conditions (i.e., cloudiness).

In the initial stage and in the crop development stage (first 50 days after transplant), before canopy is fully developed, the estimation of ETp in AquaCrop is higher than the one determined by IRRISAT. This is due to the conceptually different calculation schemes: Whilst IRRISAT has a resistance-based approach in the Penman–Monteith linked to the LAI at the pixel scale (according to the "big leaf" schematization), AquaCrop has a crop coefficient approach for soil evaporation depending on soil properties. Diversely, with growing LAI and soil cover, the two estimates converge toward very similar values of ETp, hence irrigation requirements.

In the senescence stage, after about the 85th day from transplant, the values of ETp derived by means of the two methods slightly diverge again, with higher ETp for IRRISAT respect to AquaCrop, again due to the difference between the LAI-based approach of the first one and the crop coefficient of the second one. Furthermore, AquaCrop explicitly consider the transpiration reduction associated with the senescence, which cannot be fully predicted by the reduction of the observed LAI applied into the Penman–Monteith equation.

The canopy cover fc maps retrieved from Sentinel-2 imagery can be more effectively employed for constraining AquaCrop predictions to the actual observed crop growth and for assessing its spatial variability. In this study we applied a simple assimilation technique, known in the literature as "direct" insertion [27], consisting in replacing the model state variable CC with the remotely retrieved variable fc. As displayed in Figure 6, AquaCrop ETp estimates after direct insertion are essentially equal to the calibrated AquaCrop, except for the development and senescence stages, when the average fc sensibly deviates from the model calibrated growth curve. As illustrated in Table 3, fc direct insertion does not affect the cumulative ETp, rather it affects its partitioning in transpiration (Tr) an evaporation (E), with a slight increase in Tr and a decrease in E. In 2018, instead, fc direct insertion implies a slight reduction of ETp, mainly associated with a reduction of Tr. Correspondingly, tomato yield was overestimated compared with the observed and the calibrated values in 2017 (+14.3%), while slightly underestimated in 2018 (−1.03%). This result confirmed that sequential assimilation of one state variable does not necessarily improve the prediction performance of all model state variables. In this sense, more complex sequential assimilation techniques are needed, in order to account for the structure of the observation and prediction errors [28], as well of the atmospheric forcing [29].

The impact of the direct insertion in terms of irrigation water requirement (IWR) is null in 2017, while it determines a 5% decrease in 2018. As illustrated in Figure 7, also the temporal pattern of IWR is almost unchanged in the two simulated seasons. A potential water saving of 70 mm could be achieved in 2018 according to AquaCrop, by accounting for the effective contribution of the summer rainfall (125 mm) to the soil water deficit.

Looking again at Table 3, it is interesting to note that IRRISAT, despite the significant reduction of the cumulative ETp both in 2017 (16%) and 2018 (19%) compared with the calibrated AquaCrop, predicts cumulative IWR just 3% smaller in 2017 and 11% smaller in 2018 than the calibrated AquaCrop. As illustrated in Figure 7, this reduction is essentially due to the initial and development stage, when AquaCrop, differently form IRRISAT, accounts for the soil evaporation in the IWR assessment. After the crop is fully developed, the cumulative IWR curves are almost parallel, testifying the good agreement of the remotely assessed daily IWR with AquaCrop.

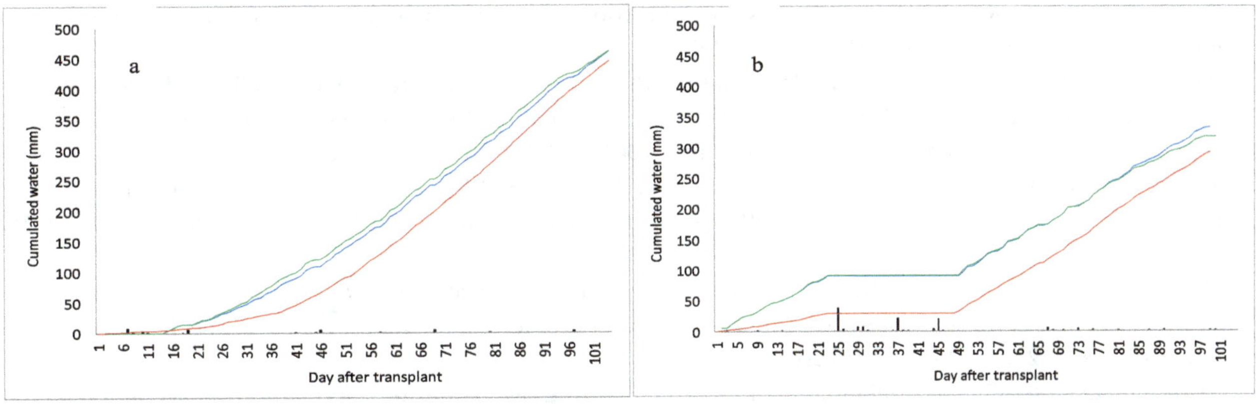

Figure 7. Cumulated irrigation water requirement (mm) estimated by the three methods: AquaCrop (blue), IRRISAT (red) and direct insertion (green) of fractional cover retrieved by Sentinel-2 imagery into AquaCrop, in 2017 (**a**) and 2018 (**b**). Rainfall is also plotted (histograms).

4. Conclusions

Sentinel-2 imagery can be effectively exploited for monitory canopy growth of tomato crops in open field. Irrigation advisory services, such as IRRISAT, which are only based on crop data retrieved by Sentinel-2 and weather data, can provide a reliable assessment of crop water requirements of tomato field crops especially when crop canopy is fully developed. It should be considered that IRRISAT does not require input data concerning the soil or crop phenology, since it is entirely based on crop growth monitoring from space. Hence, integrating Sentinel-2 imagery with a crop growth model such as AquaCrop can be an effective strategy for assessing crop water requirement in the initial and development stage of the crop, as well as for identifying the senescence stage. Further, being the satellite imagery a spatial information, the integration into a crop model can help in assessing crop water requirement at field or higher scales, i.e., at territorial level. Thus, a sequential assimilation can be used to support irrigation planning by Irrigation and Land Reclamation consortia. In this study a simple direct insertion method has been applied for assimilating canopy cover retrieved by Sentinel-2 imagery into AquaCrop, which does not guarantee an optimal model-data integration. Additional studies are required for testing more advanced data assimilation techniques, accounting for the structure of the model state and observation errors.

Author Contributions: Conceptualization, A.D.M., G.B.C. and F.A.; Data curation, S.F.B.; Formal analysis, A.D.M. and G.B.C.; Investigation, A.D.M., M.M. and F.A.; Methodology, A.D.M., G.B.C., S.F.B., M.M., G.D., C.D.M. and F.A.; Software, S.F.B.; Supervision, G.D., S.O. and F.A.; Validation, G.B.C.; Writing—original draft, A.D.M., G.B.C. and F.A.; Writing—review & editing, M.M.

Acknowledgments: The authors wish to thank OPERA project being carried out within the Water-JPI ERA-NET Cofund WaterWorks2015 Call. We wish to thank also the team of the Irrigation and Land Reclamation Consortium of "Bacino Inferiore del Volturno", and Francesco D'Amore, for their availability, help and support in organizing field campaigns.

References

1. Altobelli, F.; Cimino, O.; Natali, F.; Orlandini, S.; Gitz, V.; Meybeck, A.; Marta, A.D. Irrigated farming systems: Using the water footprint as an indicator of environmental, social and economic sustainability. *J. Agric. Sci.* **2018**, *156*, 711–722. [CrossRef]
2. Tsakmakis, I.D.; Zoidou, M.; Gikas, G.D.; Sylaios, G.K. Impact of Irrigation Technologies and Strategies on Cotton Water Footprint Using AquaCrop and CROPWAT Models. *Environ. Process.* **2018**, *5*, 181–199. [CrossRef]

3. Kustas, W.; Anderson, M. Advances in thermal infrared remote sensing for land surface modeling. *Agric. For. Meteorol.* **2009**, *149*, 2071–2081. [CrossRef]
4. Anderson, M.C.; Norman, J.M.; Mecikalski, J.R.; Otkin, J.A.; Kustas, W.P. A climatological study of evapotranspiration and moisture stress across the continental United States based on thermal remote sensing: 2. Surface moisture climatology. *J. Geophys. Res.* **2007**, *112*, D11112. [CrossRef]
5. Karimi, P.; Bastiaanssen, W.G.M.; Molden, D. Water Accounting Plus (WA+)—A water accounting procedure for complex river basins based on satellite measurements. *Hydrol. Earth Syst. Sci.* **2013**, *17*, 2459–2472. [CrossRef]
6. Drusch, M.; Del Bello, U.; Carlier, S.; Colin, O.; Fernandez, V.; Gascon, F.; Hoersch, B.; Isola, C.; Laberinti, P.; Martimort, P. Sentinel-2: ESA's optical high-resolution mission for GMES operational services. *Remote Sens. Environ.* **2012**, *120*, 25–36. [CrossRef]
7. Vuolo, F.; D'Urso, G.; De Michele, C.; Bianchi, B.; Cutting, M. Satellite-based irrigation advisory services: A common tool for different experiences from Europe to Australia. *Agric. Water Manag.* **2015**, *147*, 82–95. [CrossRef]
8. D'Urso, G. Current Status and Perspectives for the Estimation of Crop Water Requirements from Earth Observation. *Ital. J. Agron.* **2010**, *5*, 107–120.
9. Casa, R.; Varella, H.; Buis, S.; Guerif, M.; de Solan, B.; Frederic, B. Forcing a wheat crop model with LAI data to access agronomic variables: Evaluation of the impact of model and LAI uncertainties and comparison with an empirical approach. *Eur. J. Agron.* **2012**, *37*, 1–10. [CrossRef]
10. Steduto, P.; Hsiao, T.C.; Raes, D.; Fereres, E. Aquacrop-the FAO crop model to simulate yield response to water: I. concepts and underlying principles. *Agron. J.* **2009**, *101*, 426–437. [CrossRef]
11. Raes, D.; Steduto, P.; Hsiao, T.C.; Fereres, E. AquaCropThe FAO Crop Model to Simulate Yield Response to Water: II. Main Algorithms and Software Description. *Agron. J.* **2009**, *101*, 438–447. [CrossRef]
12. Linker, R.; Ioslovich, I. Assimilation of canopy cover and biomass measurements in the crop model AquaCrop. *Biosyst. Eng.* **2017**, *162*, 57–66. [CrossRef]
13. Jin, X.; Kumar, L.; Li, Z.; Xu, X.; Guijun, Y.; Wang, J. Estimation of Winter Wheat Biomass and Yield by Combining the AquaCrop Model and Field Hyperspectral Data. *Remote Sens.* **2016**, *8*, 972. [CrossRef]
14. Silvestro, P.C.; Pignatti, S.; Pascucci, S.; Yang, H.; Li, Z.; Guijun, Y.; Huang, W.; Casa, R. Estimating Wheat Yield in China at the Field and District Scale from the Assimilation of Satellite Data into the Aquacrop and Simple Algorithm for Yield (SAFY) Models. *Remote Sens.* **2017**, *9*, 509. [CrossRef]
15. Jin, X.; Liu, S.; Baret, F.; Hemerlé, M.; Comar, A. Estimates of plant density of wheat crops at emergence from very low altitude UAV imagery. *Remote Sens. Environ.* **2017**, *198*, 105–114. [CrossRef]
16. Jin, X.; Kumar, L.; Li, Z.; Feng, H.; Xu, X.; Guijun, Y.; Wang, J. A review of data assimilation of remote sensing and crop models. *Eur. J. Agron.* **2018**, *92*, 141–152. [CrossRef]
17. Zhao, Y.; Chen, S.; Shen, S. Assimilating remote sensing information with crop model using Ensemble Kalman Filter for improving LAI monitoring and yield estimation. *Ecol. Model.* **2013**, *270*, 30–42. [CrossRef]
18. Trombetta, A.; Iacobellis, V.; Tarantino, E.; Gentile, F. Calibration of the AquaCrop model for winter wheat using MODIS LAI images. *Agric. Water Manag.* **2016**, *164*, 304–316. [CrossRef]
19. Pelosi, A.; Medina, H.; Villani, P.; D'Urso, G.; Chirico, G.B. Probabilistic forecasting of reference evapotranspiration with a limited area ensemble prediction system. *Agric. Water Manag.* **2016**, *178*, 106–118. [CrossRef]
20. Chirico, G.B.; Pelosi, A.; De Michele, C.; Bolognesi, S.F.; D'Urso, G. Forecasting potential evapotranspiration by combining numerical weather predictions and visible and near-infrared satellite images: An application in southern Italy. *J. Agric. Sci.* **2018**, *19*, 702–710. [CrossRef]
21. Garrigues, S.; Lacaze, R.; Baret, F.J.; Morisette, J.T.; Weiss, M.; Nickeson, J.E.; Fernandes, R.; Plummer, S.; Shabanov, N.V.; Myneni, R.B.; et al. Validation and intercomparison of global Leaf Area Index products derived from remotesensing data. *Geophys. Res.* **2008**, *113*. [CrossRef]
22. Jacquemoud, S.; Baret, F. PROSPECT: A model of leaf optical properties spectra. *Remote Sens. Environ.* **1990**, *34*, 75–91. [CrossRef]
23. Verhoef, W. Light scattering by leaf layers with application to canopy reflectance modeling: The SAIL model. *Remote Sens. Environ.* **1984**, *16*, 125–141. [CrossRef]
24. Atzberger, C.; Berger, K. Spatially constrained inversion of radiative transfer models for improved LAI mapping from future Sentinel-2 imagery. *Remote Sens. Environ.* **2012**, *120*, 208–218. [CrossRef]

25. D'Urso, G.; Calera Belmonte, A. Operative approaches to determine crop water requirements from earth observation data: Methodologies and applications. *AIP Conf. Proc.* **2006**, *852*, 14–25.
26. Thuillier, G.; Hersé, M.; Labs, D.; Foujols, T.; Peetermans, W.; Gillotay, D.; Simon, P.C.; Mandel, H. The solar spectral irradiance from 200 to 2400 nm as measured by the solspec spectrometer from the atlas and Eureca missions. *Sol. Phys.* **2003**, *214*, 1–22. [CrossRef]
27. Chirico, G.B.; Medina, H.; Romano, N. Kalman filters for assimilating near-surface observations in the Richards equation. I: Retrieving state profiles with linear and nonlinear numerical schemes. *Hydrol. Earth Syst. Sci.* **2014**, *18*, 2503–2520. [CrossRef]
28. Medina, H.; Romano, N.; Chirico, G.B. Kalman filters for assimilating near-surface observations into the Richards equation—Part 2: A dual filter approach for simultaneous retrieval of states and parameters. *Hydrol. Earth Syst. Sci.* **2014**, *18*, 2521–2541. [CrossRef]
29. Pelosi, A.; Medina, H.; Van den Bergh, J.; Vannitsem, S.; Chirico, G.B. Adaptive Kalman filtering for postprocessing ensemble numerical weather predictions. *Mon. Weather Rev.* **2017**, *145*, 4837–4854. [CrossRef]

11

Multi-Temporal Unmanned Aerial Vehicle Remote Sensing for Vegetable Mapping using an Attention-Based Recurrent Convolutional Neural Network

Quanlong Feng [1,2], Jianyu Yang [2,*], Yiming Liu [2], Cong Ou [2], Dehai Zhu [2], Bowen Niu [2], Jiantao Liu [3] and Baoguo Li [2]

1 College of Resources and Environmental Sciences, China Agricultural University, Beijing 100193, China; fengql@cau.edu.cn

2 College of Land Science and Technology, China Agricultural University, Beijing 100083, China; liuym0086@cau.edu.cn (Y.L.); oucong@cau.edu.cn (C.O.); zhudehai@cau.edu.cn (D.Z.); s20193081417@cau.edu.cn (B.N.); libg@cau.edu.cn (B.L.)

3 School of Surveying and Geo-Informatics, Shandong Jianzhu University, Jinan 250101, China; liujiantao18@sdjzu.edu.cn

* Correspondence: ycjyyang@cau.edu.cn

Abstract: Vegetable mapping from remote sensing imagery is important for precision agricultural activities such as automated pesticide spraying. Multi-temporal unmanned aerial vehicle (UAV) data has the merits of both very high spatial resolution and useful phenological information, which shows great potential for accurate vegetable classification, especially under complex and fragmented agricultural landscapes. In this study, an attention-based recurrent convolutional neural network (ARCNN) has been proposed for accurate vegetable mapping from multi-temporal UAV red-green-blue (RGB) imagery. The proposed model firstly utilizes a multi-scale deformable CNN to learn and extract rich spatial features from UAV data. Afterwards, the extracted features are fed into an attention-based recurrent neural network (RNN), from which the sequential dependency between multi-temporal features could be established. Finally, the aggregated spatial-temporal features are used to predict the vegetable category. Experimental results show that the proposed ARCNN yields a high performance with an overall accuracy of 92.80%. When compared with mono-temporal classification, the incorporation of multi-temporal UAV imagery could significantly boost the accuracy by 24.49% on average, which justifies the hypothesis that the low spectral resolution of RGB imagery could be compensated by the inclusion of multi-temporal observations. In addition, the attention-based RNN in this study outperforms other feature fusion methods such as feature-stacking. The deformable convolution operation also yields higher classification accuracy than that of a standard convolution unit. Results demonstrate that the ARCNN could provide an effective way for extracting and aggregating discriminative spatial-temporal features for vegetable mapping from multi-temporal UAV RGB imagery.

Keywords: vegetable mapping; multi-temporal UAV; recurrent convolutional neural network; attention mechanism

---

1. Introduction

Accurate vegetable mapping is of great significance for modern precision agriculture. The spatial distribution map for different kinds of vegetables is the basis for automated agricultural activities such as unmanned aerial vehicle (UAV)-based fertilizer and pesticide spraying. Traditional vegetable mapping is usually based on field survey or visual interpretation of remote sensing imagery, which is time-consuming and inconvenient. Hence, it is of great importance to study the automatic methods for precise vegetable classification. However, previous studies mainly focused on the staple crop (e.g., corn, paddy rice) classification [1], in this regard, we are highly motivated to propose an effective method for vegetable classification based on UAV observations, which could provide a useful reference for future studies on vegetable mapping.

In previous studies, optical satellite imagery was firstly utilized for vegetable and crop mapping. Wikantika et al. applied linear spectral mixture analysis to map vegetable parcels from mountainous regions based on Landsat Enhanced Thematic Mapper (ETM) data [2]. Belgiu et al. utilized multi-temporal Sentinel-2 imagery for crop mapping based on a time-weighted dynamic time warping method and achieved comparable accuracy with random forest (RF) [3]. Rupasinghe et al. adopted Pleiades data and a support vector machine (SVM) to classify the coastal vegetation cover and also yielded a good classification performance [4]. Wan et al. also used SVM and single-date WorldView-2 for crop type classification and justified the role of texture features in improving the classification accuracy [5]. Meanwhile, as the new generation sensor of Landsat satellite, data acquired by the Operational Land Imager (OLI) from Landsat-8 has also been used for vegetable and crop type classification. For instance, Asgarian et al. used multi-date OLI imagery for vegetable and crop mapping in central Iran based on decision tree and SVM and achieved good results [6].

However, when compared with staple crops, the land parcel of vegetables is small in size, resulting in a large amount of mixed pixels in space-borne images. Different from space-borne observations, a UAV could obtain images with very high or ultra-high spatial resolution where the mixed pixel is no longer a problem. Meanwhile, a UAV could be deployed whenever necessary, which makes it an efficient tool for rapid monitoring of land resources [7–11]. Due to payload capacity limitations, off-the-shelf digital cameras have been equipped in small-sized UAVs [7,8,11]. Under this circumstance, the images acquired only have three bands (i.e., red, green, blue, RGB), resulting in a low spectral resolution which limits the performance of differentiating various vegetable categories. To reduce this impact, we introduce multi-temporal UAV data to obtain useful phenological information to enhance the inter-class separability. Afterwards, a robust classification model, the attention-based recurrent convolutional neural network (ARCNN), is constructed to further improve the classification accuracy.

Compared with mono-temporal or single-date observation, multi-temporal datasets could provide useful phenological information, which aids for plant and vegetation classification during growing season [10–15]. Pádua et al. adopted multi-temporal UAV-based RGB imagery to differentiate grapevine vegetation from other plants in a vineyard [11], and indicates that although RGB images have a low spectral resolution, the inclusion of multi-temporal observations makes it possible for accurate plant classification. Van Iersel conducted river floodplain vegetation classification using multi-temporal UAV data and a hybrid method, which is based on the combination of random forest and object-based image analysis (OBIA) [14]. In our previous research, we also incorporated multi-temporal Landsat imagery and a series of classifiers (i.e., decision trees, SVM and RF) for cropland mapping of the Yellow River delta [15], which justified the effectiveness of multi-temporal observations in enhancing the classification performance.

The above mentioned studies are mainly based on low-level, manually designed features (i.e., spectral indices, textures) and machine-learning classifiers, which might show poor performance in obtaining the high-level and representative features for accurate vegetable mapping. Besides, lots of domain expertise together with engineering skills are always involved in these manually designed features [16,17]. Meanwhile, deep learning offers a novel way for discovering informative features through a hierarchical learning framework [18], which shows promising performance in several computer vision (CV) applications such as sematic segmentation [19,20], object detection [21] and image classification [22–24]. Recently, deep learning models have been widely studied in the field of remote sensing [25–29], including cloud detection [30], building extraction [31–33], land object detection [34,35], scene classification [36,37] and land cover mapping [38–43]. Specifically, Kussul et al. utilized both one- and two-dimensional CNN to classify crop and land cover types from both synthetic aperture radar (SAR) and optical satellite imagery [41]. Rezaee et al. [42] also adopted an AlexNet [22] pre-trained on ImageNet for wetland classification based on mono-temporal RapidEye optical data. It should be noted that the vegetable land parcels in China are always mixed with other crops and the landscape is rather fragmented, leading to a large variability in the shape and scale of land parcels [44]. However, previous studies usually mirror deep learning models from computer vision field while neglecting the complex nature of agricultural landscapes. Therefore, how to build an effective deep learning model to account for the fragmented landscape is a key issue in vegetable mapping.

Meanwhile, the introduction of multi-temporal UAV data calls for effective methods of temporal feature extraction and spatial-temporal fusion to further boost the classification accuracy. Early studies [11,12,14] usually stacked or concatenated multi-temporal data without considering the hidden temporal dependencies. With the development of recurrent neural network (RNN), models such as long-short term memory (LSTM) [45] have been adopted to establish the relationship between sequential remote sensing data [46–50]. Ndikumana et al. utilized multi-temporal SAR Sentinel-1 imagery and a RNN for crop type classification, and indicated that the RNN showed a higher accuracy than several popular machine learning models (i.e., RF and SVM) [46]. Mou et al. cascaded a CNN and RNN to aggregate spectral, spatial and temporal features for change detection [47]. In addition, when it comes to vegetable or crop mapping, it should be noted that the importance or contribution of each mono-temporal dataset to classification may vary during the growing season. Therefore, how to model the sequential relationship between multi-temporal UAV data hence to further boost the classification accuracy is of great significance.

To tackle the above issues, this study proposes an attention-based recurrent convolutional neural network (ARCNN) for accurate vegetable mapping from multi-temporal UAV data. The proposed model integrates a multi-scale deformable CNN and an attention-based RNN into a trainable end-to-end network. The former is to learn and extract the representative spatial features from UAV data to account for the scale and shape variations under fragmented agricultural landscape, while the latter is to model the dependency across multi-temporal images to obtain useful phenological information. The proposed network yields an effective solution for spatial-temporal feature fusion, based on which the vegetable mapping accuracy could be boosted.

2. Materials and Methods

2.1. Study Area

Both the study area and multi-temporal UAV imagery used in this research are illustrated in Figure 1.

The study area includes a vegetable field which is located in Xijingmeng Village of Shenzhou City, Hebei province, China. There are various kinds of vegetables, such as Chinese cabbage, carrot, leaf mustard, etcetera. Meanwhile, the study area also locates in the North China Plain, which belongs to a continental monsoon climate, where summer is humid and hot, while winter is dry and cold. The annual temperature is about 13.4 °C and the annual precipitation is about 486 mm. Vegetables are usually planted in late August and harvested in early November.

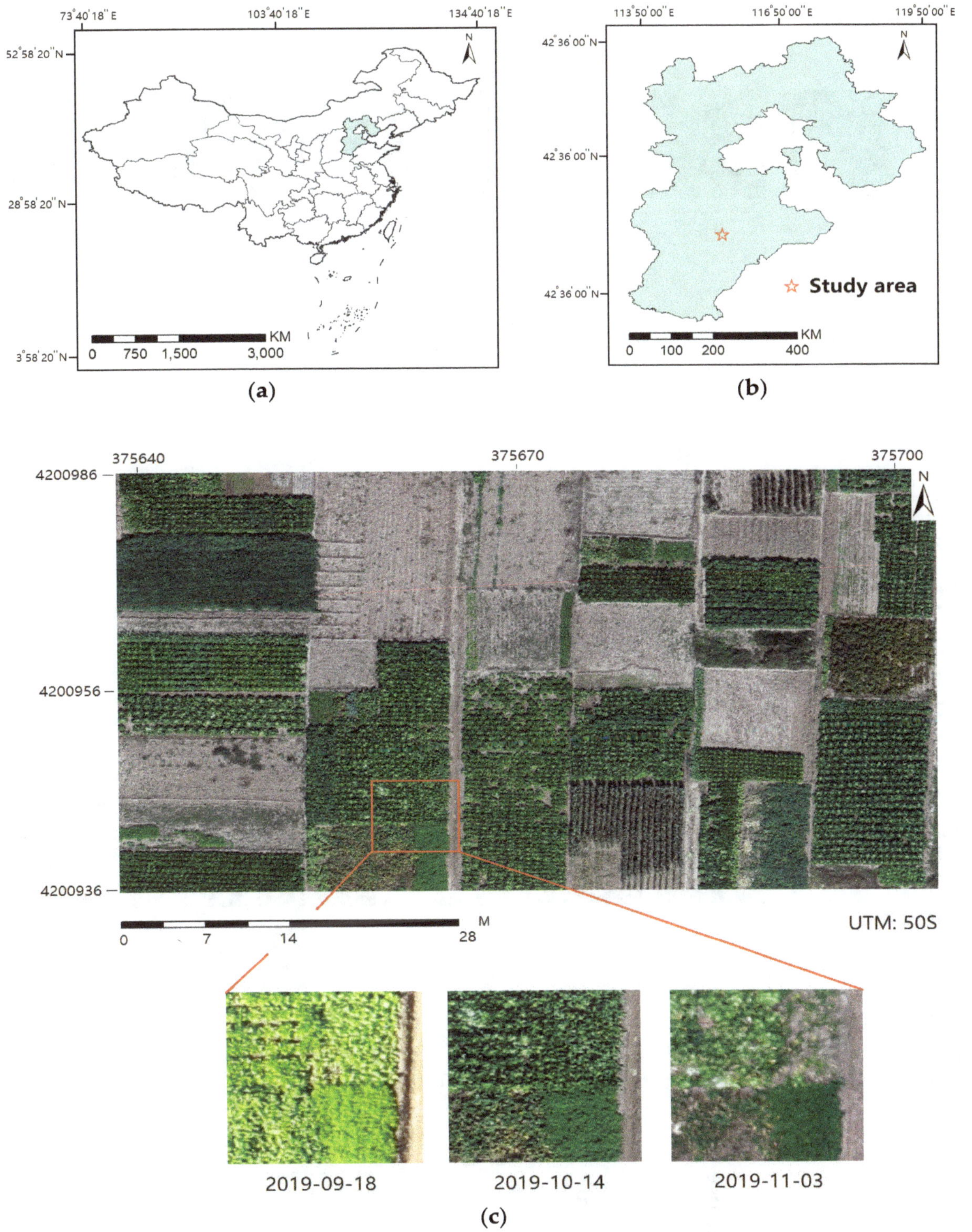

Figure 1. Study area. (**a**) China; (**b**) Hebei Province; and (**c**) Xijingmeng Village and multi-temporal UAV images.

A field survey was conducted along with the UAV flight. Vegetable and crop types, locations measured by global positioning system (GPS) and photographs were recorded for every land parcel. According to the results of the field survey, there were a total of fourteen land cover categories, including eight vegetable types (i.e., carrot, Chinese cabbage, leaf mustard, turnip, spinach, kohlrabi, potherb and scallion), four crop types (i.e., millet, sweet potato, corn and soybean), weed and bare soil (Table 1).

Table 1. Classification scheme of this study.

No.	Class Name	Training/Testing	Ground Image	No.	Class Name	Training/Testing	Ground Image
1	Carrot	200/200		8	Millet	200/200	
2	Chinese cabbage	400/400		9	Weed	100/100	
3	Leaf mustard	200/200		10	Bare soil	200/200	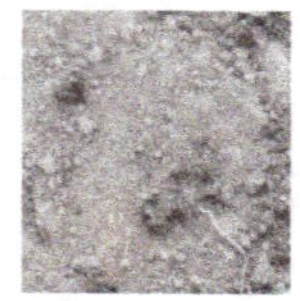
4	Turnip	200/200		11	Sweet potato	200/200	
5	Spinach	50/50		12	Corn	50/50	
6	Kohlrabi	50/50		13	Soybean	200/200	
7	Potherb	100/100		14	Scallion	100/100	

Training and testing datasets were obtained from UAV imagery by visual inspection based on the sampling sites' GPS coordinates and the corresponding land cover categories. Numbers of both training and testing datasets are shown in Table 1. Besides, Table 1 also shows the ground image taken during the field work to depict the detailed appearance of various vegetables and crops.

Meanwhile, Figure 2 illustrates the spatial distribution of both training and testing samples. It indicates that all the samples are randomly distributed and no overlap exists between training and testing regions. Besides, because we adopted patch-based per-pixel classification, all the training and testing samples are pixels from the region of interest (ROI). In this study, the number of training and testing samples are both 2250, respectively, which accounts for a small area (0.03%) of the total study region (7,105,350 pixels).

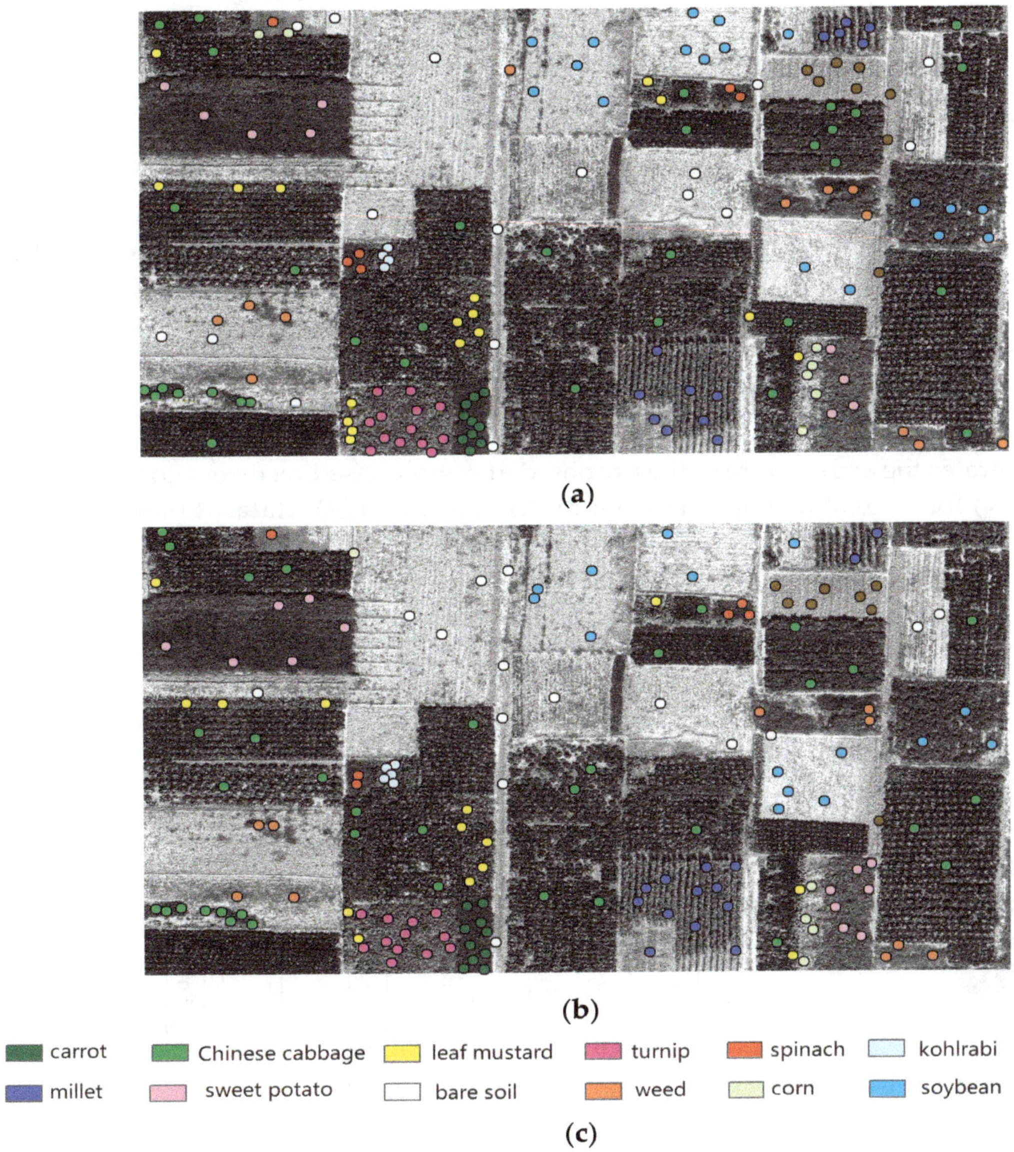

Figure 2. Spatial distribution of (**a**) training samples; (**b**) testing samples; (**c**) legend.

2.2. *Dataset Used*

We utilized a small-sized UAV, DJI-Inspire 2 [51], for the image data acquisition. The camera onboard is an off-the-shelf, light-weight digital camera with only three RGB bands. Therefore, the low spectral resolution would make it difficult to separate various vegetable categories if only considering single-date UAV data. To tackle this issue, we introduce multi-temporal UAV observations, which could obtain the phenological information during the growing season to increase the inter-class separability.

We conducted three flights in the autumn of 2019 (Table 2). During each flight, the flying height was set to be 80 m, achieving a very high spatial resolution of 2.5 cm/pixel. Besides, the width and height of the study area is 3535 and 2010 pixels (88.4 m and 50.3 m), respectively. Actually, the extent of the study area is at the limit of UAV data coverage. Although the study area may still seem small, it is limited by the operation range of the mini-UAV used. In future study, we would try high altitude long endurance (HALE) UAV to acquire images of a larger study region.

Table 2. Multi-temporal UAV images utilized in this study.

	Season	Date	Data Source
T1	Autumn	18 September 2019	UAV RGB data
T2	Autumn	14 October 2019	UAV RGB data
T3	Autumn	3 November 2019	UAV RGB data

The raw images acquired during each flight were orthorectified firstly and then mosaicked to an entire image by Pix4D [52]. Specifically, several key parameters in Pix4D are set as follows. "Aerial Grid or Corridor" is chosen for matching image pairs, "Automatic" is selected for targeted number of key points, matching window size is 7 × 7 and 1 GSD is used for resolution. The rest of the parameters are set to default values. Afterwards, image registration was performed among the multi-temporal UAV data by ENVI (the Environment for Visualizing Images) [53].

2.3. Overview of the ARCNN

Figure 3 illustrates the architecture of the proposed attention-based recurrent convolutional neural network (ARCNN) for vegetable mapping from multi-temporal UAV data. It mainly contains two parts, (1) a spatial feature extraction module based on a multi-scale deformable convolutional network (MDCN), and (2) a spatial-temporal feature fusion module based on a bi-directional RNN and attention mechanism. The former is to learn representative spatial features while the latter is to aggregate spatial and temporal features for the final vegetable classification.

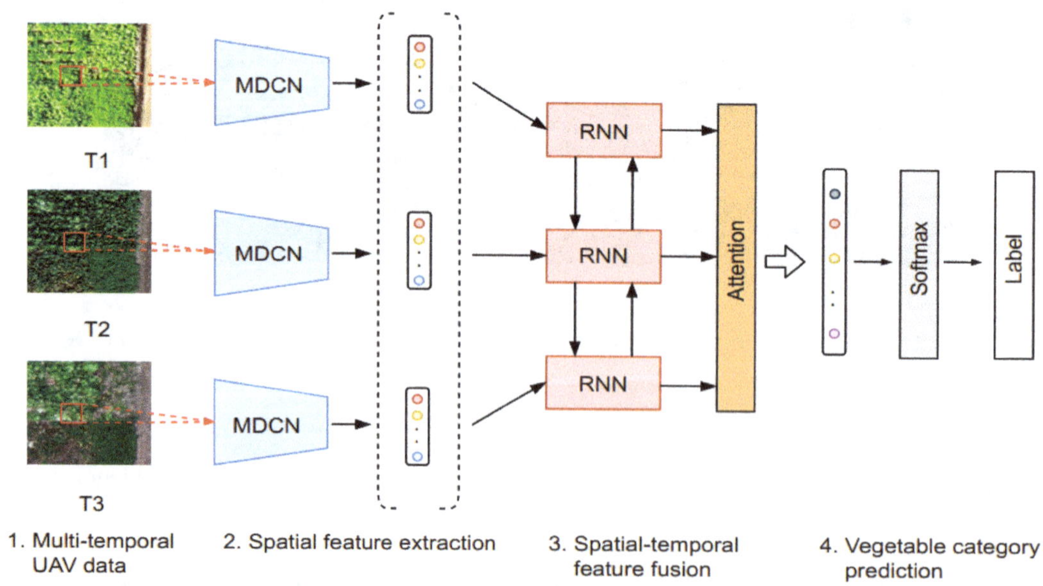

Figure 3. The overview of the proposed attention-based recurrent convolutional neural network (ARCNN) for vegetable mapping based on multi-temporal UAV data.

2.4. Spatial Feature Extraction Based on MDCN

Accurate vegetable classification requires discriminative features. In this section, a multi-scale deformable convolutional network (MDCN) is proposed to learn and extract rich spatial features from UAV imagery, which is to account for the scale and shape variations of land parcels. Specifically, MDCN is an improved version of our previous study [44], and the network structure is depicted as Figure 4.

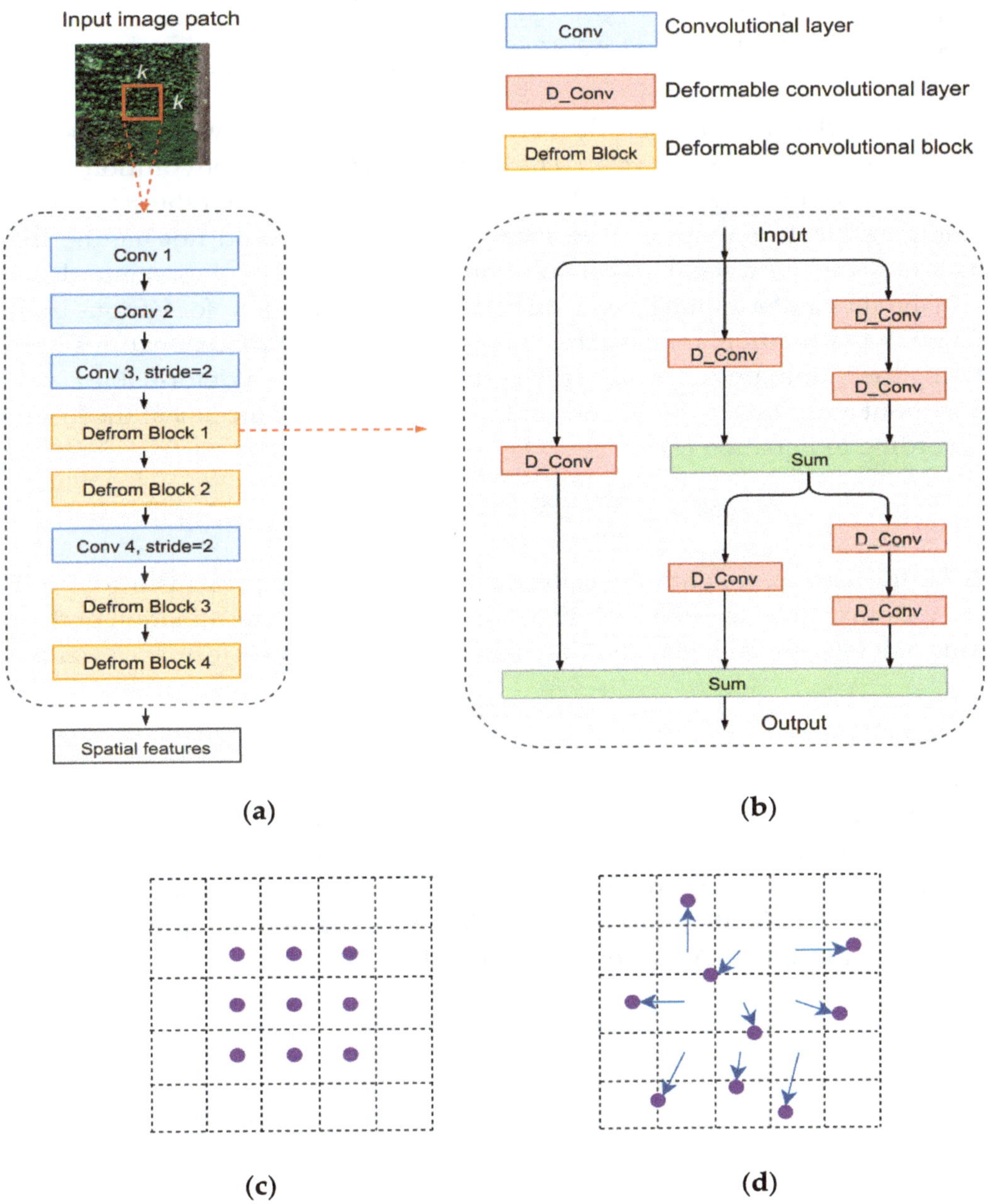

Figure 4. (**a**) Network structure of the proposed multi-scale deformable convolutional network (MDCN); (**b**) deformable convolutional block; (**c**) standard convolution; and (**d**) deformable convolution.

Same as our previous work, the input of MDCN is an image patch which is located at the center of the labeled pixel. The dimension of the patch is $k \times k \times c$ [44], where k stands for the patch size while c refers to the channel number. Specifically, MDCN includes four regular convolutional layers and four deformable convolutional blocks. Table 3 shows the detailed configuration of the MDCN.

Table 3. Detailed configuration of the MDCN.

Layer Name	Input Size	Output Size	Kernel Size	Filter Number	Stride
Input	11 × 11 × 3	–	–	–	–
Conv1	11 × 11 × 3	11 × 11 × 64	3 × 3	64	1
Conv2	11 × 11 × 64	11 × 11 × 128	3 × 3	128	1
Conv3	11 × 11 × 128	6 × 6 × 128	3 × 3	128	2
Deform Block 1	6 × 6 × 128	6 × 6 × 128	–	–	–
Deform Block 2	6 × 6 × 128	6 × 6 × 128	–	–	–
Conv4	6 × 6 × 128	3 × 3 × 128	3 × 3	256	2
Deform Block 3	3 × 3 × 256	3 × 3 × 256	–	–	–
Deform Block 4	3 × 3 × 256	3 × 3 × 256	–	–	–

The deformable block contains multiple streams of deformable convolution [54], which could learn hierarchical and multi-scale features. The role of deformable convolution is to model the shape variations under complex agricultural landscapes. Considering that the standard convolution only samples the given feature map at fixed locations [54,55], it could not handle the geometric transformations. Compared with standard convolution, deformable convolution introduces additional offsets along with the standard sampling grid [54,55], which could account for various transformations for scale, aspect ratio and rotation, making it an ideal tool to extract robust features under complex landscapes. During the training process, both the kernel and offsets of a deformable convolution unit can be learned without additional supervision. In this situation, the output y at the location p_0 could be calculated according to Equation (1):

$$y(p_0) = \sum w(p_i) * x(p_0 + p_i + \Delta p_i) \quad (1)$$

where w stands for the learned weights, p_i means the i th location, x represents the input feature map and Δp_i refers to the offset to be learned [54]. In addition, as for the determination of the patch size k, we referred to our previous research [44] and the highest classification performance was reached when k equaled 11.

2.5. Spatial-Temporal Feature Fusion

After the extraction of spatial features from every mono-temporal UAV image, it is essential to establish the relationship between these sequential features to yield a complete feature representation for boosting the vegetable classification performance. In this section, we exploit an attention based bi-directional LSTM (Bi-LSTM-Attention) for the fusion of spatial and temporal features (Figure 5). The network structure of Bi-LSTM-Attention is illustrated as follows.

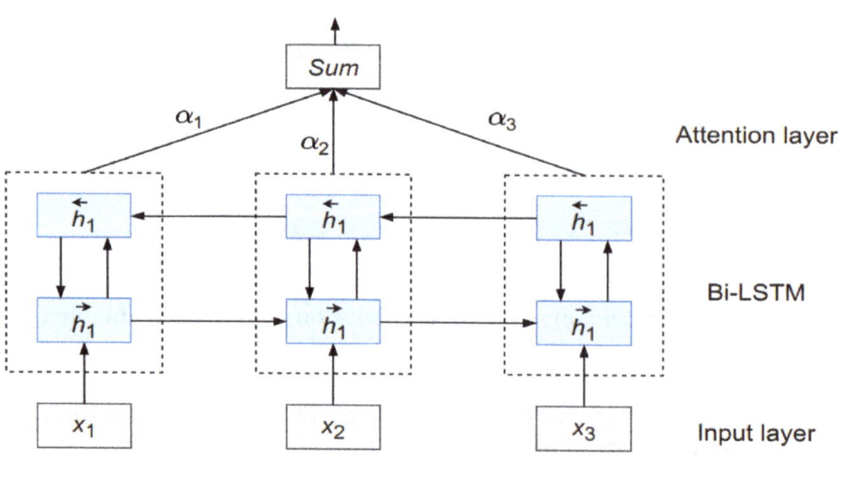

Figure 5. Architecture of the attention-based bi-directional LSTM.

Specifically, LSTM is a variant of RNN, which contains one input layer, one or several hidden layers and one output layer [45]. It should be noted that LSTM is more specialized in capturing long-range dependencies between sequential signals than other RNN models. LSTM utilizes a vector (i.e., memory cell) to store the long-term memory and adopts a series of gates to control the information flow [45] (Figure 6). The hidden layer is updated as follows:

$$i_t = \sigma(W_{ix}x_t + W_{ih}h_{t-1} + b_i) \tag{2}$$

$$f_t = \sigma(W_{fx}x_t + W_{fh}h_{t-1} + b_f) \tag{3}$$

$$c_t = f_t c_{t-1} + i_t \tanh(W_{cx}x_t + W_{ch}h_{t-1} + b_c) \tag{4}$$

$$o_t = \sigma(W_{ox}x_t + W_{oh}h_{t-1} + b_o) \tag{5}$$

$$h_t = o_t \tanh(c_t) \tag{6}$$

where i refers to the input gate, f stands for the forget gate, o refers to the output gate, c is the memory cell and σ stands for the logistic sigmoid function [45].

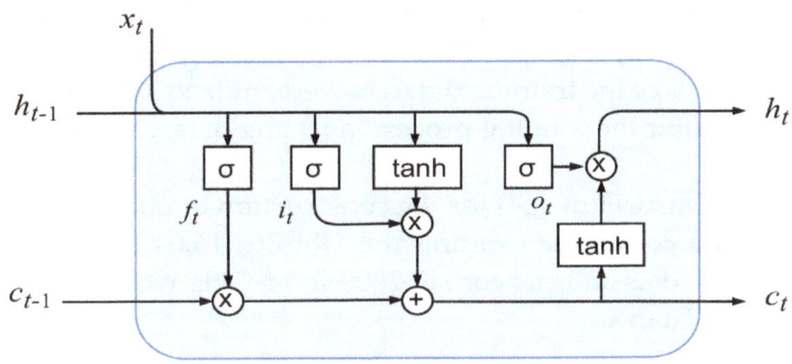

Figure 6. Structure of the LSTM.

LSTM has long been utilized in natural language processes (NLP) [56,57]. Recently, it has been introduced in the remote sensing field for change detection and land cover mapping. In this section, we exploit a bi-directional LSTM [57] to learn the relationship between multi-temporal spatial features extracted from the UAV image. As shown in Figure 5, two LSTMs are stacked together while the hidden state of first LSTM is fed into the second one, and the second LSTM follows a reverse order of the former to fully understand the dependencies of the sequential signals in a bi-directional way. In addition, to further improve the performance, we append an attention layer to the output of the second LSTM. Actually, attention mechanism is widely studied in the field of CV and NLP [58–60], which could automatically adjust the weight of input feature vectors according to their importance to the current task. Therefore, we also incorporate an attention layer to re-weight the sequential features to boost the classification performance.

Let H be a matrix containing a series of vectors $[h_1, h_2, \ldots, h_T]$ that are produced by the bi-directional LSTM, where T denotes the length of the input features. The output of the attention layer is formed by a weighted sum of vectors described as follows:

$$M = \tanh(H) \tag{7}$$

$$\alpha = softmax(w^T M) \tag{8}$$

$$R_{att} = H\alpha^T \tag{9}$$

where α is the attention vector and while R_{att} denotes the fused and attention-weighted spatial-temporal features. Additionally, the features outputted from the Bi-LSTM-Attention are re-weighted or re-calibrated adaptively, which could enhance the informative feature vectors and suppress the noisy and useless ones.

Finally, all the reweighted features were firstly sent to a fully-connected layer and then to a softmax classifier to predict the final vegetable category.

2.6. Details of Network Training

When training started, all the weights of the neural network were initialized through He normalization [61], and biases were all set to be zero. We adopt cross-entropy loss (Equation (10)) [62] as the loss function to train the proposed ARCNN:

$$CE = -\sum_{i} y_i^p \log(y_i) \tag{10}$$

where CE is short for cross-entropy loss, y^p is the predicted result and y is one-hot representation of the ground-truth label. Adam [63] was utilized as the optimization method with a learning rate of 1×10^{-4}. In the training procedure, the model with the lowest validation loss was saved.

We have conducted data augmentation to reduce the impact of limited labeled data in this study. Specifically, all training image patches were flipped and rotated by a random angle from 90°, 180° and 270°. Afterwards, we split 90% of the training datasets for the optimization of parameters. The remaining 10% of the training datasets were utilized as validation sets for performance evaluation during training. After the training process, a testing dataset was adopted to obtain the final classification accuracy.

Furthermore, we used TensorFlow [64] for the construction of our proposed model. The training process was performed on a computer running the Ubuntu 16.04 operation system. The central processing unit (CPU) involved as an Intel core i7-7800 @ 3.5 GHz while the graphics processing unit (GPU) was an NVIDIA GTX TitanX.

2.7. Accuracy Assessment

In this study, we utilized both qualitative and quantitative methods to verify the effectiveness of the proposed ARCNN for vegetable mapping. Specifically, as for the former, we used visual inspection to check for classification errors. While for the latter, a confusion matrix (CM) was obtained from the testing dataset. A series of metrics were calculated from the CM, including overall accuracy (OA), producer's accuracy (PA), user's accuracy (UA) and the Kappa coefficient.

As for the numbers of points chosen for each class, they were actually determined by the area ratio. For instance, the class of Chinese cabbage had the largest area ratio, therefore, the number of training/testing sample points was set to 400, which was the biggest among all the categories. Furthermore, other land cover types, such as spinach and kohlrabi, which only accounted for a small area on the entire study region, had a small number of sample points (only 50).

To further justify the effectiveness of the proposed method, we adopted both ablation analysis and comparison experiments with classic machine learning methods. Specifically, as for the ablation study, we justified the role of both attention-based RNN on vegetable mapping using the following setups. (1) Feature-stacking: concatenating or stacking the spatial features derived from each single-date data for classification; (2) Bi-LSTM: using a bi-directional LSTM for classification; (3) Bi-LSTM-Attention: using the attention-based bi-directional LSTM for classification and (4) standard convolution: using common, non-deformable convolution operations for classification. Besides, ablation study has also been done to justify the impact of deformable convolution when compared with the standard convolution operations.

Meanwhile, classic machine learning methods such as MLC, RF and SVM were also included for comparison experiments. In specific, MLC has long been studied in remote sensing image classification, where the predicted labels are generated based on the maximum likelihood when compared with the training samples. The basic assumption of MLC is that the training samples should follow the normal distribution, which is hard to satisfy in reality, resulting in a limited classification performance. RF belongs to an ensemble of decision trees and the predicted results are determined by the average output of each decision tree [65]. RF has no restrictions on training data distribution and has outperformed MLC in many remote sensing studies. As for SVM, it is based on the Vapnik–Chervonenkis (VC) dimension theory which aims at the minimization of structure risk, resulting in good performance, especially under the situation of limited data [66]. Parameters involved in SVM usually contain kernel type, penalty coefficient, etcetera.

3. Results

3.1. Results of Vegetable Mapping Based on Multi-Temporal Data

Figure 7 shows the vegetable mapping result generated from the proposed ARCNN. It manifests that the distribution of each vegetable category is close to field surveyed when compared with the ground truth (GT) map of Figure 8. The classification errors mainly lie between Chinese cabbage and leaf mustard, potherb, turnip and spinach.

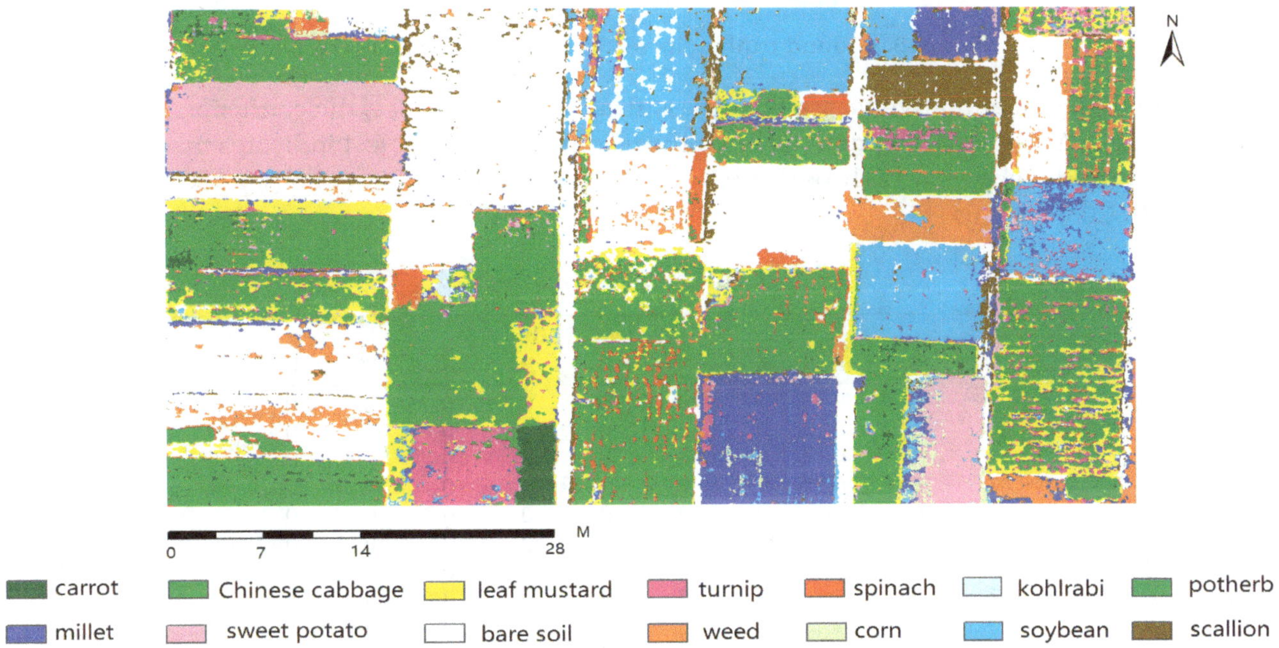

Figure 7. Vegetable map generated from the proposed ARCNN and multi-temporal UAV RGB datasets.

In order to further visually justify the classification results, Figure 8 shows the ground truth map which is manually vectorized from the UAV data. Actually, when compared with the GT map, the classification map of Figure 7 shows a salt and pepper effect. On one side, the classification model in this research belongs to a per-pixel method, which does not consider the boundary information of each land parcel, resulting in a more scattered classification result. On the other hand, the GT map is an ideal description of the spatial extent of every land cover category, neglecting the variations within each land parcel. For instance, several weed regions are missing in the GT map due to small areas. Additionally the bare soil regions in some land parcels have also been neglected. However, in practice, based on the classification map of Figure 7, we could easily generate a land cover map which is more accurate and concise just like Figure 8, which would justify the value of the proposed method in vegetable mapping. To make the boundaries of each land parcel more accurate, in future study we will research semantic segmentation models such as fully convolutional neural networks [19] to improve the visual effect.

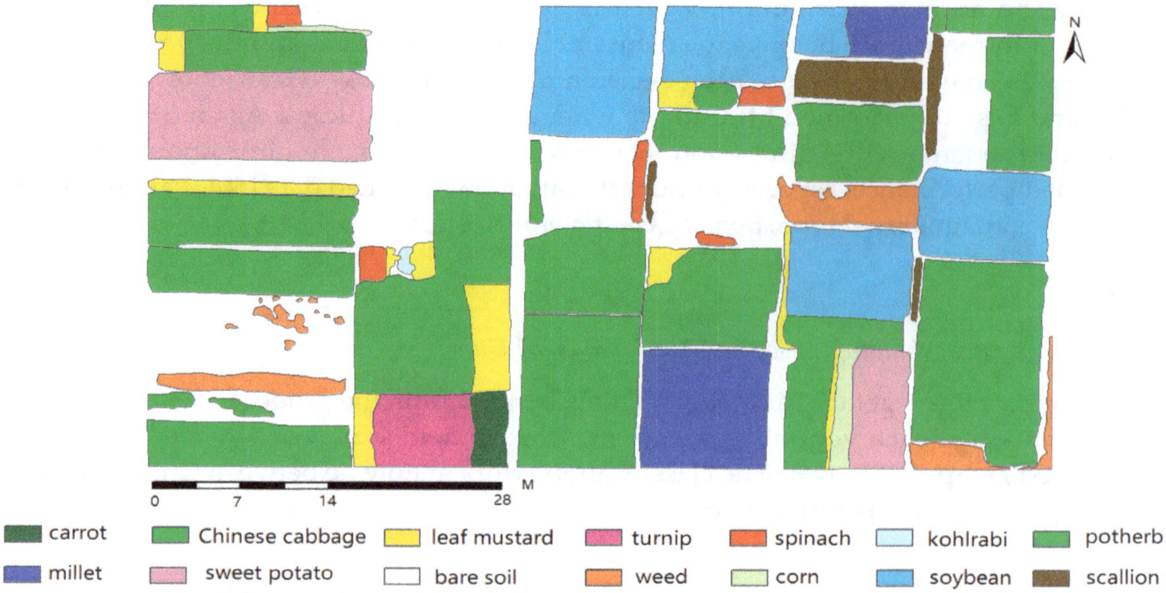

Figure 8. Ground truth map vectorized from the UAV data.

Meanwhile, to quantitatively assess the classification performance, the confusion matrix, Kappa coefficient, OA, PA and UA were derived from the testing dataset. Table 4 indicates that the proposed classification model shows a high performance with both a high OA (92.80%) and a high Kappa coefficient (0.9206).

Table 4. Confusion matrix.

Class	Ground Truth														
	1	2	3	4	5	6	7	8	9	10	11	12	13	14	UA
1	200	9	0	0	0	0	0	0	0	0	0	0	0	0	95.7
2	0	353	21	3	0	0	0	0	0	0	0	0	0	0	93.6
3	0	36	149	12	0	0	0	0	0	0	0	0	0	0	75.6
4	0	0	7	182	0	0	0	16	0	0	0	0	4	0	87.1
5	0	0	0	0	50	0	0	0	0	0	0	0	0	0	100
6	0	0	0	0	0	50	0	0	0	0	0	0	0	0	100
7	0	0	16	0	0	0	100	0	0	0	0	0	0	0	86.2
8	0	0	2	0	0	0	0	184	0	0	0	8	0	0	94.8
9	0	0	0	0	0	0	0	0	100	0	0	0	0	0	100
10	0	0	0	0	0	0	0	0	0	200	0	2	0	9	94.8
11	0	0	0	0	0	0	0	0	0	0	200	0	0	0	200
12	0	0	0	1	0	0	0	0	0	0	0	33	0	0	97.1
13	0	2	5	2	0	0	0	0	0	0	0	7	196	0	92.5
14	0	0	0	0	0	0	0	0	0	0	0	0	0	91	100
PA	100	88.3	74.5	91.0	100	100	100	92.0	100	100	100	66.0	98.0	91.0	
OA		92.80		Kappa						0.9206					

1: carrot; 2: Chinese cabbage; 3: leaf mustard; 4: turnip; 5: spinach; 6: kohlrabi; 7: potherb; 8: millet; 9: weed; 10: bare soil; 11: sweet potato; 12: corn; 13: soybean; 14: scallion; PA: producer's accuracy; UA: user's accuracy and OA: overall accuracy.

Table 4 indicates that the omissions and commissions mainly exist among leaf mustard and Chinese cabbage, potherb and turnip. For instance, several leaf mustard pixels were misclassified as Chinese cabbage and vice versa. This was understandable, since both color and shape of these leafy green vegetables (Chinese cabbage, leaf mustard, potherb, etc.) are very similar, especially at the early growth stage. Meanwhile, the RGB image has the drawback of a low spectral resolution, making it hard to differentiate these vegetable categories when using only color and shape information.

In addition, only a few mistakes occurred among the other categories, which verifies the effectiveness of the proposed vegetation mapping method.

3.2. Results of Vegetable Mapping Based on Mono-Temporal UAV Data

Figure 9 shows the vegetable map generated from both mono- and multi-temporal classification.

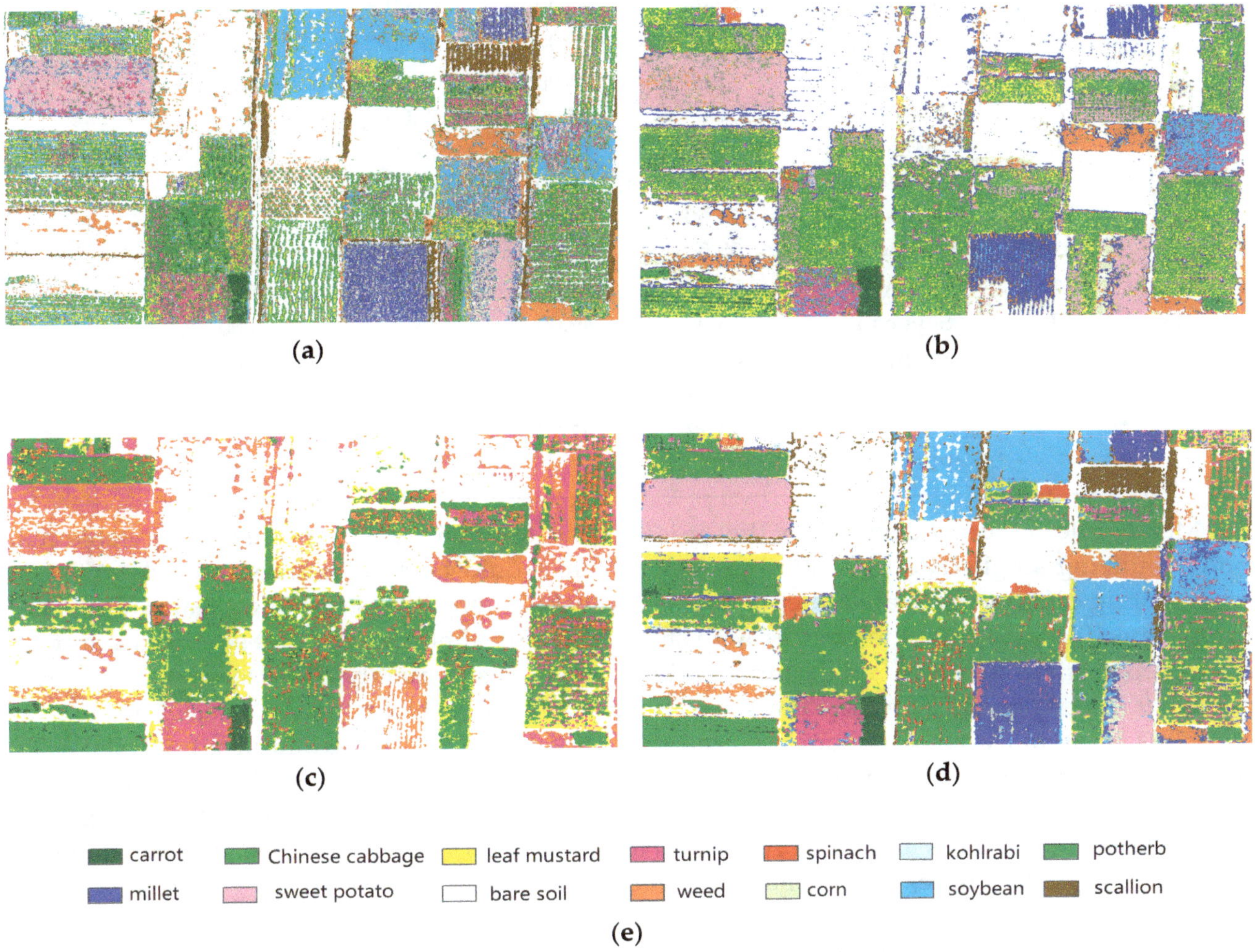

Figure 9. Vegetable map generated from **(a)** T1/2019-09; **(b)** T2/2019-10; **(c)** T3/2019-11 and **(d)** multi-temporal datasets using the proposed method. **(e)** Legend.

As mentioned above, one hypothesis of this study is that the inclusion of multi-temporal UAV data could provide additional phenological information, which would enhance the inter-class separability to cover the shortage of low spectral resolution caused by off-the-shelf digital cameras. Therefore, in this section, a contrast experiment was conducted to compare the performance between multi-temporal and mono-temporal classification. It should be noted that when using single-date UAV data for classification, the spatial-temporal feature fusion module (i.e., Bi-LSTM-Attention) would be non-functional during the training and testing procedure.

Figure 9 indicates that the incorporation of multi-temporal UAV images could significantly improve the classification performance when compared with mono-temporal data, which shows fewer obvious errors from visual inspection. This is in accordance with quantitative assessment (Table 5). It indicates that the overall classification accuracy improved by 19.76%–28.13%, with an average increase of 24.49%, after the inclusion of multi-temporal data.

Table 5. Class-level accuracy for mono- and multi-temporal classification.

No.	Class Name	T1 (%)	T2 (%)	T3 (%)	Proposed (%)
1	Carrot	89.00	93.00	84.00	100
2	Chinese cabbage	46.50	68.50	77.25	88.25
3	Leaf mustard	29.00	39.00	50.00	74.50
4	Turnip	59.00	43.50	75.00	91.00
5	Spinach	–	38.00	18.00	100
6	Kohlrabi	22.00	34.00	–	100
7	Potherb	–	25.00	57.00	100
8	Millet	73.50	76.50	–	92.00
9	Weed	75.00	95.00	85.00	100
10	Bare soil	92.50	94.50	79.00	100
11	Sweet potato	55.50	84.00	–	100
12	Corn	78.00	76.00	–	66.00
13	Soybean	75.00	49.00	–	98.00
14	Scallion	95.00	–	–	91.00
	OA (%)	64.67	67.22	73.04	92.80
	Kappa	0.6067	0.6314	0.6744	0.9206

Meanwhile, Figure 9 also shows that it is difficult to obtain a high-precision vegetable map if only utilizing single-date UAV RGB images. There would be a large amount of classification errors among different vegetable categories, especially between Chinese cabbage, leaf mustard and turnip. Specifically, during the early growth stage (T1), large amounts of Chinese cabbage and leaf mustard pixels are misclassified as turnip (Figure 9a). This is mainly because these leafy green vegetables share very similar appearances (e.g., color, shape and texture patterns), which leads to a low inter-class separability hence a poor classification accuracy (64.67%). In the middle growth stage (T2), the classification accuracy of Chinese cabbage has been greatly improved due to its shape change due to the growth process. However, it still remains difficult to separate leaf mustard from Chinese cabbage (Figure 9b). When it comes to the ripe stage (T3), the leaf mustard could finally be differentiated from Chinese cabbage (Figure 9c). This is mainly because the Chinese cabbage shows a round head in the ripe stage (Table 1), which is greatly different to leaf mustard.

Table 5 shows the class-level accuracy for each vegetable category and other land cover types. It indicates that there is a significant accuracy gap between mono- and multi-temporal classification when using UAV RGB imagery. This is understandable because if using single-date UAV data alone, the similarity of color and texture patterns between various vegetables would yield a low inter-class separability. This is even more so at the early growth stage (T1), when vegetable seedlings share very similar appearances, resulting in the lowest classification accuracy with an OA of 64.67%. However, with the inclusion of multi-temporal UAV images, the additional phenological information would increase the separability among various vegetables, which could boost the final classification performance.

3.3. *Results of Ablation Analysis*

To justify the effectiveness of the proposed ARCNN model, a series of ablation experiments are conducted and the results are shown as follows.

3.3.1. Results of Different Fusion Methods

In this section, we consider the following methods for the fusion of spatial-temporal features: (1) feature-stacking; (2) Bi-LSTM and (3) Bi-LSTM-Attention. The description of these methods is in Section 2.7. The experimental results are shown in Table 6.

Table 6. Comparison between different spatial-temporal feature fusion methods.

Method	OA (%)	Kappa
Feature-stacking	89.56	0.8849
Bi-LSTM	90.93	0.8999
Bi-LSTM-Attention	92.80	0.9206

Table 6 indicates that the Bi-LSTM-Attention module used in this study outperforms both feature-stacking and Bi-LSTM, which increases the OA by 3.24% and 1.87%, respectively. The role of Bi-LSTM-Attention will be discussed in Section 4.1.

3.3.2. Results of Standard Convolution

In this section, we replaced all the deformable convolution operations by standard convolution units in the proposed network to justify the role of deformable convolution in vegetable mapping. Table 7 shows the comparison results.

Table 7. Comparison of standard and deformable convolution.

Method	OA (%)	Kappa
Standard convolution	91.96	0.9111
Deformable convolution	92.80	0.9206

Table 7 implies that the inclusion of deformable convolution could improve the vegetable mapping accuracy. The detailed discussion will be presented in Section 4.2.

3.4. Results of Comparison with Other Methods

To further justify the effectiveness of the proposed classification model, we compared it with several machine learning classifiers and other deep learning models. As for the former, we conducted comparison experiments using MLC, RF and SVM based on the same training and testing datasets. We used grid search for the parameterization of both RF and SVM. It turns out that an RF with 300 decision trees and a max depth of 15, and a SVM with radial basis kernel, a gamma [66] of 0.001 and a penalty coefficient (C) [66] of 100 has the best performance, respectively.

Table 8 shows the comparison results between the proposed model and other classical machine learning methods. It indicates that the deep learning based model has an advantage over the classical methods. A detailed discussion of this will follow in Section 4.4.

Table 8. Comparison with classical machine learning classifiers.

Method	OA (%)	Kappa
MLC	46.04	0.4095
RF	63.96	0.6023
SVM	84.76	0.8317
Proposed	92.80	0.9206

In addition, we have conducted comparison experiments with several previous studies, mainly including Ndikumana et al. (stacked LSTMs) [46], Mou et al. (CNN-RNN) [47] and Ji et al. (3D-CNN) [43]. Because the dimension of input and output of these models are different from ours, we have made necessary changes accordingly when reproducing these DL models. The experimental results are shown as follows.

Table 9 indicates that the proposed ARCNN in the research has a better performance when compared with several previous deep learning models. The OA is boosted by an increase of 2.53%

to 8.36% while the Kappa has risen by 2.78% to 9.23%. The detailed discussion will be presented in Section 4.4.

Table 9. Comparison with other deep learning methods.

Method	OA (%)	Kappa
Ndikumana et al.	84.44	0.8283
Mou et al.	90.27	0.8928
Ji et al.	90.18	0.8915
Proposed	92.80	0.9206

4. Discussion

4.1. Impact of Attention-Based RNN on Vegetable Mapping

In this section, we will discuss the impact of attention mechanisms in the RNN for vegetable mapping. Specifically, according to the ablation study results of Section 3.3.1, the comparisons were made between different methods for spatial-temporal feature fusion, including feature-stacking, Bi-LSTM and Bi-LSTM-Attention. Results show that the feature-stacking yields the lowest accuracy with an OA of 89.56% and a Kappa of 0.8849. The reason is that feature-stacking just concatenates all the multi-temporal features without considering the relationship and temporal dependencies across the sequential UAV data. Meanwhile, since Bi-LSTM could understand the dependencies of the sequential features in a bi-directional way, therefore, it shows a better performance than the simple feature-stacking method with an OA improvement of 1.37%. In this study, we added an attention layer on the top of Bi-LSTM to further improve its performance. The attention based Bi-LSTM could enhance the important features while suppressing the less informative ones, outperforming both feature-stacking and Bi-LSTM with an OA increase of 3.24% and 1.87%, respectively, which verifies its effectiveness in spatial-temporal feature fusion.

4.2. Impact of Deformable Convolution on Vegetable Mapping

Another hypothesis of this study is that the scale and shape variations could be accounted for using the deformable convolution. According to Table 7, it indicates that the inclusion of deformable convolution could boost the classification performance. The OA has been improved from 91.96% to 92.80% with a rise of about 1%, justifying the role of deformable convolution. The reason for the lower accuracy of standard convolution is that it has a fixed kernel shape, which lacks the capability to model the geometric transformations of complex landscapes. On the other hand, deformable convolution has a flexible receptive field, which could be adaptive to the variability of shape and scale of remotely sensed imagery [44]. Therefore, the deformable convolution shows a better performance, especially under the complex and fragmented agricultural landscape in this study.

4.3. Impact of Multi-Temporal UAV Data on Vegetable Mapping

In addition, we will further discuss the role of multi-temporal UAV data on vegetable mapping. In fact, one of the main objectives of this study is to explore whether the incorporation of multi-temporal UAV RGB images could improve the vegetable classification accuracy. The initial motivation lies in the fact that RGB images acquired by UAV have a low spectral resolution, which would make it hard for the fine-grained classification of various vegetable categories. Therefore, in this study we have selected images from three important periods, i.e., the sowing period, the growing period and the harvesting period, to capture the phenological characteristics of different vegetables. Although the number of three dates may seem limited, all of them fall into the distinct periods of the vegetable and crop growth stage, which could still provide additional and useful time-series features for classification.

Meanwhile, in previous studies, images from only three dates have been studied for remote sensing image classification and they outperform the single-date dataset. For instance, Palchowdhuri et al. used three images from both multi-temporal Sentinel-2 and WordView-3 imagery for crop classification in Coalville in the United Kingdom and achieved an accuracy of 91% [67]. Similar findings were also reported in Yang et al., where three images from summer, autumn and winter were integrated for coastal land cover mapping [68]. In our previous study [15], we also utilized only three images during the whole crop growing season for the cropland classification in the Yellow River Delta of China, which yielded an average accuracy of 89%, justifying the role of three dates for classification. In future research, images from a longer temporal range could be included to further improve the classification performance.

4.4. Comparison with Other Methods

In this section, we focused on the detailed discussion between the proposed ARCNN and other classical machine learning methods and several previous deep learning methods. Specifically, Table 8 indicates that our proposed method outperforms machine learning methods such as MLC, RF and SVM with an OA increase of 27.29%, 21.58% and 8.64%, respectively. The results are in accordance with [46] and our previous studies [39,44]. The reason could be that classical machine learning methods lack the ability to capture the high-level representative features when compared to deep learning models, leading to a performance gap in vegetable mapping.

In addition, there is a need to compare the proposed ARCNN with other methods for multi-temporal UAV image classification. Recent researches such as van Iersel et al. [14] and Michez et al. [12], they both utilized object-oriented image analysis (OBIA) and random forest for plant classification from multi-date UAV data. Manually designed features such as band ratio and vegetation indices were used for classification. Compared with their studies, we replace the manually designed features with high-level and discriminative features that are automatically learned from deep neural networks, (i.e., CNN and RNN), which could enhance the feature's representativeness. To the best of our knowledge, this study is the first case to introduce deep learning methods in multi-temporal UAV image classification. Therefore, the proposed method in this research might provide useful reference for future studies.

Meanwhile, it is also necessary to compare the proposed ARCNN with other deep learning models for remote sensing image classification. Early studies mainly utilized LSTM for multi-temporal classification. One representative research is Ndikumana et al., where five LSTMs were stacked for the classification using multi-temporal SAR Sentinel-1 data [46]. The input data in Ndikumana's study are a single pixel with a time curve, which neglects the rich, contextual relationship hidden in the spatial features, showing a relatively lower accuracy (84.44%). Different from Ndikumana et al. [46], we have added a CNN in front of LSTM to enrich the representative spatial feature extraction.

Mou et al. also cascaded a CNN and RNN for change detection from two optical remote sensing images [47]. Compared with Mou et al. [47], our model makes two significant improvements. Firstly, from the perspective of CNN, we incorporate the multi-scale deformable convolutions, which could aggregate multi-level contextual features. Secondly, we used the attention mechanism with a bi-directional LSTM to further enhance the modeling of sequential signals in multi-temporal remote sensing data. All the above modifications have improved the classification from 90.18% by Mou et al. to 92.80%.

Besides, Ji et al. adopted a 3D-CNN to extract spatial-temporal features for crop type mapping from multi-temporal satellite imagery [43]. Compared with Ji et al. [43], our method has also gained a more accurate result. The reason lies in that 3D CNN cannot explicitly establish the relationship between the sequential signals, which has flaws in the generation of spatial-temporal feature fusion and integration. Furthermore, our Bi-LSTM-Attention module is more straightforward in mining the relationship across multi-temporal data than a 3D-CNN.

5. Conclusions

This study proposed an attention-based recurrent convolutional neural network (ARCNN) for accurate vegetable mapping based on multi-temporal unmanned aerial vehicle (UAV) red-green-blue

(RGB) data. The proposed ARCNN first leverages a multi-scale deformable CNN to learn and extract the rich spatial features from each mono-temporal UAV image, which aims to account for the shape and scale variations under complex and fragmented agricultural landscapes. Afterwards, an attention-based bi-directional long-short term memory (LSTM) is introduced to model the relationship between the sequential features, from which spatial and temporal features are fused and aggregated. Finally, the fused features are fed to a fully connected layer and a softmax classifier to determine the vegetable category.

Experimental results showed that the proposed ARCNN yields a high classification performance with an overall accuracy (OA) of 92.08% and a Kappa coefficient of 0.9206. When compared with mono-temporal classification, the incorporation of multi-temporal UAV data could boost the OA significantly by an average increase of 24.49%, which verifies the hypothesis that multi-temporal UAV observations could enhance the inter-class separability and thus reduce the drawback of low spectral resolution of off-the-shelf digital cameras. The Bi-LSTM-Attention module outperforms other fusion methods such as feature-staking and bi-directional LSTM with an OA increase of 3.24% and 1.87%, respectively, justifying its effectiveness in modeling the dependency across the sequential features. Meanwhile, the introduction of deformable convolution could also improve the OA by about 1% when compared with standard convolution. In addition, the proposed ARCNN also shows a higher performance than other classical machine learning classifiers such as maximum likelihood classifier, random forest and support vector machine, and several previous deep learning methods for remote sensing classification.

This study demonstrates that the proposed ARCNN could yield an accurate vegetable mapping result from multi-temporal UAV RGB data. The drawback of low spectral resolution of RGB images could be compensated by introducing additional phenological information and robust deep learning models. Although images from only three dates were included, a good classification result could still be achieved providing all three dates fall into the distinct growing periods of vegetables. Finally, the proposed model could be viewed as a general framework for multi-temporal remote sensing image classification. As for future work, more study cases should be considered to justify the effectiveness of the proposed method. Additionally, semantic segmentation models should be incorporated to get a more accurate vegetable map.

Author Contributions: Methodology, Q.F.; validation, Q.F. and J.Y.; data curation, Y.L. and C.O.; writing—original draft preparation, Q.F.; writing—review and editing, J.Y., D.Z., B.N., J.L. and B.L. All authors have read and agreed to the published version of the manuscript.

Acknowledgments: Special thanks to the anonymous reviewers and editors for their very useful comments and suggestions to help improve the quality of this paper. The authors give thanks to Zhe Liu from the China Agricultural University for assisting in the field work. We would also like to thank Beijing IRIS Remote Sensing Technology Limited, Inc. for their help in preprocessing UAV raw data.

References

1. Torres-Sánchez, T.; Peña, J.M.; de Castro, A.I.; López-Granados, F. Multi-temporal mapping of the vegetation fraction in early-season wheat fields using images from UAV. *Comput. Electron. Agric.* **2014**, *103*, 104–113. [CrossRef]
2. Wikantika, K.; Uchida, S.; Yamamoto, S. Mapping vegetable area with spectral mixture analysis of the Landsat-ETM. In Proceedings of the IEEE International Geoscience and Remote Sensing Symposium, Toronto, ON, Canada, 24–28 June 2002; pp. 1965–1967.
3. Belgiu, M.; Csillik, O. Sentinel-2 cropland mapping using pixel-based and object-based time-weighted dynamic time warping analysis. *Remote Sens. Environ.* **2018**, *204*, 509–523. [CrossRef]

4. Rupasinghe, P.A.; Milas, A.S.; Arend, K.; Simonson, M.A.; Mayer, C.; Mackey, S. Classification of shoreline vegetation in the Western Basin of Lake Erie using airborne hyperspectral imager HSI2, Pleiades and UAV data. *Int. J. Remote Sens.* **2019**, *40*, 3008–3028. [CrossRef]
5. Wan, S.; Chang, S.H. Crop classification with WorldView-2 imagery using Support Vector Machine comparing texture analysis approaches and grey relational analysis in Jianan Plain, Taiwan. *Int. J. Remote Sens.* **2019**, *40*, 8076–8092. [CrossRef]
6. Asgarian, A.; Soffianian, A.; Pourmanafi, S. Crop type mapping in a highly fragmented and heterogeneous agricultural landscape: A case of central Iran using multi-temporal Landsat 8 imagery. *Comput. Electron. Agric.* **2016**, *127*, 531–540. [CrossRef]
7. Feng, Q.; Liu, J.; Gong, J. UAV remote sensing for urban vegetation mapping using random forest and texture analysis. *Remote Sens.* **2015**, *7*, 1074–1094. [CrossRef]
8. Feng, Q.; Liu, J.; Gong, J. Urban flood mapping based on unmanned aerial vehicle remote sensing and random forest classifier—A case of Yuyao, China. *Water* **2015**, *7*, 1437–1455. [CrossRef]
9. Dai, Y.; Gong, J.; Li, Y.; Feng, Q. Building segmentation and outline extraction from UAV image-derived point clouds by a line growing algorithm. *Int. J. Digit. Earth* **2017**, *10*, 1077–1097. [CrossRef]
10. Böhler, J.E.; Schaepman, M.E.; Kneubühler, M. Optimal timing assessment for crop separation using multispectral unmanned aerial vehicle (UAV) data and textural features. *Remote Sens.* **2019**, *11*, 1780. [CrossRef]
11. Pádua, L.; Marques, P.; Hruška, J.; Adão, T.; Peres, E.; Morais, R.; Sousa, J.J. Multi-temporal vineyard monitoring through UAV-based RGB imagery. *Remote Sens.* **2018**, *10*, 1907. [CrossRef]
12. Michez, A.; Piégay, H.; Lisein, J.; Claessens, H.; Lejeune, P. Classification of riparian forest species and health condition using multi-temporal and hyperspatial imagery from unmanned aerial system. *Environ. Monit. Assess.* **2016**, *188*, 146. [CrossRef] [PubMed]
13. Moeckel, T.; Dayananda, S.; Nidamanuri, R.R.; Nautiyal, S.; Hanumaiah, N.; Buerkert, A.; Wachendorf, M. Estimation of vegetable crop parameter by multi-temporal UAV-borne images. *Remote Sens.* **2018**, *10*, 805. [CrossRef]
14. Van Iersel, W.; Straatsma, M.; Middelkoop, H.; Addink, E. Multitemporal Classification of river floodplain vegetation using time series of UAV images. *Remote Sens.* **2018**, *10*, 1144. [CrossRef]
15. Feng, Q.; Gong, J.; Liu, J.; Li, Y. Monitoring cropland dynamics of the yellow river delta based on multi-temporal Landsat imagery over 1986 to 2015. *Sustainability* **2015**, *7*, 14834–14858. [CrossRef]
16. Chen, L.; Yang, W.; Xu, K.; Xu, T. Evaluation of local features for scene classification using VHR satellite images. In Proceedings of the 2011 Joint Urban Remote Sensing Event, Munich, Germany, 11–13 April 2011; pp. 385–388.
17. Zhu, Q.; Zhong, Y.; Zhao, B.; Xia, G.; Zhang, L. Bag-of-visual-words scene classifier with local and global features for high spatial resolution remote sensing imagery. *IEEE Geosci. Remote Sens. Lett.* **2016**, *13*, 747–751. [CrossRef]
18. LeCun, Y.; Bengio, Y.; Hinton, G. Deep learning. *Nature* **2015**, *521*, 436–444. [CrossRef]
19. Shelhamer, E.; Long, J.; Darrell, T. Fully convolutional networks for semantic segmentation. *IEEE Trans. Pattern Anal. Mach. Intell.* **2017**, *39*, 640–651. [CrossRef]
20. LaLonde, R.; Bagci, U. Capsules for Object Segmentation. *arXiv* **2018**, arXiv:1804.04241. Available online: https://arxiv.org/abs/1804.04241.pdf (accessed on 17 April 2020).
21. Lin, T.Y.; Goyal, P.; Girshick, R.; He, K.; Dollar, P. Focal loss for dense object detection. In Proceedings of the IEEE International Conference on Computer Vision (ICCV), Venice, Italy, 22–29 October 2017; pp. 2999–3007.
22. Krizhevsky, A.; Sutskever, I.; Hinton, G.E. Imagenet classification with deep convolutional neural networks. *Proc. Adv. Neural Inf. Process. Syst.* **2012**, 1097–1105. [CrossRef]
23. He, K.; Zhang, X.; Ren, S.; Sun, J. Deep residual learning for image recognition. In Proceedings of the IEEE Conference on Computer Vision and Pattern Recognition, Las Vegas, NV, USA, 27–30 June 2016; pp. 770–778.
24. Hu, J.; Shen, L.; Sun, G. Squeeze-and-Excitation Networks. *arXiv* **2017**, arXiv:1709.01507. Available online: https://arxiv.org/pdf/1709.01507.pdf (accessed on 17 April 2020).
25. Zhu, X.X.; Tuia, D.; Mou, L.; Xia, G.; Zhang, L.; Xu, F.; Fraundorfer, F. Deep learning in remote sensing: A comprehensive review and list of resources. *IEEE Geosci. Remote Sens. Mag.* **2017**, *5*, 8–36. [CrossRef]
26. Zhang, L.; Zhang, L.; Du, B. Deep learning for remote sensing data: A technical tutorial on the state of the art. *IEEE Geosci. Remote Sens. Mag.* **2016**, *4*, 22–40. [CrossRef]

27. Kellenberger, B.; Marcos, D.; Tuia, D. Detecting mammals in UAV images: Best practices to address a substantially imbalanced dataset with deep learning. *Remote Sens. Environ.* **2018**, *216*, 139–153. [CrossRef]
28. Carrio, A.; Sampedro, C.; Rodriguez-Ramos, A.; Campoy, P. A review of deep learning methods and applications for unmanned aerial vehicles. *J. Sens.* **2017**, 3296874. [CrossRef]
29. Kamilaris, A.; Prenafeta-Boldú, F.X. Deep learning in agriculture: A survey. *Comput. Electron. Agric.* **2018**, *147*, 70–90. [CrossRef]
30. Chen, Y.; Fan, R.; Bilal, M.; Yang, X.; Wang, J.; Li, W. Multilevel cloud detection for high-resolution remote sensing imagery using multiple convolutional neural networks. *ISPRS Int. J. Geo. Inf.* **2018**, *7*, 181. [CrossRef]
31. Alshehhi, R.; Marpu, P.R.; Woon, W.L.; Mura, M.D. Simultaneous extraction of roads and buildings in remote sensing imagery with convolutional neural networks. *ISPRS J. Photogramm. Remote Sens.* **2017**, *130*, 139–149. [CrossRef]
32. Xu, Y.; Wu, L.; Xie, Z.; Chen, Z. Building Extraction in Very High Resolution Remote Sensing Imagery Using Deep Learning and Guided Filters. *Remote Sens.* **2018**, *10*, 144. [CrossRef]
33. Li, X.; Yao, X.; Fang, Y. Building-A-Nets: Robust Building Extraction From High-Resolution Remote Sensing Images With Adversarial Networks. *IEEE J. Sel. Top. Appl. Earth Obs. Remote Sens.* **2018**, *11*, 3680–3687. [CrossRef]
34. Deng, Z.; Sun, H.; Zhou, S.; Zhao, J.; Zou, H. Toward Fast and Accurate Vehicle Detection in Aerial Images Using Coupled Region-Based Convolutional Neural Networks. *IEEE J. Sel. Top. Appl. Earth Obs. Remote Sens.* **2017**, *10*, 3652–3664. [CrossRef]
35. Ammour, N.; Alhichri, H.; Bazi, Y.; Benjdira, B.; Alajlan, N.; Zuair, M. Deep Learning Approach for Car Detection in UAV Imagery. *Remote Sens.* **2017**, *9*, 312. [CrossRef]
36. Han, W.; Feng, R.; Wang, L.; Cheng, Y. A semi-supervised generative framework with deep learning features for high-resolution remote sensing image scene classification. *ISPRS J. Photogramm. Remote Sens.* **2018**, *145*, 23–43. [CrossRef]
37. Wang, Q.; Liu, S.; Chanussot, J.; Li, X. Scene classification with recurrent attention of VHR remote sensing images. *IEEE Trans. Geosci. Remote Sens.* **2019**, *57*, 1155–1167. [CrossRef]
38. Rußwurm, M.; Körner, M. Multi-temporal land cover classification with sequential recurrent encoders. *ISPRS Int. J. Geo. Inf.* **2018**, *7*, 129. [CrossRef]
39. Feng, Q.; Zhu, D.; Yang, J.; Li, B. Multisource hyperspectral and LiDAR data fusion for urban land-use mapping based on a modified two-branch convolutional neural network. *ISPRS Int. J. Geo. Inf.* **2019**, *8*, 28. [CrossRef]
40. Huang, B.; Zhao, B.; Song, Y. Urban land-use mapping using a deep convolutional neural network with high spatial resolution multispectral remote sensing imagery. *Remote Sens. Environ.* **2018**, *214*, 73–86. [CrossRef]
41. Kussul, N.; Lavreniuk, M.; Skakun, S.; Shelestov, A. Deep learning classification of land cover and crop types using remote sensing data. *IEEE Geosci. Remote Sens. Lett.* **2017**, *14*, 778–782. [CrossRef]
42. Rezaee, M.; Mahdianpari, M.; Zhang, Y.; Salehi, B. Deep convolutional neural network for complex wetland classification using optical remote sensing imagery. *IEEE J. Sel. Top. Appl. Earth Obs. Remote Sens.* **2018**, *11*, 3030–3039. [CrossRef]
43. Ji, S.; Zhang, C.; Xu, A.; Shi, Y.; Duan, Y. 3D convolutional neural networks for crop classification with multi-temporal remote sensing images. *Remote Sens.* **2018**, *10*, 75. [CrossRef]
44. Feng, Q.; Yang, J.; Zhu, D.; Liu, J.; Guo, H.; Bayartungalag, B.; Li, B. Integrating multitemporal sentinel-1/2 data for coastal land cover classification using a multibranch convolutional neural network: A case of the Yellow River Delta. *Remote Sens.* **2019**, *11*, 1006. [CrossRef]
45. Hochreiter, S.; Schmidhuber, J. Long short-term memory. *Neural Comput.* **1997**, *9*, 1735–1780. [CrossRef] [PubMed]
46. Ndikumana, E.; Ndikumana, E.; Ho Tong Minh, D.; Baghdadi, N.; Courault, D.; Hossard, L. Deep recurrent neural network for agricultural classification using multitemporal SAR Sentinel-1 for Camargue, France. *Remote Sens.* **2018**, *10*, 1217. [CrossRef]
47. Mou, L.; Bruzzone, L.; Zhu, X.X. Learning spectral-spatial-temporal features via a recurrent convolutional neural network for change detection in multispectral imagery. *IEEE Trans. Geosci. Remote Sens.* **2019**, *57*, 924–935. [CrossRef]
48. Song, A.; Choi, J.; Han, Y.; Kim, Y. Change detection in hyperspectral images using recurrent 3D fully convolutional networks. *Remote Sens.* **2018**, *10*, 1827. [CrossRef]

49. Liu, Q.; Zhou, F.; Hang, R.; Yuan, X. Bidirectional-convolutional LSTM based spectral-spatial feature learning for hyperspectral image classification. *Remote Sens.* **2017**, *9*, 1330.
50. Mou, L.; Ghamisi, P.; Zhu, X.X. Deep recurrent neural networks for hyperspectral image classification. *IEEE Trans. Geosci. Remote Sens.* **2017**, *55*, 3639–3655. [CrossRef]
51. DJI-Inspire 2. Available online: https://www.dji.com/cn/inspire-2/ (accessed on 17 March 2020).
52. Pix4D. Available online: http://pix4d.com/ (accessed on 17 March 2020).
53. ENVI. Available online: http://www.enviidl.com/ (accessed on 17 March 2020).
54. Dai, J.; Qi, H.; Xiong, Y.; Li, Y.; Zhang, G.; Hu, H.; Wei, Y. Deformable convolutional networks. *arXiv* **2017**, arXiv:1703.06211. Available online: https://arxiv.org/abs/1703.06211 (accessed on 17 March 2020).
55. Jin, Q.; Meng, Z.; Pham, T.D.; Chen, Q.; Wei, L.; Su, R. DUNet: A Deformable Network for Retinal Vessel Segmentation. *arXiv* **2018**, arXiv:1811.01206. Available online: https://arxiv.org/pdf/1811.01206.pdf (accessed on 17 April 2020). [CrossRef]
56. Pan, D.; Yuan, J.; Li, L.; Sheng, D. Deep neural network-based classification model for Sentiment Analysis. *arXiv* **2019**, arXiv:1907.02046. Available online: https://arxiv.org/abs/1907.02046 (accessed on 17 April 2019).
57. Melamud, O.; Goldberger, J.; Dagan, I. Context2vec: Learning generic context embedding with bidirectional LSTM. In Proceedings of the 20th SIGNLL Conference on Computational Natural Language Learning (CoNLL), Berlin, Germany, 11–12 August 2016; pp. 51–61.
58. Cui, W.; Wang, F.; He, X.; Zhang, D.; Xu, X.; Yao, M.; Wang, Z.; Huang, J. Multi-scale semantic segmentation and spatial relationship recognition of remote sensing images based on an attention model. *Remote Sens.* **2019**, *11*, 1044. [CrossRef]
59. Xu, R.; Tao, Y.; Lu, Z.; Zhong, Y. Attention-mechanism-containing neural networks for high-resolution remote sensing image classification. *Remote Sens.* **2018**, *10*, 1602. [CrossRef]
60. Zhao, B.; Wu, X.; Feng, J.; Peng, Q.; Yan, S. Diversified visual attention networks for fine-grained object classification. *IEEE Trans. Multimedia* **2017**, *19*, 1245–1256. [CrossRef]
61. He, K.; Zhang, X.; Ren, S.; Sun, J. Delving deep into rectifiers: Surpassing human-level performance on ImageNet classification. *arXiv* **2015**, arXiv:1502.01852. Available online: https://arxiv.org/pdf/1502.01852.pdf (accessed on 17 March 2020).
62. Cox, D. The Regression Analysis of Binary Sequences. *J. Royal Stat. Soc. Ser. B* **1958**, *20*, 215–242. [CrossRef]
63. Kingma, D.P.; Ba, J. Adam: A Method for Stochastic Optimization. *arXiv* **2014**, arXiv:1412.6980. Available online: https://arxiv.org/abs/1412.6980 (accessed on 17 March 2020).
64. TensorFlow. Available online: https://tensorflow.google.cn/ (accessed on 17 March 2020).
65. Breiman, L. Random forests. *Mach. Learn.* **2001**, *45*, 5–32. [CrossRef]
66. Chapelle, O.; Vapnik, V.; Bousquet, O.; Mukherjee, S. Choosing multiple parameters for support vector machines. *Mach. Learn.* **2002**, *46*, 131–159. [CrossRef]
67. Palchowdhuri, Y.; Valcarce-Diñeiro, R.; King, P.; Sanabria-Soto, M. Classification of multi-temporal spectral indices for crop type mapping: A case study in Coalville, UK. *J. Agric. Sci.* **2018**, *156*, 24–36. [CrossRef]
68. Yang, X.; Chen, L.; Li, Y.; Xi, W.; Chen, L. Rule-based land use/land cover classification in coastal areas using seasonal remote sensing imagery: A case study from Lianyungang City, China. *Environ. Monit. Assess.* **2015**, *187*, 449. [CrossRef]

12

Multi-Temporal Agricultural Land-Cover Mapping using Single-Year and Multi-Year Models based on Landsat Imagery and IACS Data

Isaac Kyere *, Thomas Astor, Rüdiger Graß and Michael Wachendorf

Grassland Science and Renewable Plant Resources, Universität Kassel, Steinstraße 19, D-37213 Witzenhausen, Germany; thastor@uni-kassel.de (T.A.); grass@wiz.uni-kassel.de (R.G.); mwach@uni-kassel.de (M.W.)

* Correspondence: isaac.kyere@uni-kassel.de

Abstract: The spatial distribution and location of crops are necessary information for agricultural planning. The free availability of optical satellites such as Landsat offers an opportunity to obtain this key information. Crop type mapping using satellite data is challenged by its reliance on ground truth data. The Integrated Administration and Control System (IACS) data, submitted by farmers in Europe for subsidy payments, provide a solution to the issue of periodic field data collection. The present study tested the performance of the IACS data in the development of a generalized predictive crop type model, which is independent of the calibration year. Using the IACS polygons as objects, the mean spectral information based on four different vegetation indices and six Landsat bands were extracted for each crop type and used as predictors in a random forest model. Two modelling methods called single-year (SY) and multiple-year (MY) calibration were tested to find out their performance in the prediction of grassland, maize, summer, and winter crops. The independent validation of SY and MY resulted in a mean overall accuracy of 71.5% and 77.3%, respectively. The field-based approach of calibration used in this study dealt with the 'salt and pepper' effects of the pixel-based approach.

Keywords: agricultural land-cover; multi-spectral; generalized model; machine learning; crop type mapping; Integrated Administration and Control System; remote sensing

---

1. Introduction

The increasing world population coupled with the high demand for agricultural resources [1] require reliable data on agricultural lands for decision making and planning towards the future [2]. The knowledge on available croplands is fundamental to food security [3], sustainable cropping [4] and the maximization of food production [5]. Information about the spatial distribution of crops and the spatial extent of croplands are also essential to ascertain the impact of any human activity on croplands [6].

Reliable and accurate information about agricultural lands requires an efficient and precise approach, which remote sensing (RS) can offer [7–9]. RS-based methods can be used to obtain various crop information, such as crop type [10], biomass [11], or yield [12]. The advent of satellite-based optical RS has revolutionized large-scale cropland mapping and has been used in many local, regional, and global agricultural projects [4,13–15].

The free availability of some of these images adds to the many advantages of satellite-based optical remote sensing in agriculture [16]. Such data, which is also available for historic time periods back to the early 1970s, provides a means to study the present landscapes in relation to how they were in the past. Landsat, which is the oldest running earth monitoring program, provides a 47-year archive of satellite data of the entire earth at a 30-meter resolution. As a result, most of the crop types and other agricultural mapping studies have used Landsat images as the main data source. For instance, Lui et al., [16] used multi-temporal Landsat-8 to successfully map winter wheat in China. Maxwell et al., [17] demonstrated an effective corn classification from Landsat images through an automated process in south-central Nebraska of USA, and Yin et al., [18] used dense Landsat time series data to map agricultural and land abandonment with a high level of accuracy in the Caucasus, covering parts of Russia and Georgia. Many of these studies either employed supervised or unsupervised classification to ascertain the needed land-cover information [19,20], either at the pixel or object-based levels. Despite their accurate performances, they have some limitations [21,22]. Supervised learning always requires field information, also known as training or ground-truth data [23]. Even with the unsupervised learning, knowledge about the study area is required to assign the correct land cover type to the classification results.

A large number of studies on crop type and cropland mapping used field data from the same mapping year [24–26]. This way of mapping is limited in situations, where there are no ground truth data available, or data collection is impossible for the period of interest. Due to the yearly and periodic changes in cultivated crops, continuous collection of ground truth information is necessary to reliably map crop types. However, given the labor-intensive, expensive and difficult nature of ground truth data collection [27], studies such as Botkin et al., [28] and Sonobe et al., [29] have recommended research into the development of training and classification methods, which is applicable to years where field information is not available (i.e., a generalized classifier).

Given the rotation of crops on fields at different seasons and the fast changes in biomass and phenology of crops, the use of temporal information is very crucial in the discrimination of crops. Prediction of land-cover based on multi-temporal data involves the use of data from several different seasons and has proven to be effective in many studies [30,31], as it integrates the varying phenological characteristics among vegetation. Leaf pigment, water, and canopy structure are proven to relate with spectral reflectance of crops but varies at different growing seasons [32]. The use of data from a single date is known to inefficiently capture the differences among the many crops which share similar spectral characteristics [10]. Manfron et al., [33] for instance, analyzed time series of satellite images to efficiently estimate the inter-annual variability of the sowing dates of winter wheat Many other studies have employed vegetation indices such as the normalized difference vegetation index and the enhanced vegetation index to capture the seasonal dynamics of crops and other land-cover characteristics [19,34].

In Europe, there exists a remarkably rich agricultural land cover data body within the Integrated Administration and Control System (IACS), which are regularly collected by farmers as part of the subsidy payment scheme in the common agricultural policy [35]. A similar agricultural data in the United States of America is the reference data collected by the Department of Agriculture (USDA) to produce the annual crop data layer (CDL) [36]. These reference data are not available to the public, which may be the reason why the CDL has served as validation data in many crop type mapping studies [8,37]. Conversely, the IACS data can be freely obtained by scientists and research institutions for scientific purposes upon an official request. However, not much has been done with the IACS data in crop type mapping. Griffiths et al., [38] created a national single-year wall-to-wall land-cover map of Germany and used IACS data as a reference to validate some part of the study area. The study of Vuolo et al., [30] demonstrated how multi-temporal Sentinel-2 data can improve the accuracy of crop prediction when IACS data was used to independently validate the classified map. To the best of the authors' knowledge, there is no research that has used multi-temporal IACS data as training data to develop a generalized model to predict crop types from satellite data at the field level.

Therefore, we hypothesize that the IACS data can be used to train a multi-temporal field-based model, which can predict crop types from a satellite image that is independent of the model's training year. Hence, the calibration data, as well as the data used for testing the models, are from different years. In addressing the stated hypothesis, two different modelling approaches, i.e., multiple-year

(MY) and single-year (SY) calibrations were tested in the present study. While SY models are calibrated using data from just one year, MY modelling involves model training based on data from two or more years. The SY and MY approaches have been applied in some crop mapping studies, e.g., [8], but were done at the pixel level, which is characterized by the problem of 'salt and pepper' effects (i.e., a misclassification of neighbouring pixels despite large similarities). On the other side, the object-based method of land cover classification, that has recently attracted considerable attention [39] as a replacement for the pixel-based [40], suffers from difficulties in the segmentation scale selection. Further, it was shown to depend on the size of the objects being mapped [41] and tend to misclassify small land-cover objects in low to medium satellite images, such as Landsat [22]. Therefore, this study employs a field/polygon-based calibration approach using the exact crop field shapes from the IACS database. Our study addresses the following questions:

(1) How well do models based on a single year's spectral information predict crops when tested on years not included in the model calibration process?
(2) What is the prediction performance of models calibrated on spectral information from multiple years?
(3) Is the accuracy of the classification models affected by field size?

2. Materials and Methods

2.1. Study Area

The study was done in the Northern Hesse region of Germany (Figure 1) comprising the districts of Kassel, Waldeck-Frankenberg, Schwalm-Eder, Hersfeld Rotenburg, and Werra-Meissner. The study area comprises ca. 6900 km^2 and is characterized by diverse landscapes and sites with favorable and less favorable environmental conditions for farming. The favourable arable lands are mostly found in flat valleys and on plateaus with moderate slopes, which are often covered by loess of substantial thickness mainly in the western and northern parts [42]. The less favourable arable sites show shallow soils with less native water and nutrient availability.

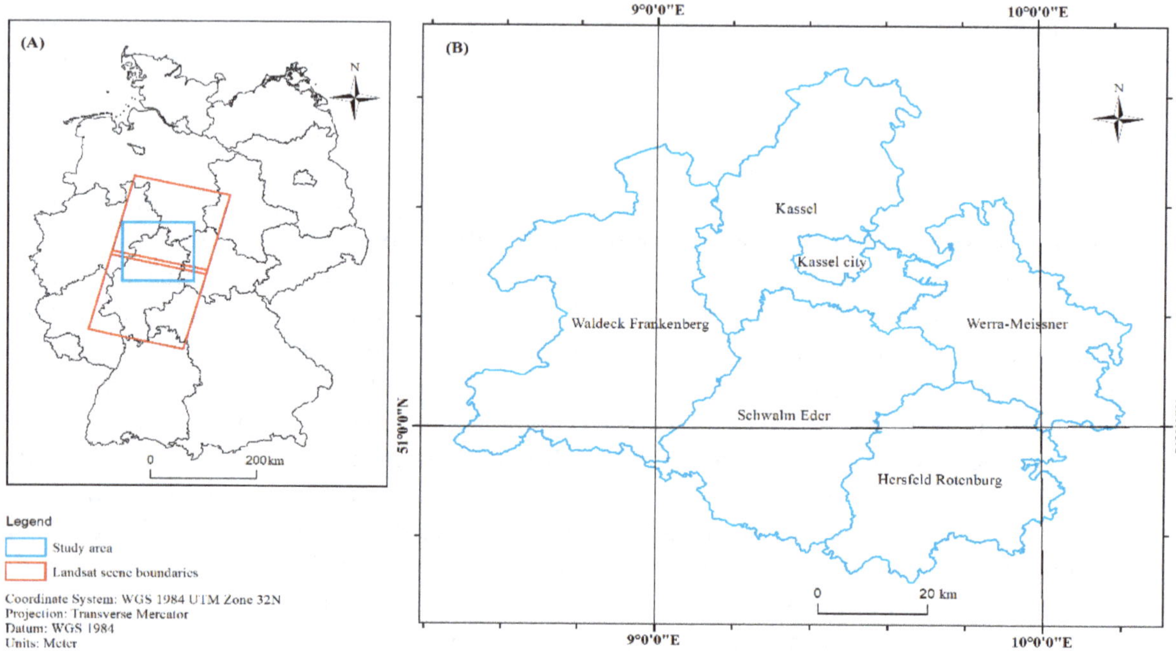

Figure 1. Map of the study area. (**A**) shows a map of Germany and the location of the study area, with the boundaries of the Landsat scenes; (**B**) shows the five districts, where the study was done.

Elevation ranges from 101 to 754 m with mean annual temperatures of 9–10 °C in the lowlands and 5–6 °C in the highlands. The mean annual rainfall ranges from 500–1300 mm [43]. The calendar of the crop types considered in this study can be seen in Table 1.

Table 1. Generalized calendar of the four crop types in the study area.

Crop Types	Sowing Window	Peak Greenness	Harvesting Window
Grassland	Depending on the grassland management system		
Maize	Late April	Mid-August	Mid-late September
Summer crops	Late March-Mid April	Mid-Late June	July-September
Winter crops	September-October	Mid-June	July-August

2.2. Data

2.2.1. Satellite Data

A total of 63 satellite images from the period April to October between 2005 and 2015 were used for this study. Surface reflectance Landsat scenes (Level-2) as summarized in Table 2 were downloaded from USGS's Earth Explorer [44]. Images from only six years were used because of little to no clouds. The images of Landsat 5 TM and 7 ETM+ had been atmospherically corrected using Landsat Ecosystem Disturbances Adaptive Processing System (LEDAPS) by NASA [45]. Surface reflectance of Landsat-8 was produced using Landsat Surface Reflectance Code [46]. Table 3 shows detailed information about the six spectral bands of the Landsat data used. Despite the small differences in the spectral ranges of the Landsat types, which have been well studied [47,48] to be smaller than 1 standard deviation of time-series of the spectral curve had no significant effect on classification results. Additionally, according to USGS [49], the Level-2 product (surface reflectance) of the Landsat images are similar, therefore, the Landsat data was not normalized. The atmospherically corrected Landsat images were accompanied by cloud mask layers. The images were categorized according to the dates when the images were captured, i.e., early summer (ES, April to May) and late summer (LS, July to October). ES and LS seasons cover the growing period of crop types in the study area; hence their use can help capture the different phenology of the crops at different stages of their development.

Table 2. Summary of satellite images used. (TM: Thematic Mapper, ETM+: Enhanced Thematic Mapper plus). The numbers in brackets represent the number of images used per date.

		Date of Image Acquisition	
Year	Satellite	Early Summer	Late Summer
2005	Landsat 5 TM Landsat 7 ETM+	03-Apr. (1), 21-Apr. (2) 4-Apr. (2)	18-Aug. (2)
2007	Landsat 5 TM Landsat 7 ETM+	02-Apr. (2), 25-Apr. (1), 27-Apr. (1) 26-Apr (2)	16-Jul. (2), 01-Aug. (1), 24-Aug. (2)
2009	Landsat 5 TM Landsat 7 ETM+	07-Apr. (2), 14-Apr. (2), 16-Apr. (2), 02-May (1), 25-May (1)	06-Aug. (2), 20-Aug. (2) 05-Aug. (1), 22-Sep. (1)
2010	Landsat 5 TM Landsat 7 ETM+	17- Apr. (2), 19-Apr. (2) 18-Apr. (2)	08-Jul. (1), 31-Jul. (1), 07-Aug. (2)
2011	Landsat 5 TM Landsat 7 ETM+	20-Apr. (1), 22-Apr. (2), 08-May (1) 21-Apr. (2), 07-May(2)	03-Aug. (1), 15-Oct. (1), 22-Oct. (1) 20-Aug. (1), 03-Sep. (2), 21-Sep (1), 28-Sep. (2)
2015	Landsat 8	24-Apr. (2)	30-Aug. (2)

Table 3. Summary of the six spectral bands of the Landsat 5, 7 and 8. NIR = Near infra-red, SWIR = Shortwave infra-red, TM = Thematic Mapper, ETM+ = Enhanced Thematic Mapper plus.

Landsat 5 TM and 7 ETM+			Landsat 8		
Band Number	**Band Name**	**Wavelength (μm)**	**Band Number**	**Band Name**	**Wavelength (μm)**
Band 1	Blue	0.441–0.514	Band 2	Blue	0.452–0.512
Band 2	Green	0.519–0.601	Band 3	Green	0.533–0.590
Band 3	Red	0.631–0.692	Band 4	Red	0.636–0.673
Band 4	NIR	0.772–0.898	Band 5	NIR	0.851–0.879
Band 5	SWIR-1	1.547–1.749	Band 6	SWIR-1	1.566–1.651
Band 7	SWIR-2	2.064–2.345	Band 7	SWIR-2	2.107–2.294

2.2.2. Reference Data

The IACS data were used as ancillary data in this study. These are spatial data collected by farmers as part of the subsidy support system within the EU. It is made up of the shapes of agricultural fields and the crop types planted in each cropping season. The models were initially developed to predict individual crop species in the study area, but tests (not shown) exhibited incorrect predictions among crops species of similar spectral characteristics and growing periods. Therefore, several crops were grouped into four crop types. They were grassland, maize, summer, and winter crops with their vegetation profiles shown in Figure 2. These vegetation profiles depict the spectral characteristics of the crop types at different stages of their development and show a similar trend across all years. Farmers are not obliged to register their fields; except for farmers who apply for subsidies. Therefore, the reference data used in this study is limited to the declared fields as submitted to the responsible agency. Fallow fields were not considered in this study since the reference data (i.e., the IACS) used for modelling consists of only cultivated fields.

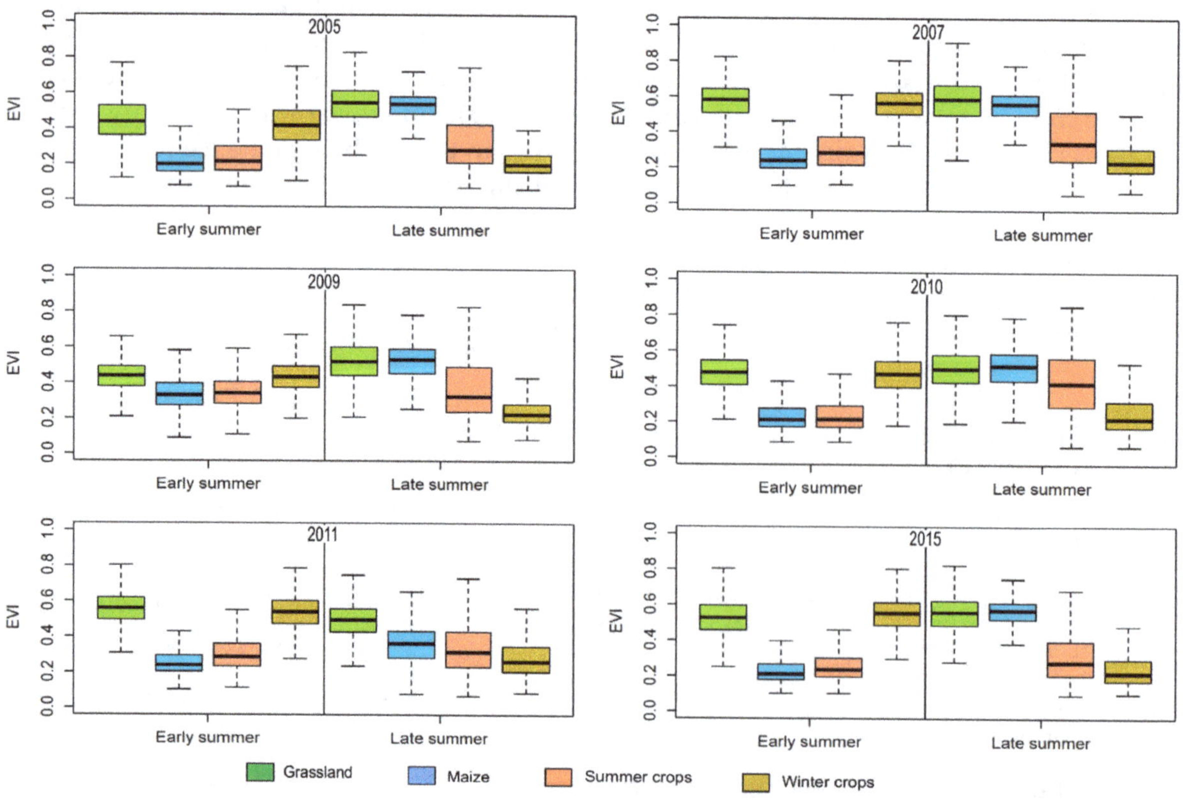

Figure 2. Vegetation profiles of the four crops based on enhanced vegetation index (EVI). Early summer = April to May, Late summer = July to October.

2.3. *Data Processing*

Image Pre-processing

The different bands (i.e., blue, green, red, near infra-red and shortwave infra-red 1 and 2) of the satellite images were stacked together, based on the years and acquisition time (Figure 3). Some of the images had small areas of clouds since cloud cover of less than 10% was considered appropriate for the study purpose. As a result, the cloud masks that came with the images were used to mask out all clouds. The masked areas were replaced with non-cloudy data from other images of the same area around the same time frame (i.e., May for ES and September, October for LS). With respect to the Landsat 7, the scan lines which resulted from the failure of the scan line corrector of the ETM+ sensor were also replaced with cloud-free satellite data from other images of the same area using the "cover" function [50] from the "raster" package in R software [51]. As our study area includes more than one Landsat image, some images were mosaicked to cover the entire area of interest. Mean values were used for overlapping layers during the mosaicking process. Figure 3 shows a complete workflow of the data analysis of this study.

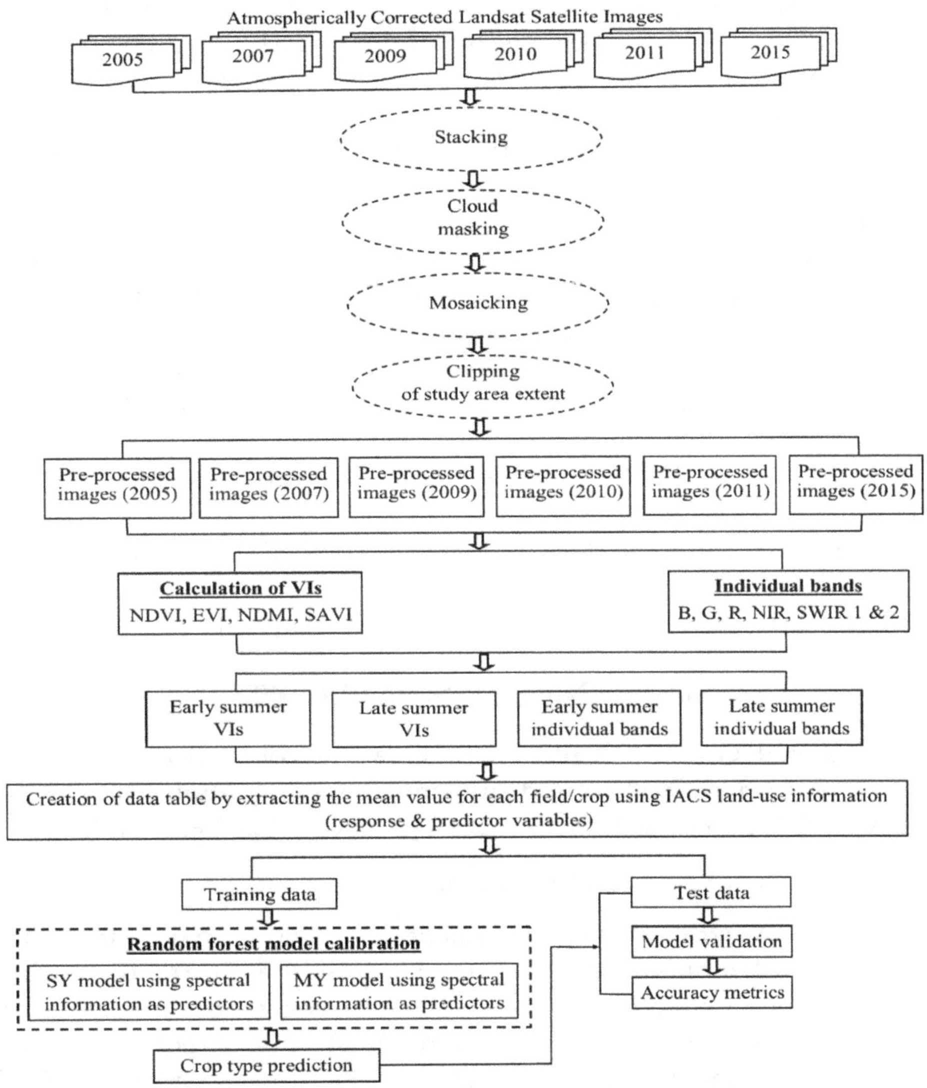

Figure 3. The workflow of the data analysis. SY-Single-year, MY-Multiple-years, VIs-Vegetation Indices, B-Blue, G-Green, R-Red, NIR-Near Infra-red, SWIR-Shortwave Infra-red, NDVI-Normalized Difference Vegetation Index, EVI-Enhanced Vegetation Index, SAVI-Soil Adjusted Vegetation Index, NDMI-Normalized Difference Moisture Index.

2.4. Model Calibration and Validation

Crop type prediction models were built based on random forest (RF) algorithm, which is an ensemble supervised machine learning classifier that creates numerous decision trees for prediction by randomly selecting subsets of the training data through the process of bagging [52]. Higher accuracies have been achieved with RF as compared to other machine learning algorithms in many crop mapping studies [7,8]. It can effectively function with only two main parameters, i.e., the number of trees to grow (*Ntree*) and the number of predictor variables selected for the best splitting of each tree node (*Mtry*) [53]. In this study, *Ntree* was set at 500 for all models since the error steadies before this number is reached, while *Mtry* was set to the square root of the input variables as reviewed by Belgiu and Drăgu [53].

2.4.1. Input Variables Used in the Model

Spectral data obtained from the satellite data and used as predictors in the crop type predictive models consisted of blue, green, red, near infra-red (NIR), shortwave infra-red 1 (SWIR 1) and shortwave infra-red 2 (SWIR 2). Additionally, four widely used spectral vegetation indices (VIs), i.e., normalized difference vegetation index (NDVI) [54], enhanced vegetation index (EVI) [34], soil adjusted vegetation index (SAVI) [55] and normalized difference moisture index (NDMI) [56], were computed from a ratio of different satellite bands (see Equations (1)–(4)) and included as explanatory variables. These VIs capture the dynamics of vegetation like greenness and vigor among others at different phenological stages. The potential of NDVI to assess vegetation dynamics of crops has been demonstrated by a number of studies [10,19]. However, it has shortcomings of sensitivity to saturation, soil background effects, or atmospheric effects. In dealing with these limitations, EVI and SAVI were added. SAVI deals with the soil background effects, while EVI uses the blue band to deal with the atmospheric influences by aerosols. EVI, NDVI, and SAVI, as shown in Equations (1)–(3), respectively, use the NIR and red bands in their computation, and they complement each other when used in vegetation analysis. NDMI, which uses NIR and SWIR for measuring the water content in vegetation, was also included in the analysis (Equation (4)), as it adds some complementary information to the other VIs.

$$\text{NDVI} = (\text{NIR} - \text{Red})/(\text{NIR} + \text{Red}) \tag{1}$$

$$\text{EVI} = \text{G} \times (\text{NIR} - \text{Red})/(\text{NIR} + \text{C}_1 \times \text{Red} - \text{C}_2 \times \text{Blue} + \text{L}) \tag{2}$$

$$\text{SAVI} = [(\text{NIR} - \text{Red})/(\text{NIR} + \text{Red} + \text{L})] \times (1 + \text{L}) \tag{3}$$

$$\text{NDMI} = (\text{NIR} - \text{SWIR})/(\text{NIR} + \text{SWIR}) \tag{4}$$

where NIR = Near Infra-red, G = gain factor, C_1 and C_2 are aerosol resistance term coefficients, L in Equation (2) is non-linear canopy background adjustment, L in Equation (3) is soil brightness factor and SWIR = Shortwave Infra − red (values: G = 2.5, $C_1 = 6$, $C_2 = 7.5$, $L_{EVI} = 1$, $L_{SAVI} = 0.5$).

2.4.2. Field-based Extraction of Spectral Information.

The extraction of the spectral information was done at field base with the exact crop fields as objects. Mean values of each crop's field were extracted from the spectral data, which consisted of six individual bands of the satellite image, as well as four vegetation indices; grouped into early summer (ES) and late summer (LS) spectral information. In all, 20 different spectral information were used as predictors in the RF models. Since the IACS data (Figure 5) represent field information and, crops are cultivated with unequal distribution of fields for each crop type, almost equal numbers of polygon/field samples for each crop type were selected. Thus, crop types with many fields were always undersampled compared to those which were represented by fewer fields. Six different data tables were built for the six respective years, which were later used to calibrate and validate the crop type prediction model, with the spectral information and crop types as predictor and response variables, respectively.

2.4.3. Crop Type Prediction Modelling

Two different modelling approaches called same-year (SY) and multiple-year (MY) training were employed. With respect to SY models, an RF algorithm was trained using only the spectral information of one year and cross-validated using the remaining years as shown in Figure 4A. This was repeated six times, where for each repetition a different year was used to train the model, and the model's performance in predicting crop types was assessed using the remaining single-year data.

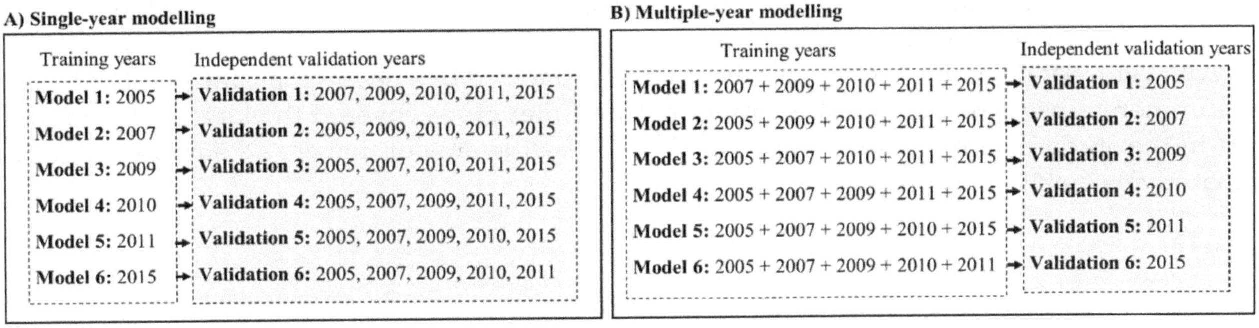

Figure 4. An illustration of the two modelling approaches.

MY models were trained by combining the extracted spectral information from five different years during the training phase and tested on an independent year. The training combination with multiple years was done six times, and with each repetition a single year was left out to validate the efficacy of the MY models (Figure 4B). The contribution of the predictors was assessed based on the internal mean decrease Gini of RF, which is the average of all Gini impurity recorded for each input variable when selected for splitting at each tree node [57]. Graphs showing the six most important predictor variables (based on percentages of the mean decrease Gini Index) of the best modelling method were created for visualization.

2.4.4. Accuracy Assessment

The performance of the models in predicting crop types was independently evaluated at field scale based on a confusion matrix. The independent validation was done by comparing the predicted crop types with the known crops using the reference data. The three most important and widely used metrics namely, overall accuracy (OA), user accuracy (UA) and producer accuracy (PA), resulting from the confusion matrix were calculated. OA assesses the overall performance of a model and is the ratio of correctly predicted crops and the total number of predicted crops. UA evaluates how well the predicted crops agree with the known reference data (i.e., the IACS field data), while PA measures the agreement between the reference data and the prediction. From the confusion matrix, the error of commission (EC) and error of omission (EO) of the respective land cover types can be obtained. Since the performance of each developed SY model was tested for all years, except for the training years, the presented accuracy measures are averages of OA, UA, and PA of the same years. Since accuracies and errors of spatial data are spatially explicit, a map was created that demonstrates our models' ability to visualize correctly and wrongly predicted fields using the best modelling method.

2.5. Relationship between Field Size and Accuracy

To check whether the accuracy of crop type prediction depended on the size of crop fields, the correctly and wrongly predicted fields along with their sizes were extracted for each of the MY models. Field sizes were rounded to the nearest multiples of 1.5 ha (i.e., the average field size) to create field size classes, with the count of the all correct (True) and wrong (False) predictions for each field size category. The percentages of correct predictions for all the field size categories were computed.

3. Results

3.1. Agricultural Land Cover Data

The IACS data used as reference data to calibrate and validate the developed models showed different field numbers, average field size, and area for grassland, maize, summer crops, and winter crops (Figure 5). The average field size for grassland was around 1 ha and did not change significantly over time. Summer crops showed a slightly higher average field size between 1.2 and 1.5 ha for all years under consideration. Maize and winter crops had the biggest average field sizes, ranging between 1.8 ha to 2 ha from 2005 to 2015.

While grassland had the highest number of fields followed by winter crops, the number of maize and summer crops were the lowest. Consequently, winter crops covered the largest area of arable lands in the study area followed by grassland, whereas the area of maize and summer crops was comparatively small.

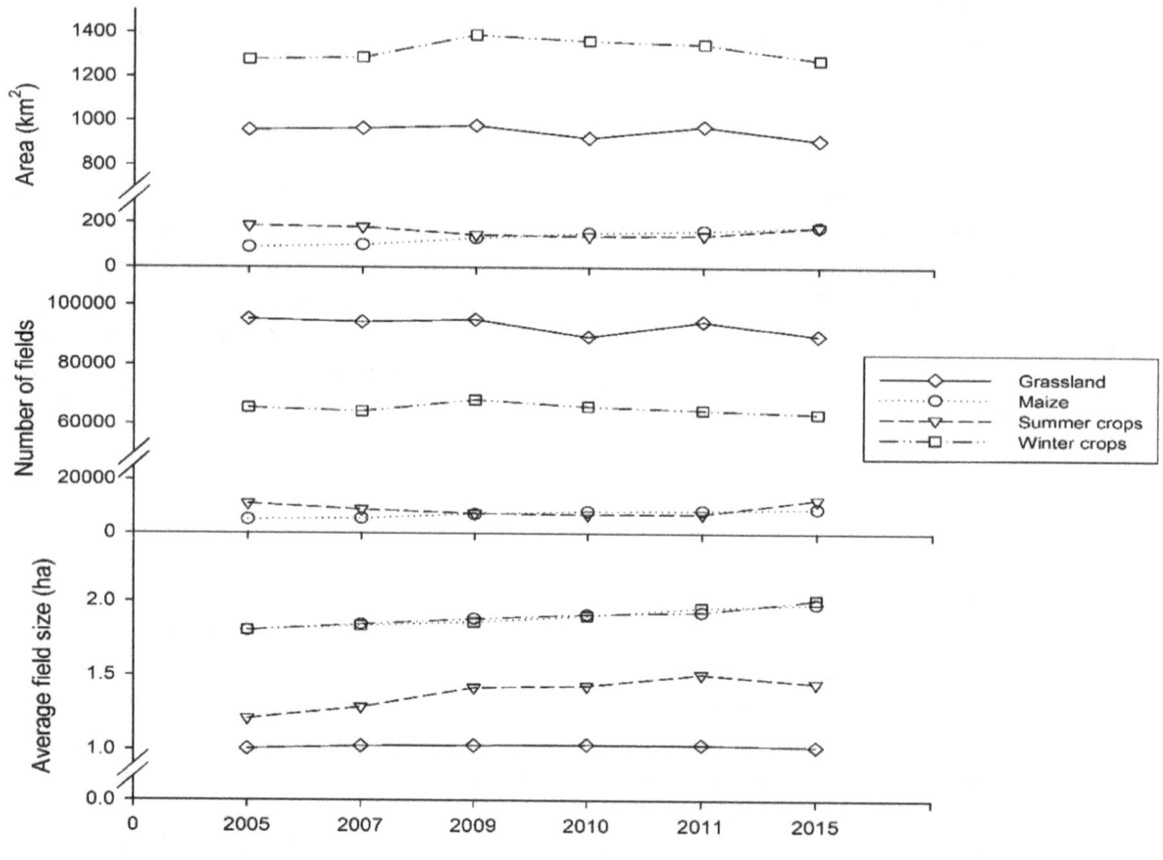

Figure 5. Characteristics of Integrated Administration and Control System (IACS) data used as reference information for the predictive crop type models.

3.2. Assessment of the Modelling Approaches

The general performance of the two modelling approaches as indicated by the OA (Figure 6) ranged from 67.7% to 73.4% with an average value of 71.5% for the SY models. The MY modelling approach showed an OA between 67.1% and 86.1% with an average of 77.3% (Figure 7). The lowest OA for the SY models was observed when they were tested with crops in the year 2015, while the highest OA value was achieved in 2011. Conversely, MY models achieved their highest OA in 2015, whereas, the lowest performance was observed in 2011.

The SY prediction of individual crop types in different years based on user (UA) and producer accuracy (PA) showed a high UA (81.6%) and PA (82.2%) values for grassland, while UA (76.7%) and PA values (67%) for maize were somewhat lower. Prediction of summer crops (UA and PA of 60.6 % and 69.5 % respectively) was less accurate than winter crops (UA 76%, PA 71.1%).

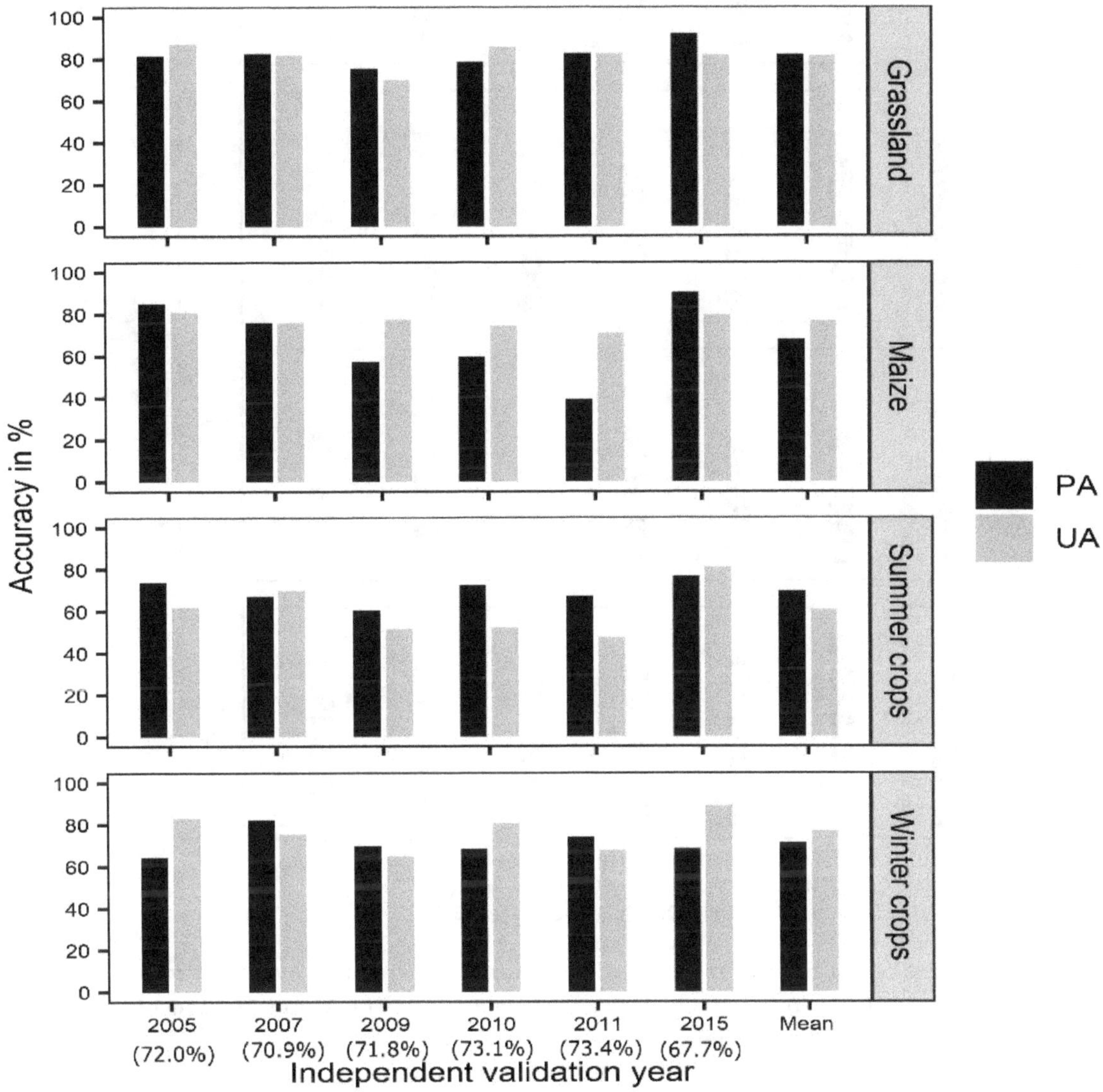

Figure 6. Accuracies of single year models. The mean bars represent average PA and UA, respectively, for each crop across years. Values in brackets represent the average overall accuracies of all models. UA = User accuracy and PA = producer accuracy.

The application of MY models to identify grassland area of different years resulted in a mean UA of 83.1% and a PA of 87.8% (Figure 7). Maize was discriminated with a mean UA and PA of 71.8% and 85.2% respectively, while winter crops were discriminated with a mean accuracy of 79% (UA) and 79.6% (PA). The average accuracy for summer crops was comparatively low (PA = 71.5% and PA = 69.3%), which was mainly a result of the confusion between summer crops and the other crops.

Overall, grasslands always exhibited higher accuracies (UA and PA > 80%) across years when predicted by the two modelling methods, whereas the arable crops were better predicted by the MY models in comparison to the SY models.

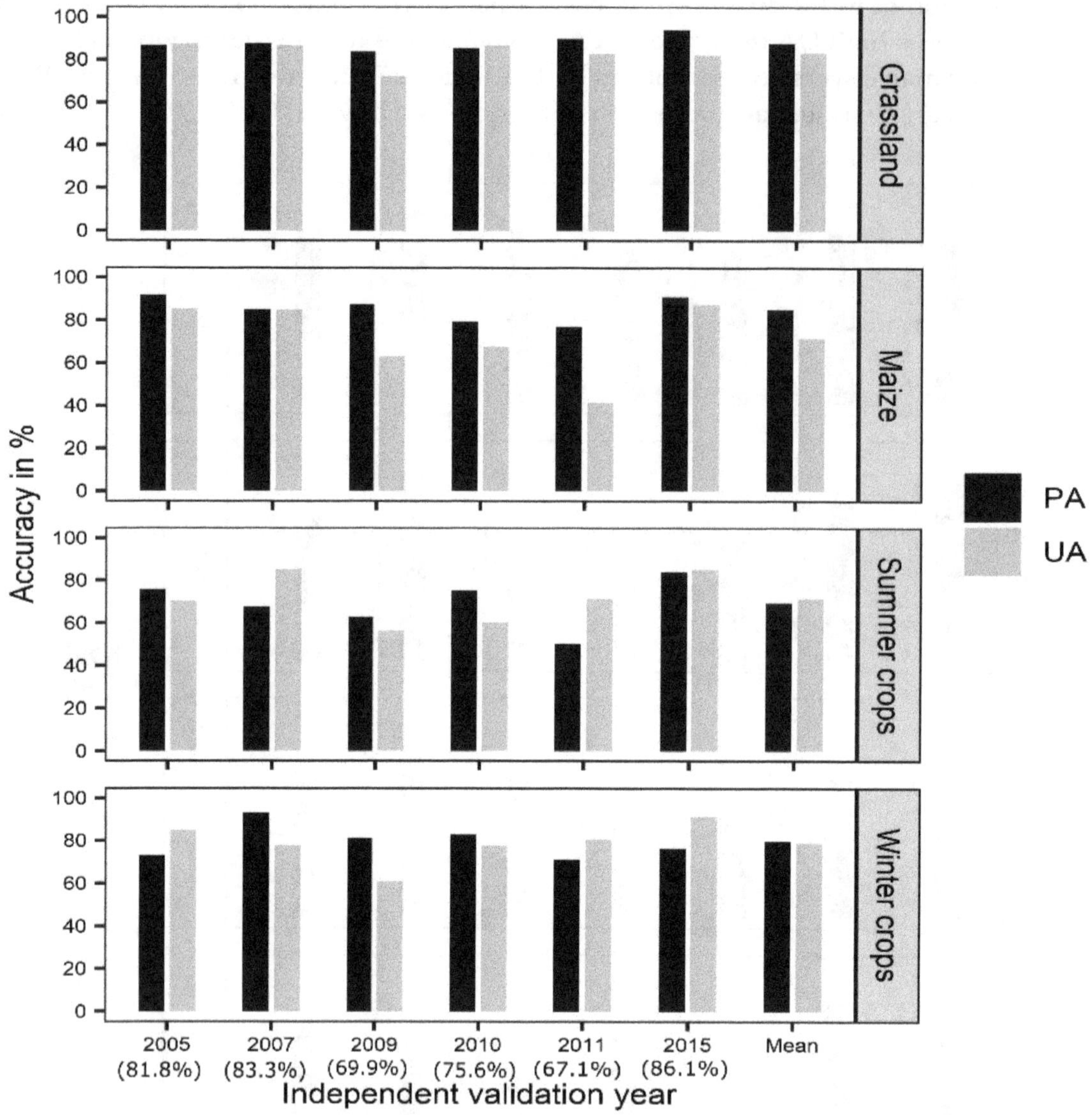

Figure 7. Accuracies of multiple-year models. The mean bars represent average producer accuracies (PA) and user accuracy (UA), respectively, for each crop across years. Values in brackets represent the average overall accuracies of all models.

3.3. Classified Maps Based on Best Modelling Method

Since MY models proved to be the best modelling approach, their capability to create accurate crop-type maps are exhibited in Figure 8 for 2015, which corresponds to an OA of 86.1% (see Appendix A, Table A6 for a confusion matrix). The map was derived with model 6 (Figure 4B), which is an MY model developed based on information from 2005, 2007, 2009, 2010, and 2011. On closer examination, regions at higher altitudes with less favourable growth conditions, which are dominated by grasslands (Figure 8A), can be clearly distinguished from fertile areas, where a multitude of arable crops is grown and where grassland is only interspersed. Moreover, the spatial distribution and patterns of crops, the shape, and edges of fields can clearly be observed. The so-called 'salt and pepper' effects, that characterize most land-cover maps at a pixel base were not experienced with the maps produced in this research, which may be a result of the fact that our models were calibrated with the mean spectral information at field scale based on the IACS field polygons. Maps of the remaining years and their respective confusion matrices are shown in the Appendix A, that is, Figures A1–A5 and Tables A1–A5, respectively.

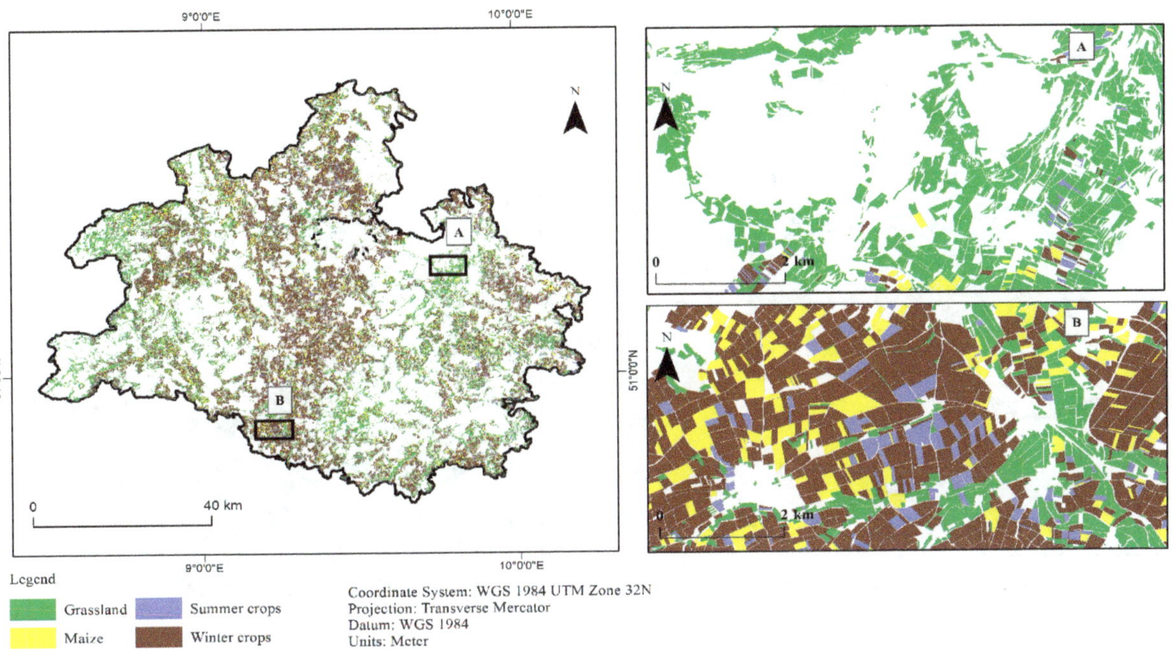

Figure 8. A classified map of 2015 resulting from a multiple-year model based on spectral information of 2005, 2007, 2009, 2010 and 2011. 'A' and 'B' show areas dominated by grassland at higher altitudes and fertile areas dominated by arable crops respectively.

3.4. *Relationship between Model Accuracy and Field Size*

An increase in mean accuracy from 74% to 87% was observed as the field size increased (i.e., 1.5–9 ha) (Figure 9). Furthermore, a slight decrease in the mean accuracy from 87% to 85% of crop type prediction was seen with increasing field size from 9 ha to 12 ha. As the field size increased further to 13.5 ha, a slight increase in accuracy was observed (from 85% to 88%). However, a marginal reduction to 87% accuracy was observed as the field size increased further. The consistent and marginal rise and fall in accuracy with increasing field sizes indicate that crop type prediction by MY models is independent of field size. But it is important to state that the majority of fields (> 60%) belong to the smallest field category (1.5 ha). The accuracy maps of the MY models can be seen in Figure 10 and the rest in the Appendix A (Figures A6–A10)

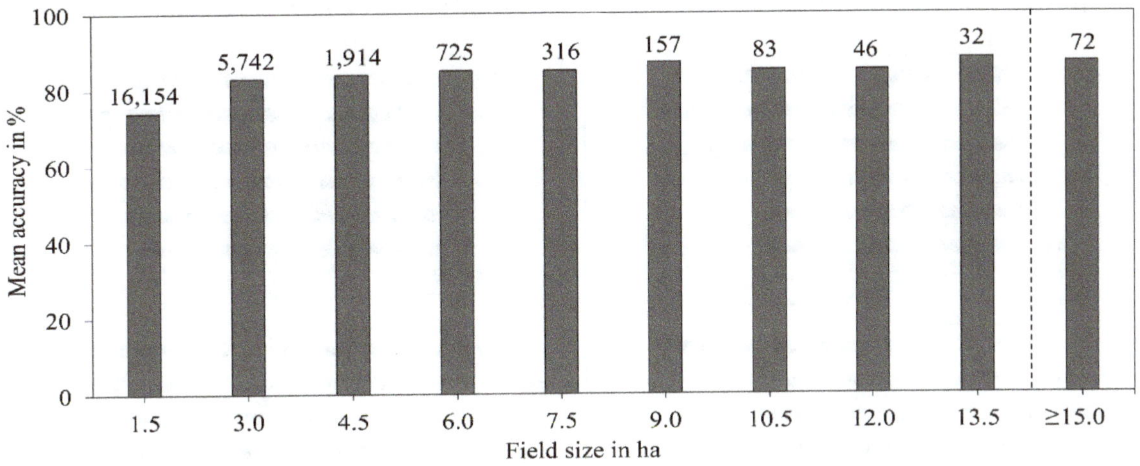

Figure 9. The relationship between the accuracy of crop type prediction and field size based on the multiple-year models. The accuracies are mean values for all years, with the last bin representing the average accuracy of all fields ranging 15–34.5 ha. The value on top of each bar represents the number of fields for each field size group.

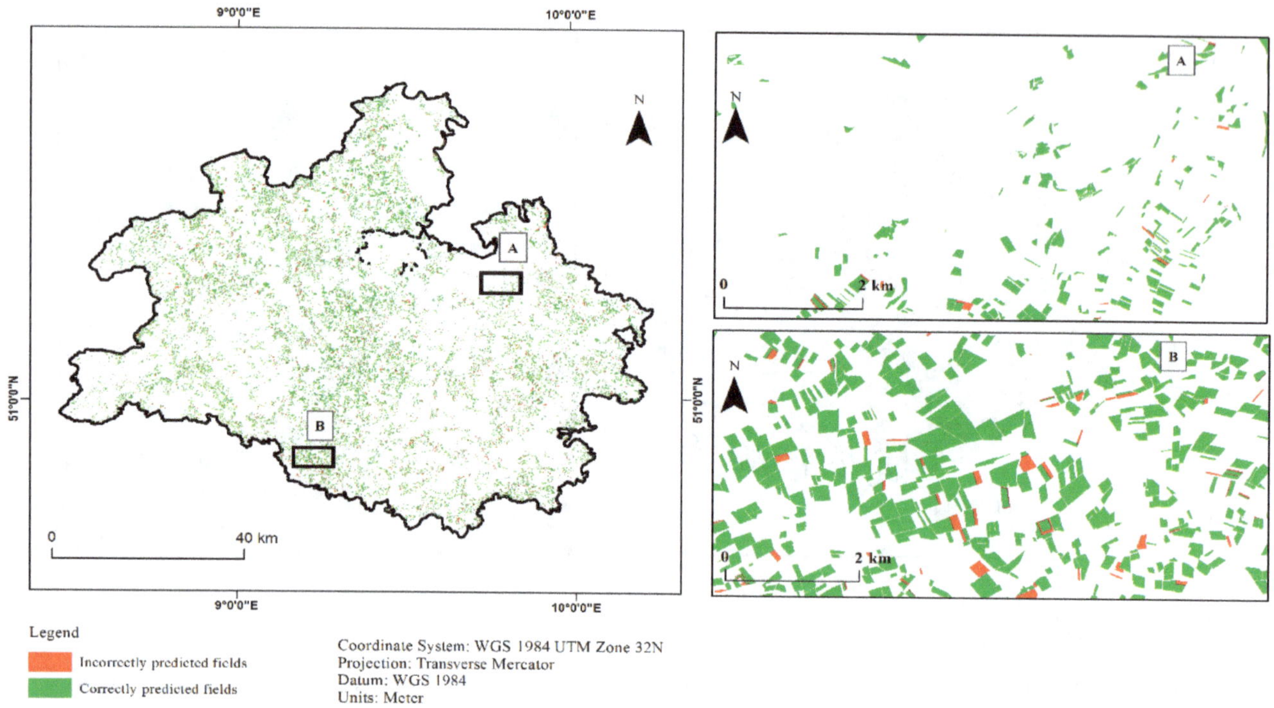

Figure 10. An IACS-based accuracy map of 2015 resulting from a multiple-year model calibrated using spectral data of 2005, 2007, 2009, 2010 and 2011 representing wrongly and correctly predicted fields. 'A' and 'B' show areas dominated by grassland at higher altitudes and fertile areas dominated by arable crops respectively.

3.5. Importance of Predictor Variables

One of the strengths of RF models is the ability to measure and assess the contribution of each predictor variable used. Figure 11 presents the importance values of the first 6 most important input variables used in the MY models. Late summer (LS) NDMI was the first most important predictor variable in most models, and only in one instance the means of early summer (ES) NDVI was ranked first (Figure 11B). Another predictor that seemed to be important across models was the early summer NDVI as shown in Figure 11A,C–F. The contribution of VIs in the prediction of crops was much stronger than the individual bands. Red and green bands are the only bands that appeared among the first six important predictors, with red being the dominant one across all models.

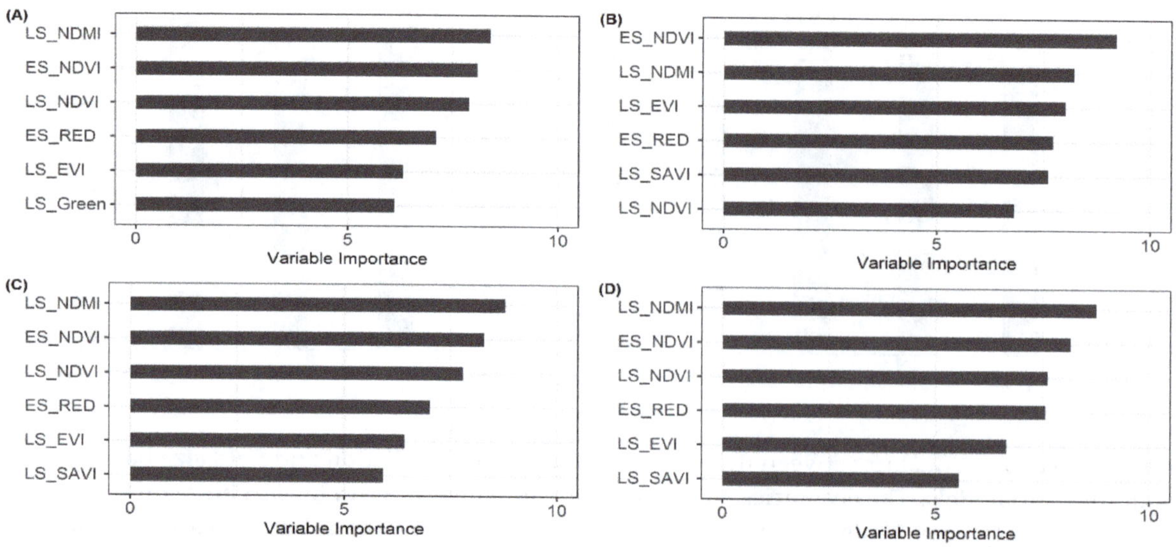

Figure 11. *Cont.*

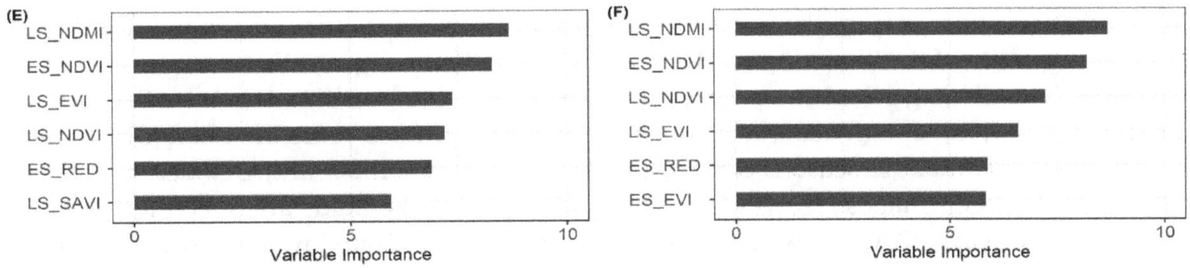

Figure 11. The 6 most important predictors used in the multi-year models expressed in decreasing order of importance from the top of the y-axis. Variable importance on the x-axis is expressed as a percentage of Mean Decrease Gini. (**A–F**) are MY models trained by a combination of (2005 + 2007 + 2009 + 2010 + 2011), (2005 + 2007 + 2009 + 2010 + 2015), (2005 + 2007 + 2009 + 2011 + 2015), (2005 + 2007 + 2010 + 2011 + 2015), (2005 + 2009 + 2010 + 2011 + 2015) and (2007 + 2009 + 2010 + 2011 + 2015) respectively. LS = Late summer, ES = Early summer.

4. Discussion

Crop type mapping in large agricultural landscapes is challenged by the daunting task of periodic training data collection. The traditional satellite-based mapping approach of using reference data from the same year impedes mapping specifically in periods where reference data is not available. The IACS data, which is a field-based crop type data presents reliable reference datasets to deal with the problem of frequent training data collection for satellite-based crop type mapping through the development of a generalized model. The issue of generalized classifiers has been raised and exploited in a few agricultural mapping studies, but less focus was put on specific crop type mapping. Thus, this study aimed at assessing the efficacy of IACS data to be used as reference data for the development of generalized SY and MY crop type models to predict grassland, maize, summer and winter crops as land-cover categories.

The accuracies achieved in our study are similar to the work of [8], where corn and soybeans were predicted based on spectral characteristics using the single-year modelling approach. While our study considers four different crop types with an acceptable average UA and PA for grassland across years (>80%), maize, summer and winter crops were predicted at somewhat lower accuracies. That means that despite the somewhat low performance of SY models in predicting the other crop types, it is able to predict grassland with an acceptable level of accuracy across years.

The overall accuracy of the SY calibration method employed in our study is rather low compared to the traditional method, where calibration and testing data are from the same year. Probable reasons may be different growing dates of the crops in different years, inter-annual differences in climate, image acquisition time as well as variation in image quality between years, as was also suggested by Laborte et al., [58] and Zhong et al., [8]. The performance of MY models in predicting crop types showed higher robustness across years than the SY calibration approach. An average increment of 6% in OA (i.e., 77.3%) was achieved by MY models across years. A similar OA of 73.1% was achieved by Massey et al., [37] when an MY calibrated model was used to predict crop types from MODIS data for an independent year. The higher robustness of MY models might be attributable to the fact that, through the use of spectral information from different years, the interannual climate variability as well as variations in image quality from different years, are reduced to a certain degree [8,58]. Thus, these factors make them generic with high prediction accuracies when applied to predict crops from data not seen by the model. Additionally, MY calibration compensates for the phenological differences of crops among years through the inclusion of many phenological situations using multiple data from different years. This makes MY models more efficient and generalized for satellite-based crop type classification when training data is not available for a period of interest.

The prediction of crops from the spectral-temporal profiles of satellite images can heavily depend on the time and quality of the images used [10,58]. Occasionally, a compromise has to be made between the quality and time of images in the same growing season, which may ultimately have the consequence that not all years will have high accuracies when data from multiple years are used [8]. In this study, for example, the 2009 and 2011 seasons comprised of very late September and October images (Table 2) due to the lack of earlier images. However, in the study area, maize fields are harvested as late as the end of September or early October, whereas summer crops are harvested much earlier to give way for winter crops, which might have developed one or two leaves at this time. Such sources of variation might explain the somewhat lower accuracies of 2009 and 2011 data as compared to the other years.

The data on the assessment of the spectral predictors used in the MY models indicates a paramount contribution of VIs in the prediction of crops as compared to the individual bands. A similar conclusion was drawn by Fletcher [59] in the discrimination of soybean and three weed species. The highest contribution of VIs in the prediction of the crops was expected since their calculation involved two or more bands and as a result, used the unique spectral characteristics of the individual bands to produce a single layer which captured the different phenological dynamics among the crops. Vegetation indices, which are based on SWIR and NIR spectral bands, are known for their contributions to plant separation [59]. Therefore, despite the contributions of the other predictor variables, the late summer NDMI is ranked as the topmost dominant predictor variables for almost all years, followed by early summer NDVI. NDMI uses a normalized ratio of the difference and sum of NIR and SWIR and is known to be sensitive to changes in water content of vegetations canopies [56]. It can, therefore, be inferred that the differences among the four crop types are better captured by the content of moisture in their leaf canopies during late summer.

The subject of object size and prediction accuracy is very crucial in the mapping of agricultural areas [39]. Our results suggest that the prediction of a particular crop type does not necessarily depend on the corresponding size of the field. It was expected that the prediction of bigger fields may be easier than with smaller fields, but the results do not confirm that. The biggest crop field category (≥15 ha) achieved a prediction accuracy of 88%, nonetheless, comparatively smaller field sizes (9 ha) also achieved the same accuracy. Thus, the present study does not support the conclusions of Castilla et al., [41], that the possibility of correct classification of land-cover type decreases with decreasing object size. The reason may be that Castilla et al., [41] employed a segmentation method, which is dependent on the land cover size, whereas the present study used the exact field polygons declared by farmers in the study area as objects for the prediction of the crop types. However, since the present study area is dominated by smaller fields with very few large fields, future research is required to further investigate the relationship between field size and accuracy of crop types prediction in agricultural areas with a relatively even distribution of field sizes.

The uniqueness of this study compared to other studies of generalized classifiers for cropland mapping is the field-based approach employed. This approach deals with some of the challenges associated with the widely used methods. The issue of segmentation scale selection of the other object-based classification [21] is avoided. Moreover, the 'salt and pepper' effects that characterize the pixel-based prediction of land-cover types are equally averted in this study. Hence, our MY modelling approach can be used to map past and present crop types which may be necessary to ascertain the impacts of any agricultural activity (e.g., biogas production) heavily dependent on croplands.

Finally, the hypothesis that IACS data can be used to calibrate models for the prediction of crop types from a satellite image differing from the calibration year has been proven through a field-based SY and MY calibration approach. However, Cai et al., [60] stated that increasing the calibration years to a maximum of 10 years can further increase accuracy. Therefore, our five-year MY models could be improved further by incorporating more years of spectral information, as more satellite data (Sentinel and EnMap) become available in the future.

5. Conclusions

For the first time, this study used a field-based approach to test the usefulness of IACS data in calibrating an RF-based model to predict crop types from satellite images, that are not from the same year as the calibration year. Thus, two modelling methods called SY (i.e., using spectral data from a single year) and MY calibration (i.e., using spectral data from multiple years) were tested in the discrimination of grassland, maize, summer and winter crops.

The results depict a superior performance of the MY approach as compared to SY model. The MY approach included a larger range of inter-annual variability in image quality, climate, and growing dates of crops from different years, thus, contributing to its robustness in predicting crop type from satellite images of different years. The approach employed in this work, unlike other object-based methods, is not dependent on field size. It is, therefore, recommended to use the field-based MY calibration approach for practical crop type mapping, particularly when reference data for the mapping year is not available. This method is useful for practical reasons and can be used to map past and present croplands for comparative analysis. However, the inclusion of soil data and phenological metrics as predictors of MY model may have a potential for future research. This might help improve performance and provide an opportunity for more specific crop type mapping, rather than generic crops like summer and winter crops as used in this study. A combination of data from different satellites like Sentinel or upcoming satellites like EnMap or HyspIRI might further improve the MY modelling approach due to higher revisiting time and thus a denser time series

Author Contributions: T.A., R.G., and M.W. conceived the idea of the research. I.K. processed the data, analyzed the results and wrote the manuscript. T.A., R.G., and M.W. supervised the study and contributed to the writing and revision of the manuscript.

Appendix A

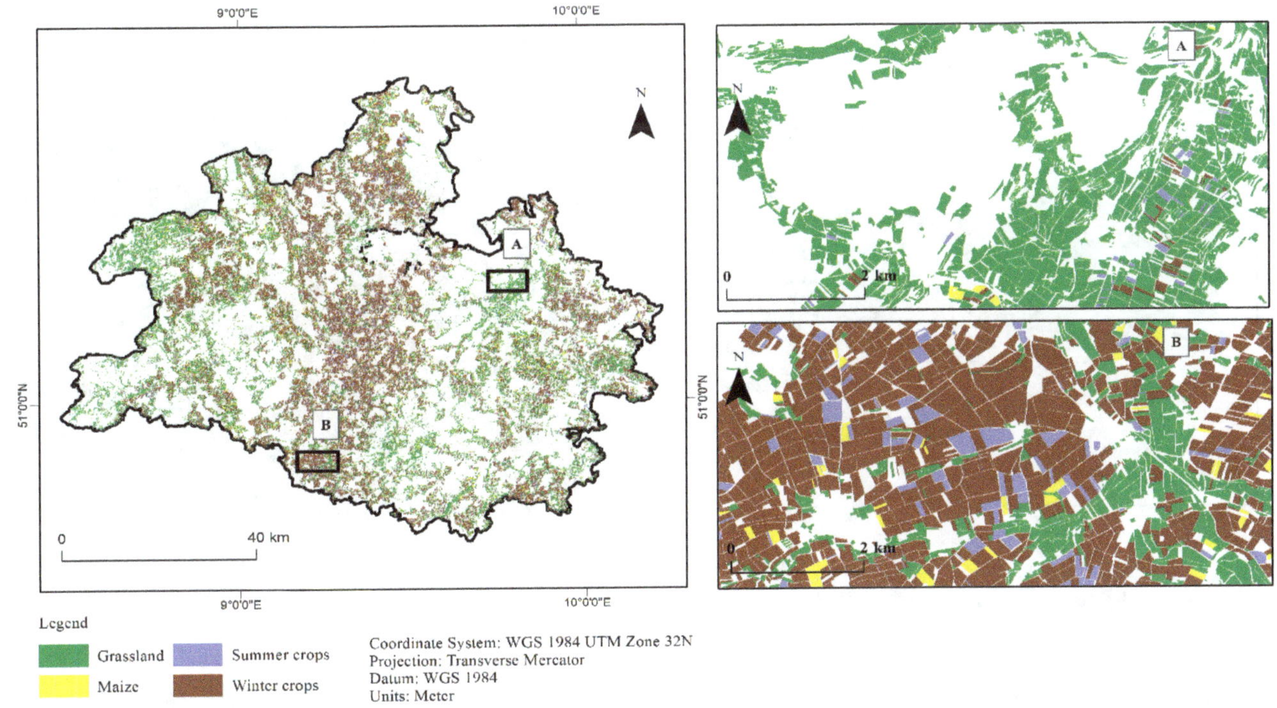

Figure A1. A classified map of 2005 resulting from a multiple-year model based on spectral information of 2007, 2009, 2010, 2011 and 2015.

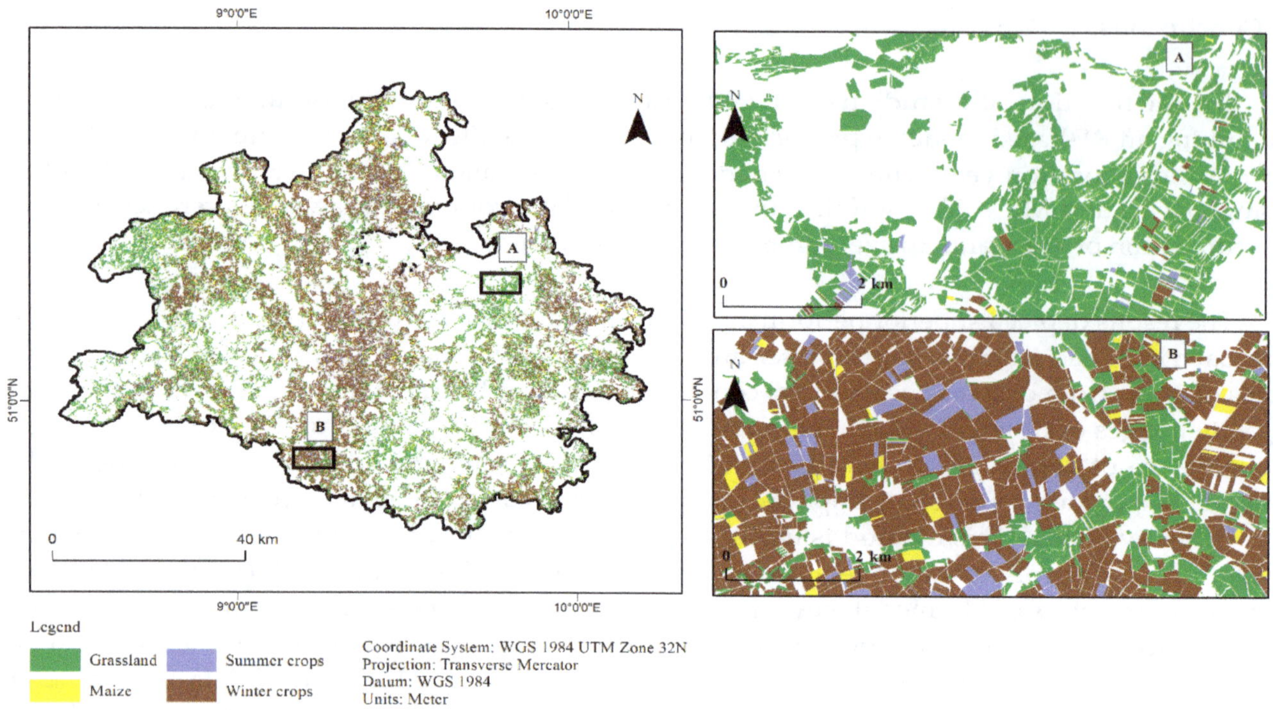

Figure A2. A classified map of 2007 resulting from a multiple-year model based on spectral information of 2005, 2009, 2010, 2011 and 2015.

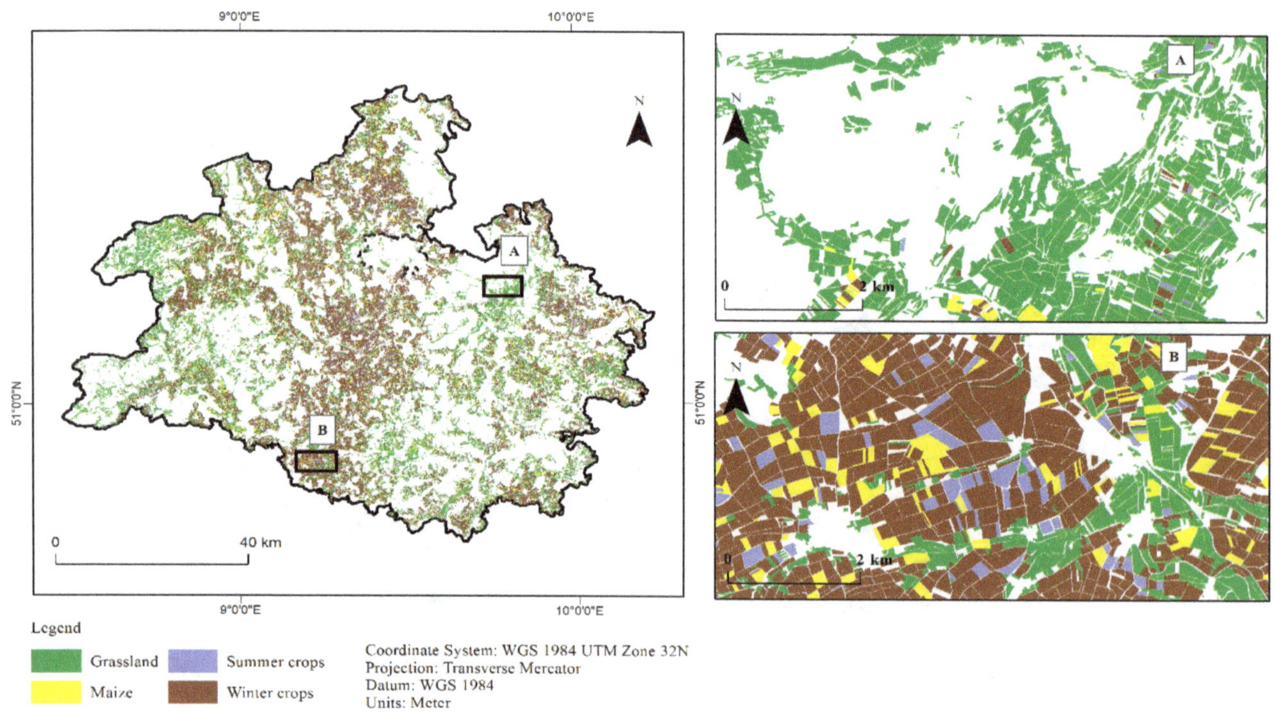

Figure A3. A classified map of 2009 resulting from a multiple-year model based on spectral information of 2005, 2007, 2010, 2011 and 2015.

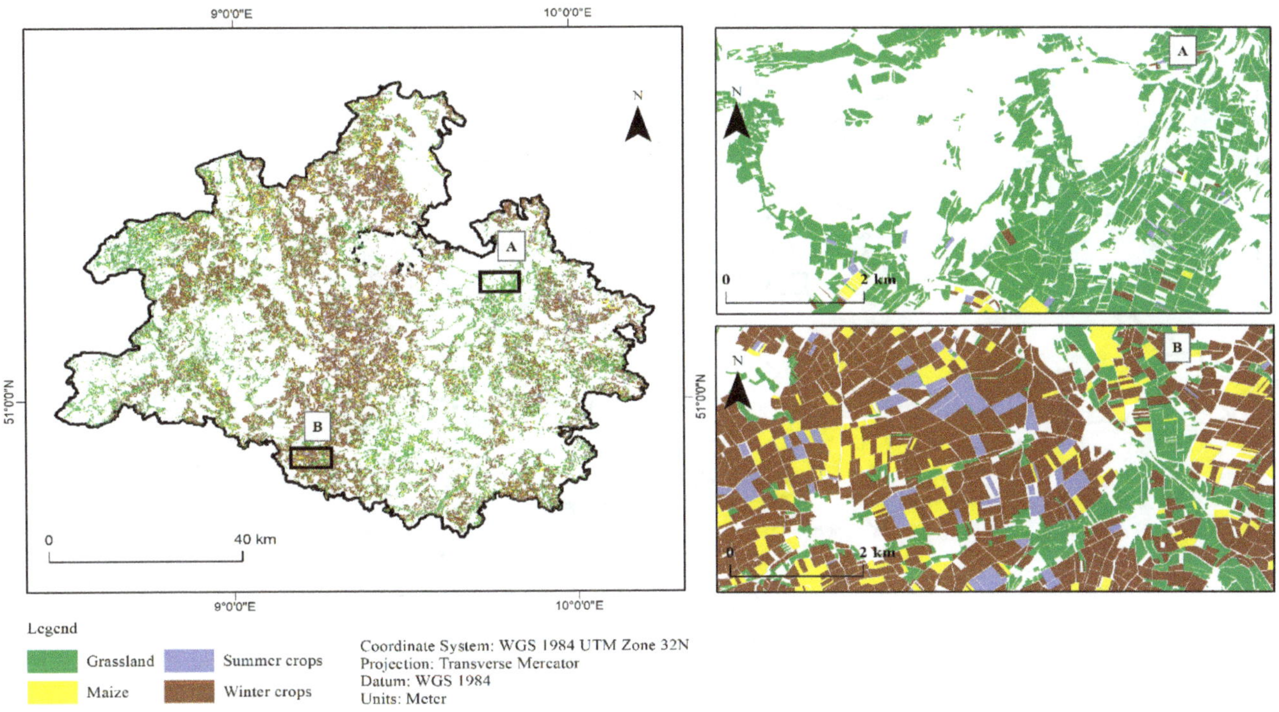

Figure A4. A classified map of 2010 resulting from a multiple-year model based on spectral information of 2005, 2007, 2009, 2011 and 2015.

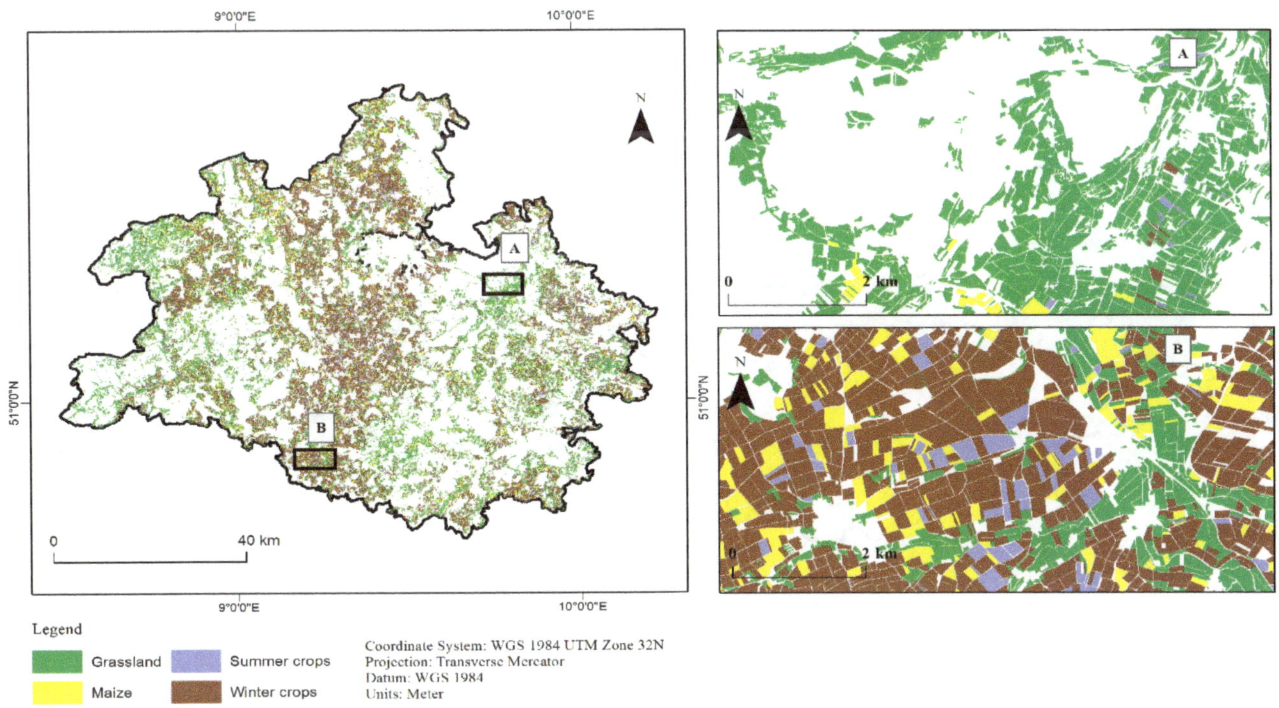

Figure A5. A classified map of 2011 resulting from a multiple-year model based on spectral information of 2005, 2007, 2009, 2010 and 2015.

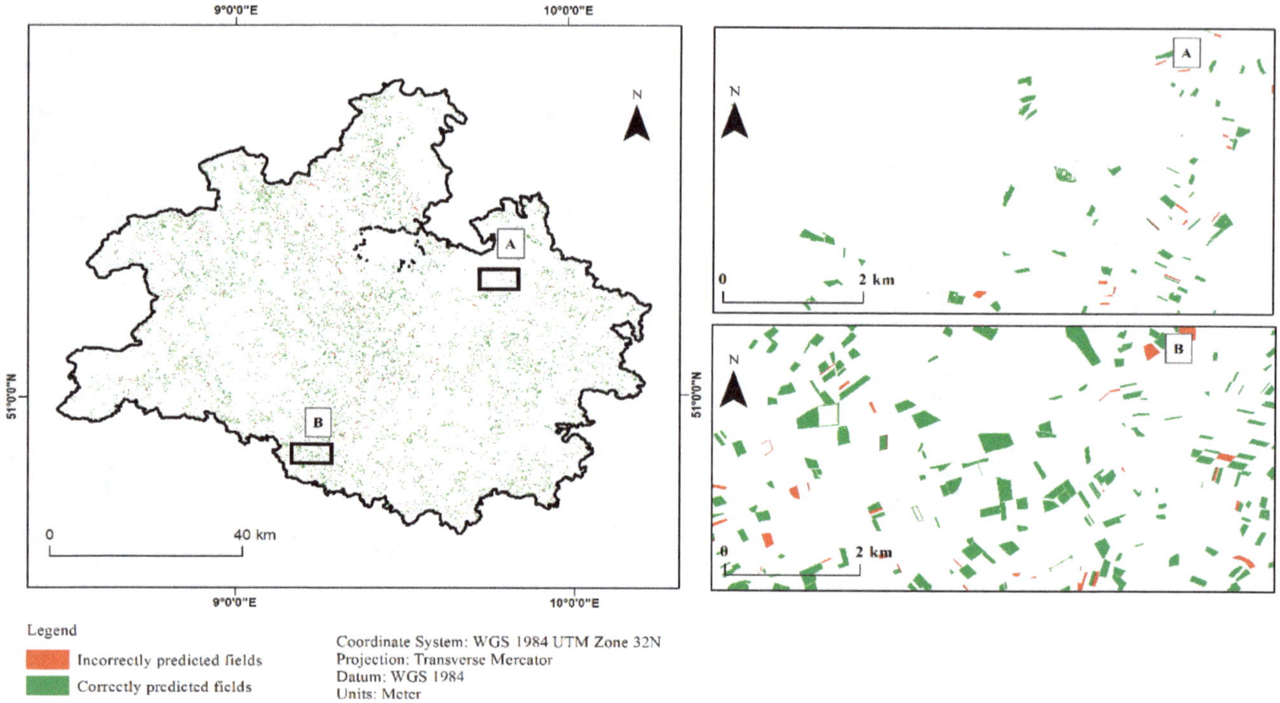

Figure A6. An IACS-based accuracy map of 2005 resulting from a multiple-year model calibrated using spectral data from 2007, 2009, 2010, 2011 and 2015.

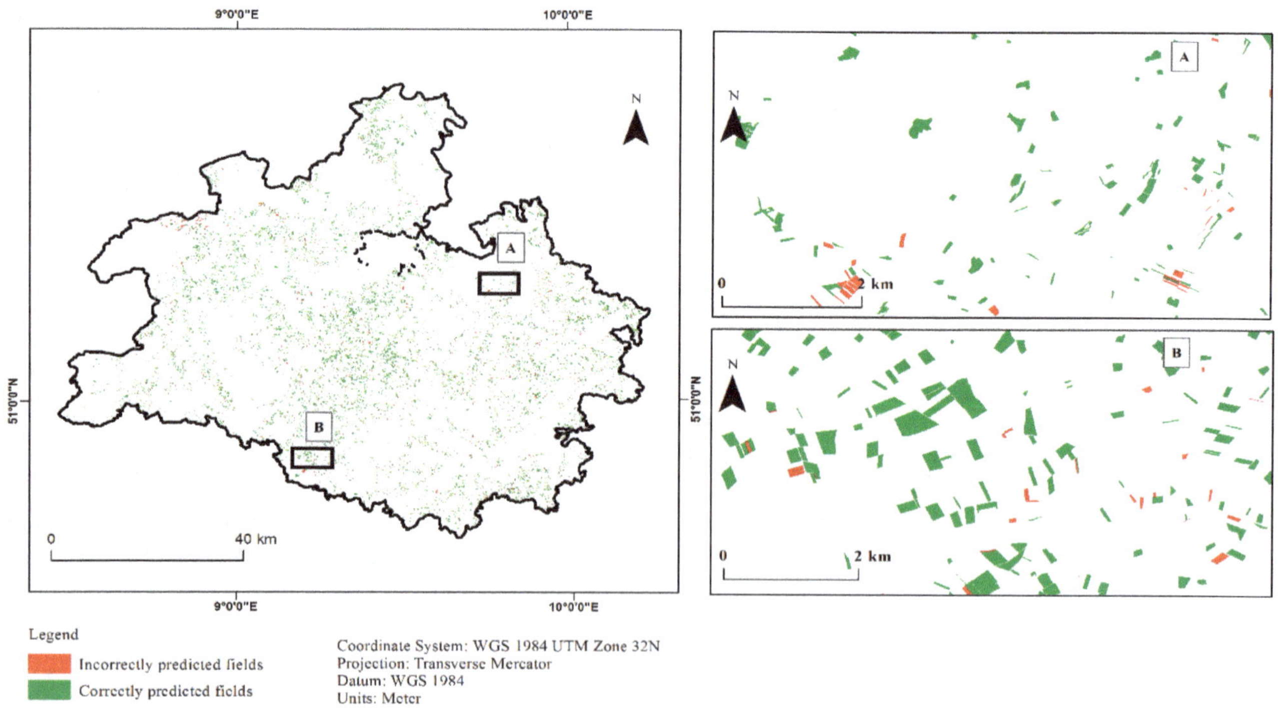

Figure A7. An IACS-based accuracy map of 2007 resulting from a multiple-year model calibrated using spectral data from 2005, 2009, 2010, 2011 and 2015.

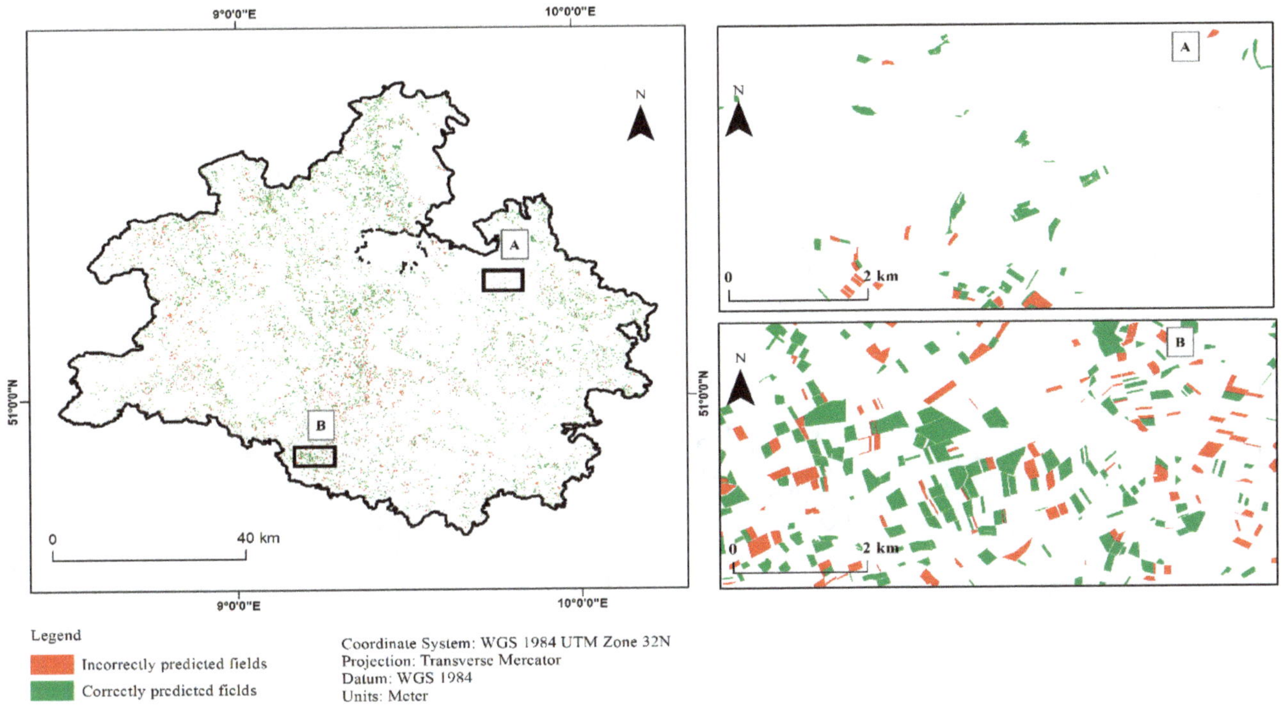

Figure A8. An IACS-based accuracy map of 2009 resulting from a multiple-year model calibrated using spectral data from 2005, 2007, 2010, 2011 and 2015.

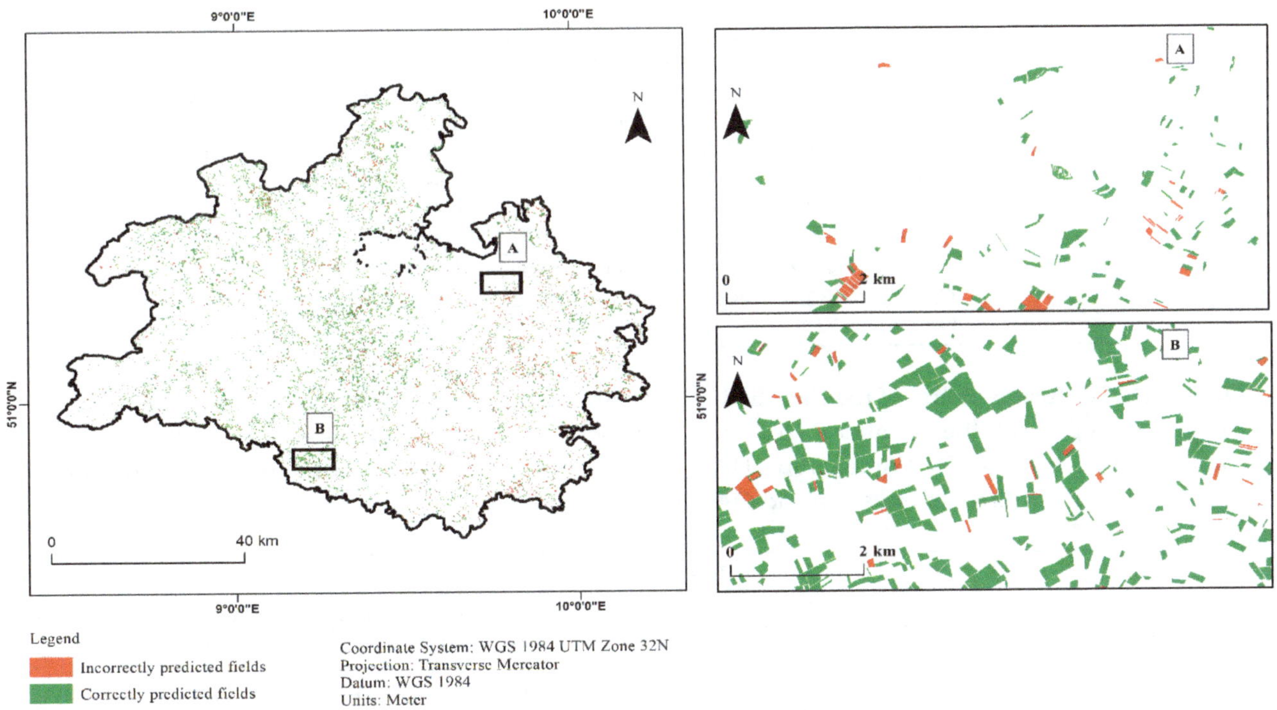

Figure A9. An IACS-based accuracy map of 2010 resulting from a multiple-year model calibrated using spectral data from 2005, 2007, 2009, 2011 and 2015.

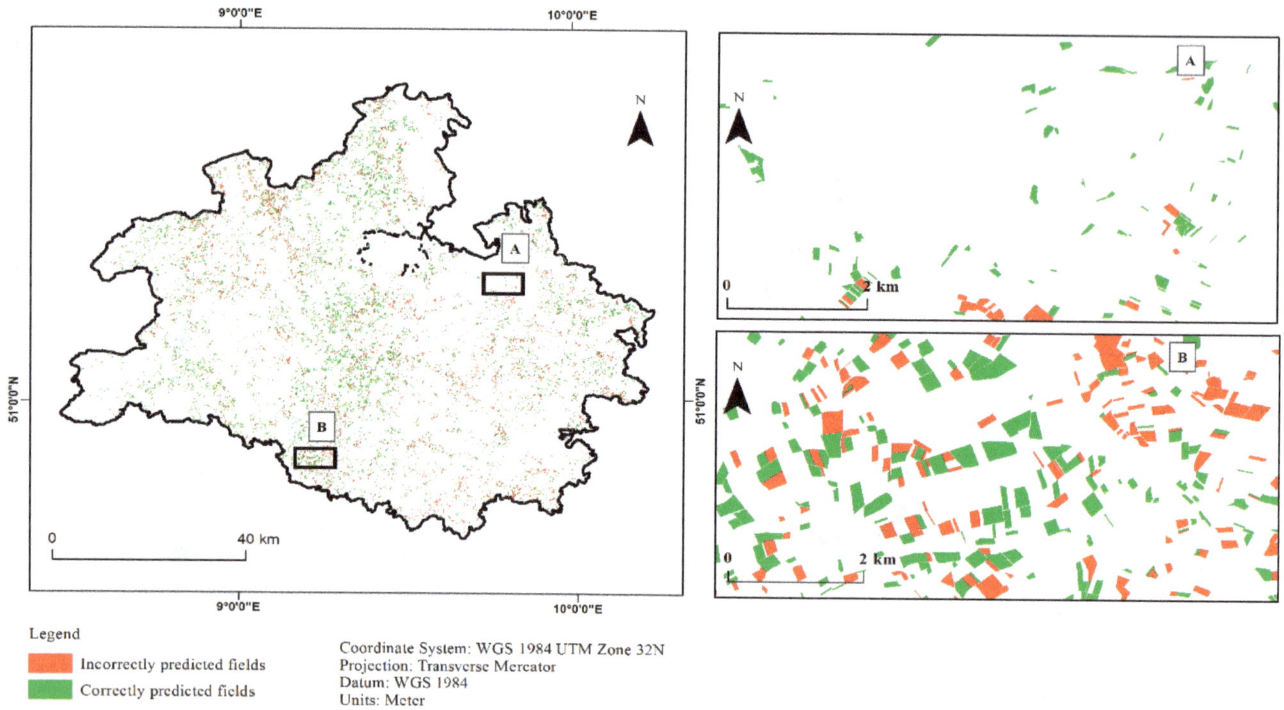

Figure A10. An IACS-based accuracy map of 2011 resulting from a multiple-year model calibrated using spectral data from 2005, 2007, 2009, 2010 and 2015.

Table A1. Confusion matrix of the classified map of 2005 using the multiple-year model. The shaded diagonals represent the number of correctly predicted crops. GL = Grassland, MZ = Maize, SC = Summer crops, WC = Winter crops, UA = User accuracy, PA = Producer accuracy, EC = Error of commission, EO = Error of omission and OA = Overall accuracy.

		Reference						
		GL	**MZ**	**SC**	**WC**	**Total**	**UA (%)**	**CE (%)**
Prediction	GL	4339	137	270	214	4960	87.48	12.52
	MZ	292	4527	442	33	5294	85.51	14.49
	SC	264	229	3792	1099	5384	70.43	29.57
	WC	105	48	496	3654	4303	84.92	15.08
	Total	5000	4941	5000	5000			
	PA (%)	86.78	91.62	75.84	73.08			
	OE (%)	13.22	8.38	24.16	26.92			
	OA (%)	81.8						

Table A2. Confusion matrix of the classified map of 2007 using the multiple-year model. The shaded diagonals represent the number of correctly predicted crops. GL = Grassland, MZ = Maize, SC = Summer crops, WC = Winter crops, UA = User accuracy, PA = Producer accuracy, EC = Error of commission, EO = Error of omission and OA = Overall accuracy.

		Reference						
		GL	**MZ**	**SC**	**WC**	**Total**	**UA (%)**	**CE (%)**
Prediction	GL	4380	203	316	156	5055	86.65	13.35
	MZ	158	3738	463	38	4397	85.01	14.99
	SC	68	358	3384	154	3964	85.37	14.63
	WC	394	90	837	4652	5973	77.88	22.12
	Total	5000	4389	5000	5000			
	PA (%)	87.60	85.17	67.68	93.04			
	OE (%)	12.40	14.83	32.32	6.96			
	OA (%)	83.32						

Table A3. Confusion matrix of the classified map of 2009 using the multiple-year model. The shaded diagonals represent the number of correctly predicted crops. GL = Grassland, MZ = Maize, SC = Summer crops, WC = Winter crops, UA = User accuracy, PA = Producer accuracy, EC = Error of commission, EO = Error of omission and OA = Overall accuracy.

		Reference						
		GL	**MZ**	**SC**	**WC**	**Total**	**UA (%)**	**CE (%)**
Prediction	GL	4185	986	430	175	5776	72.45	27.55
	MZ	166	4070	394	26	4656	87.41	12.59
	SC	345	889	3347	748	5329	62.81	37.19
	WC	304	531	1743	4051	6629	61.11	38.89
	Total	5000	6476	5914	5000			
	PA (%)	83.70	62.85	56.59	81.02			
	OE (%)	16.30	37.15	43.41	18.98			
	OA (%)	69.91						

Table A4. Confusion matrix of the classified map of 2010 using the multiple-year model. The shaded diagonals represent the number of correctly predicted crops. GL = Grassland, MZ = Maize, SC = Summer crops, WC = Winter crops, UA = User accuracy, PA = Producer accuracy, EC = Error of commission, EO = Error of omission and OA = Overall accuracy.

		Reference						
		GL	**MZ**	**SC**	**WC**	**Total**	**UA (%)**	**CE (%)**
Prediction	GL	4262	376	179	91	4908	86.84	13.16
	MZ	172	4990	956	185	6303	79.17	20.83
	SC	238	1830	4388	827	7283	60.25	39.75
	WC	328	171	307	3897	4703	82.86	17.14
	Total	5000	7367	5830	5000			
	PA (%)	85.24	67.73	75.27	77.94			
	OE (%)	14.76	32.27	24.73	22.06			
	OA (%)	75.6						

Table A5. Confusion matrix of the classified map of 2011 using the multiple-year model. The shaded diagonals represent the number of correctly predicted crops. GL = Grassland, MZ = Maize, SC = Summer crops, WC = Winter crops, UA = User accuracy, PA = Producer accuracy, EC = Error of commission, EO = Error of omission and OA = Overall accuracy.

		Reference						
		GL	**MZ**	**SC**	**WC**	**Total**	**UA (%)**	**CE (%)**
Prediction	GL	4493	209	264	460	5426	82.81	17.19
	MZ	141	3427	732	151	4451	76.99	23.01
	SC	108	4073	4644	363	9188	50.54	49.46
	WC	258	507	872	4026	5663	71.09	28.91
	Total	5000	8216	6512	5000			
	PA (%)	89.86	41.71	71.31	80.52			
	OE (%)	10.14	58.29	28.69	19.48			
	OA (%)	67.09						

Table A6. Confusion matrix of the classified map of 2015 using the multiple-year model. The shaded diagonals represent the number of correctly predicted crops. GL = Grassland, MZ = Maize, SC = Summer crops, WC = Winter crops, UA = User accuracy, PA = Producer accuracy, EC = Error of commission, EO = Error of omission and OA = Overall accuracy.

		Reference						
		GL	**MZ**	**SC**	**WC**	**Total**	**UA (%)**	**CE (%)**
Prediction	GL	10,313	461	592	1163	12,529	82.31	17.69
	MZ	183	8111	891	69	9254	87.65	12.35
	SC	305	296	9943	1134	11,678	85.14	14.86
	WC	199	53	464	7634	8350	91.43	8.57
	Total	11,000	8921	11,890	10,000			
	PA (%)	93.75	90.92	83.62	76.34			
	OE (%)	6.25	9.08	16.38	23.66			
	OA (%)	86.1						

References

1. Tilman, D.; Balzer, C.; Hill, J.; Befort, B.L. Global food demand and the sustainable intensification of agriculture. *Proc. Natl. Acad. Sci. USA* **2011**, *108*, 20260–20264. [CrossRef] [PubMed]
2. Nagy, A.; Fehér, J.; Tamás, J. Wheat and maize yield forecasting for the Tisza river catchment using MODIS NDVI time series and reported crop statistics. *Comput. Electron. Agric.* **2018**, *151*, 41–49. [CrossRef]
3. See, L.; Fritz, S.; You, L.; Ramankutty, N.; Herrero, M.; Justice, C.; Becker-Reshef, I.; Thornton, P.; Erb, K.; Gong, P.; et al. Improved global cropland data as an essential ingredient for food security. *Glob. Food Secur.* **2015**, *4*, 37–45. [CrossRef]
4. Gumma, M.K.; Thenkabail, P.S.; Teluguntla, P.; Rao, M.N.; Mohammed, I.A.; Whitbread, A.M. Mapping rice-fallow cropland areas for short-season grain legumes intensification in South Asia using MODIS 250 m time-series data. *Int. J. Digit. Earth* **2016**, *9*, 981–1003. [CrossRef]
5. Funk, C.C.; Brown, M.E. Declining global per capita agricultural production and warming oceans threaten food security. *Food Secur.* **2009**, *1*, 271–289. [CrossRef]
6. Pérez-Hoyos, A.; Rembold, F.; Kerdiles, H.; Gallego, J. Comparison of global land cover datasets for cropland monitoring. *Remote Sens.* **2017**, *9*, 1118. [CrossRef]
7. Tatsumi, K.; Yamashiki, Y.; Canales Torres, M.A.; Taipe, C.L.R. Crop classification of upland fields using Random forest of time-series Landsat 7 ETM+ data. *Comput. Electron. Agric.* **2015**, *115*, 171–179. [CrossRef]
8. Zhong, L.; Gong, P.; Biging, G.S. Efficient corn and soybean mapping with temporal extendability: A multi-year experiment using Landsat imagery. *Remote Sens. Environ.* **2014**, *140*, 1–13. [CrossRef]
9. Castillejo-González, I. Mapping of Olive Trees Using Pansharpened QuickBird Images: An Evaluation of Pixel- and Object-Based Analyses. *Agronomy* **2018**, *8*, 288. [CrossRef]
10. Foerster, S.; Kaden, K.; Foerster, M.; Itzerott, S. Crop type mapping using spectral-temporal profiles and phenological information. *Comput. Electron. Agric.* **2012**, *89*, 30–40. [CrossRef]
11. Moeckel, T.; Dayananda, S.; Nidamanuri, R.R.; Nautiyal, S.; Hanumaiah, N.; Buerkert, A.; Wachendorf, M. Estimation of vegetable crop parameter by multi-temporal UAV-borne images. *Remote Sens.* **2018**, *10*, 805. [CrossRef]
12. Wu, B.; Meng, J.; Li, Q.; Yan, N.; Du, X.; Zhang, M. Remote sensing-based global crop monitoring: Experiences with China's CropWatch system. *Int. J. Digit. Earth* **2014**, *7*, 113–137. [CrossRef]
13. Pittman, K.; Hansen, M.C.; Becker-Reshef, I.; Potapov, P.V.; Justice, C.O. Estimating global cropland extent with multi-year MODIS data. *Remote Sens.* **2010**, *2*, 1844–1863. [CrossRef]
14. Waldhoff, G.; Lussem, U.; Bareth, G. Multi-Data Approach for remote sensing-based regional crop rotation mapping: A case study for the Rur catchment, Germany. *Int. J. Appl. Earth Obs. Geoinf.* **2017**, *61*, 55–69. [CrossRef]
15. Habyarimana, E.; Piccard, I.; Catellani, M.; De Franceschi, P.; Dall'Agata, M. Towards Predictive Modeling of Sorghum Biomass Yields Using Fraction of Absorbed Photosynthetically Active Radiation Derived from Sentinel-2 Satellite Imagery and Supervised Machine Learning Techniques. *Agronomy* **2019**, *9*, 203. [CrossRef]

16. Liu, J.; Feng, Q.; Gong, J.; Zhou, J.; Liang, J.; Li, Y. Winter wheat mapping using a random forest classifier combined with multi-temporal and multi-sensor data. *Int. J. Digit. Earth* **2018**, *11*, 783–802. [CrossRef]
17. Maxwell, S.K.; Nuckols, J.R.; Ward, M.H.; Hoffer, R.M. An automated approach to mapping corn from Landsat imagery. *Comput. Electron. Agric.* **2004**, *43*, 43–54. [CrossRef]
18. Yin, H.; Prishchepov, A.V.; Kuemmerle, T.; Bleyhl, B.; Buchner, J.; Radeloff, V.C. Mapping agricultural land abandonment from spatial and temporal segmentation of Landsat time series. *Remote Sens. Environ.* **2018**, *210*, 12–24. [CrossRef]
19. Zheng, B.; Myint, S.W.; Thenkabail, P.S.; Aggarwal, R.M. A support vector machine to identify irrigated crop types using time-series Landsat NDVI data. *Int. J. Appl. Earth Obs. Geoinf.* **2015**, *34*, 103–112. [CrossRef]
20. Waldner, F.; Hansen, M.C.; Potapov, P.V.; Löw, F.; Newby, T.; Ferreira, S.; Defourny, P. National-scale cropland mapping based on spectral-temporal features and outdated land cover information. *PLoS ONE* **2017**, *12*, e0181911. [CrossRef]
21. Phiri, D.; Morgenroth, J. Developments in Landsat land cover classification methods: A review. *Remote Sens.* **2017**, *9*, 967. [CrossRef]
22. Liu, D.; Xia, F. Assessing object-based classification: Advantages and limitations. *Remote Sens. Lett.* **2010**, *1*, 187–194. [CrossRef]
23. Chen, D.; Stow, D. The Effect of Training Strategies on Supervised Classification at Different Spatial Resolutions. *Photogramm. Eng. Remote Sens.* **2002**, *68*, 1155–1162.
24. Li, Q.; Wang, C.; Zhang, B.; Lu, L. Object-based crop classification with Landsat-MODIS enhanced time-series data. *Remote Sens.* **2015**, *7*, 16091–16107. [CrossRef]
25. Zhu, L.; Radeloff, V.C.; Ives, A.R. Improving the mapping of crop types in the Midwestern U.S. by fusing Landsat and MODIS satellite data. *Int. J. Appl. Earth Obs. Geoinf.* **2017**, *58*, 1–11. [CrossRef]
26. Chen, Y.; Lu, D.; Moran, E.; Batistella, M.; Dutra, L.V.; Sanches, I.D.; da Silva, R.F.B.; Huang, J.; Luiz, A.J.B.; de Oliveira, M.A.F. Mapping croplands, cropping patterns, and crop types using MODIS time-series data. *Int. J. Appl. Earth Obs. Geoinf.* **2018**, *69*, 133–147. [CrossRef]
27. Foody, G.M.; Mathur, A.; Sanchez-Hernandez, C.; Boyd, D.S. Training set size requirements for the classification of a specific class. *Remote Sens. Environ.* **2006**, *104*, 1–14. [CrossRef]
28. Botkin, D.B.; Estes, J.E.; MacDonald, R.M.; Wilson, M.V. Studying the Earth's Vegetation from Space. *Bioscience* **1984**, *34*, 508–514. [CrossRef]
29. Sonobe, R.; Yamaya, Y.; Tani, H.; Wang, X.; Kobayashi, N.; Mochizuki, K.I. Mapping crop cover using multi-temporal landsat 8 oli imagery. *Int. J. Remote Sens.* **2017**, *38*, 4348–4361. [CrossRef]
30. Vuolo, F.; Neuwirth, M.; Immitzer, M.; Atzberger, C.; Ng, W.-T. How much does multi-temporal Sentinel-2 data improve crop type classification? *Int. J. Appl. Earth Obs. Geoinf.* **2018**, *72*, 122–130. [CrossRef]
31. Zhao, Y.; Feng, D.; Yu, L.; Wang, X.; Chen, Y.; Bai, Y.; Hernández, H.J.; Galleguillos, M.; Estades, C.; Biging, G.S.; et al. Detailed dynamic land cover mapping of Chile: Accuracy improvement by integrating multi-temporal data. *Remote Sens. Environ.* **2016**, *183*, 170–185. [CrossRef]
32. Xiao, X.; Boles, S.; Liu, J.; Zhuang, D.; Frolking, S.; Li, C.; Salas, W.; Moore, B. Mapping paddy rice agriculture in southern China using multi-temporal MODIS images. *Remote Sens. Environ.* **2005**, *95*, 480–492. [CrossRef]
33. Manfron, G.; Delmotte, S.; Busetto, L.; Hossard, L.; Ranghetti, L.; Brivio, P.A.; Boschetti, M. Estimating inter-annual variability in winter wheat sowing dates from satellite time series in Camargue, France. *Int. J. Appl. Earth Obs. Geoinf.* **2017**, *57*, 190–201. [CrossRef]
34. Huete, A.; Didan, K.; Miura, T.; Rodriguez, E.P.; Gao, X.; Ferreira, L.G. Overview of the radiometric and biophysical performance of the MODIS vegetation indices. *Remote Sens. Environ.* **2002**, *83*, 195–213. [CrossRef]
35. Bill, R.; Nash, E.; Grenzdörffer, G. Integrated Administration and Control System. In *Springer Handbook of Geographic Information System*; Kresse, W., Danko, D.M., Eds.; Springer: Berlin, Germany, 2012; pp. 801–803. ISBN 978-3-540-72678-4.
36. Boryan, C.; Yang, Z.; Mueller, R.; Craig, M. Monitoring US agriculture: The US department of agriculture, national agricultural statistics service, cropland data layer program. *Geocarto Int.* **2011**, *26*, 341–358. [CrossRef]
37. Massey, R.; Sankey, T.T.; Congalton, R.G.; Yadav, K.; Thenkabail, P.S.; Ozdogan, M.; Sánchez Meador, A.J. MODIS phenology-derived, multi-year distribution of conterminous U.S. crop types. *Remote Sens. Environ.* **2017**, *198*, 490–503. [CrossRef]

38. Griffiths, P.; Nendel, C.; Hostert, P. Intra-annual reflectance composites from Sentinel-2 and Landsat for national-scale crop and land cover mapping. *Remote Sens. Environ.* **2019**, *220*, 135–151. [CrossRef]
39. Ma, L.; Li, M.; Ma, X.; Cheng, L.; Du, P.; Liu, Y. ISPRS Journal of Photogrammetry and Remote Sensing A review of supervised object-based land-cover image classification. *ISPRS J. Photogramm. Remote Sens.* **2017**, *130*, 277–293. [CrossRef]
40. Blaschke, T.; Hay, G.J.; Kelly, M.; Lang, S.; Hofmann, P.; Addink, E.; Queiroz Feitosa, R.; van der Meer, F.; van der Werff, H.; van Coillie, F.; et al. Geographic Object-Based Image Analysis—Towards a new paradigm. *ISPRS J. Photogramm. Remote Sens.* **2014**, *87*, 180–191. [CrossRef]
41. Castilla, G.; Hernando, A.; Zhang, C.; McDermid, G.J. The impact of object size on the thematic accuracy of landcover maps. *Int. J. Remote Sens.* **2014**, *35*, 1029–1037. [CrossRef]
42. Wagner, B. Spatial analysis of loess and loess-like sediments in the Weser-Aller catchment (Lower Saxony and Northern Hesse, NW Germany). *Quarernary Sci. J.* **2011**, *60*, 27–46. [CrossRef]
43. HLNUG Hessian Agency for Nature Conservation, Environment and Geology. Available online: Atlas.umwelt.hessen.de (accessed on 9 November 2018).
44. USGS Landsat Surface Reflectance Products. Available online: https://earthexplorer.usgs.gov/ (accessed on 24 July 2018).
45. Masek, J.G.; Vermote, E.F.; Saleous, N.E.; Wolfe, R.; Hall, F.G.; Huemmrich, K.F.; Gao, F.; Kutler, J.; Lim, T.K. A landsat surface reflectance dataset for North America, 1990–2000. *IEEE Geosci. Remote Sens. Lett.* **2006**, *3*, 68–72. [CrossRef]
46. Vermote, E.; Justice, C.; Claverie, M.; Franch, B. Preliminary analysis of the performance of the Landsat 8/OLI land surface reflectance product. *Remote Sens. Environ.* **2016**, *185*, 46–56. [CrossRef]
47. Flood, N. Continuity of reflectance data between landsat-7 ETM+ and landsat-8 OLI, for both top-of-atmosphere and surface reflectance: A study in the australian landscape. *Remote Sens.* **2014**, *6*, 7952–7970. [CrossRef]
48. Li, P.; Jiang, L.; Feng, Z. Cross-comparison of vegetation indices derived from landsat-7 enhanced thematic mapper plus (ETM+) and landsat-8 operational land imager (OLI) sensors. *Remote Sens.* **2014**, *6*, 310–329. [CrossRef]
49. USGS Landsat Science Products-Landsat Surface Reflectance. Available online: https://www.usgs.gov/land-resources/nli/landsat/landsat-surface-reflectance?qt-science_support_page_related_con=0#qt-science_support_page_related_con (accessed on 10 May 2019).
50. Hijmans, R.J.; van Etten, J.; Cheng, J.; Mattiuzzi, M.; Sumner, M.; Greenberg, J.A.; Lamigueiro, O.P.; Bevan, A.; Racine, E.B.; Shortridge, A. Package "raster". *R Package* **2019**, version 2.9-5. 1–60.
51. Team, R.D.C. R Development Core Team R. *R A Lang. Environ. Stat. Comput.* **2015**, *1*, 409.
52. Breiman, L. Random forest. *Mach. Learn.* **2001**, *45*, 5–32. [CrossRef]
53. Belgiu, M.; Drăgu, L. Random forest in remote sensing: A review of applications and future directions. *ISPRS J. Photogramm. Remote Sens.* **2016**, *114*, 24–31. [CrossRef]
54. Tucker, C.J. Red and photographic infrared linear combinations for monitoring vegetation. *Remote Sens. Environ.* **1979**, *8*, 127–150. [CrossRef]
55. Huete, A. A soil-adjusted vegetation index (SAVI). *Remote Sens. Environ.* **1988**, *25*, 295–309. [CrossRef]
56. Gao, B.C. NDWI-A normalized difference water index for remote sensing of vegetation liquid water from space. *Remote Sens. Environ.* **1996**, *58*, 257–266. [CrossRef]
57. Liaw, A.; Wiener, M. Classification and Regression by randomForest. *R News* **2002**, *2*, 18–22.
58. Laborte, A.G.; Maunahan, A.A.; Hijmans, R.J. Spectral Signature Generalization and Expansion Can Improve the Accuracy of Satellite Image Classification. *PLoS ONE* **2010**, *5*, e10516. [CrossRef] [PubMed]
59. Fletcher, R.S. Using Vegetation Indices as Input into Random Forest for Soybean and Weed Classification. *Am. J. Plant Sci.* **2016**, *07*, 2186–2198. [CrossRef]
60. Cai, Y.; Guan, K.; Peng, J.; Wang, S.; Seifert, C.; Wardlow, B.; Li, Z. A high-performance and in-season classification system of field-level crop types using time-series Landsat data and a machine learning approach. *Remote Sens. Environ.* **2018**, *210*, 35–47. [CrossRef]

Permissions

All chapters in this book were first published in MDPI; hereby published with permission under the Creative Commons Attribution License or equivalent. Every chapter published in this book has been scrutinized by our experts. Their significance has been extensively debated. The topics covered herein carry significant findings which will fuel the growth of the discipline. They may even be implemented as practical applications or may be referred to as a beginning point for another development.

The contributors of this book come from diverse backgrounds, making this book a truly international effort. This book will bring forth new frontiers with its revolutionizing research information and detailed analysis of the nascent developments around the world.

We would like to thank all the contributing authors for lending their expertise to make the book truly unique. They have played a crucial role in the development of this book. Without their invaluable contributions this book wouldn't have been possible. They have made vital efforts to compile up to date information on the varied aspects of this subject to make this book a valuable addition to the collection of many professionals and students.

This book was conceptualized with the vision of imparting up-to-date information and advanced data in this field. To ensure the same, a matchless editorial board was set up. Every individual on the board went through rigorous rounds of assessment to prove their worth. After which they invested a large part of their time researching and compiling the most relevant data for our readers.

The editorial board has been involved in producing this book since its inception. They have spent rigorous hours researching and exploring the diverse topics which have resulted in the successful publishing of this book. They have passed on their knowledge of decades through this book. To expedite this challenging task, the publisher supported the team at every step. A small team of assistant editors was also appointed to further simplify the editing procedure and attain best results for the readers.

Apart from the editorial board, the designing team has also invested a significant amount of their time in understanding the subject and creating the most relevant covers. They scrutinized every image to scout for the most suitable representation of the subject and create an appropriate cover for the book.

The publishing team has been an ardent support to the editorial, designing and production team. Their endless efforts to recruit the best for this project, has resulted in the accomplishment of this book. They are a veteran in the field of academics and their pool of knowledge is as vast as their experience in printing. Their expertise and guidance has proved useful at every step. Their uncompromising quality standards have made this book an exceptional effort. Their encouragement from time to time has been an inspiration for everyone.

The publisher and the editorial board hope that this book will prove to be a valuable piece of knowledge for researchers, students, practitioners and scholars across the globe.

List of Contributors

Evangelos Anastasiou, Nikoleta Darra, Vasileios Psiroukis, George Xanthopoulos and Spyros Fountas
Department of Natural Resources Management & Agricultural Engineering, Agricultural University of Athens, Iera Odos 75, 11855 Athens, Greece

Athanasios Balafoutis
Department of Natural Resources Management & Agricultural Engineering, Agricultural University of Athens, Iera Odos 75, 11855 Athens, Greece
Institute for Bio-Economy & Agri-Technology, Centre of Research & Technology Hellas, Dimarchou Georgiadou 118, 38221 Volos, Greece

Aikaterini Biniari
Faculty of Crop Science, Agricultural University of Athens, Iera Odos 75, 11855 Athens, Greece

Maggie Mulley and Lammert Kooistra
Laboratory of Geo-Information Science and Remote Sensing, Wageningen University and Research, 6700 AA Wageningen, The Netherlands

Laurens Bierens
TEC-IB B.V., Oude Veiling 29, 2635 GK Den Hoorn (ZH), The Netherlands

Salima Yousfi, Adrian Gracia-Romero, José Luis Araus and Maria Dolores Serret
Section of Plant Physiology, University of Barcelona, 08028 Barcelona, Spain
AGROTECNIO Center, University of Lleida, 25198 Lleida, Spain

Nassim Kellas, Mohamed Kaddour and Ahmed Chadouli
Institut Technique des Grandes Cultures, Alger 16016, Algeria

Mohamed Karrou
International Center for Agricultural Research in the Dry Areas (ICARDA), Avenue Hafiane Cherkaoui, Rabat 10112, Morocco

Ke Zhang, Xiaojun Liu, Jingshan Lu, Songyang Li, Qiang Cao, Yan Zhu, Weixing Cao and Yongchao Tian
National Engineering and Technology Center for Information Agriculture, Key Laboratory for Crop System Analysis and Decision Making, Ministry of Agriculture, 1 Weigang Road, Nanjing 210095, China
Key Laboratory for Information Agriculture, Jiangsu, Collaborative Innovation Center for Modern Crop Production, Nanjing Agricultural University, 1 Weigang Road, Nanjing 210095, China

Syed Tahir Ata-Ul-Karim
National Engineering and Technology Center for Information Agriculture, Key Laboratory for Crop System Analysis and Decision Making, Ministry of Agriculture, 1 Weigang Road, Nanjing 210095, China
Key Laboratory for Information Agriculture, Jiangsu, Collaborative Innovation Center for Modern Crop Production, Nanjing Agricultural University, 1 Weigang Road, Nanjing 210095, China
Key Laboratory of Soil Environment and Pollution Remediation, Institute of Soil Science, Chinese Academy of Science, Nanjing 210008, China

Brian Krienke
Department of Agronomy and Horticulture, University of Nebraska-Lincoln, Lincoln, NE 68583, USA

Jian Chen, Zichao Zhang, Kai Zhang and Shubo Wang
College of Engineering, China Agricultural University, Beijing 100083, China

Yu Han
College of Water Resources & Civil Engineering, China Agricultural University, Beijing 100083, China

Francesco Novelli and Francesco Vuolo
Institute of Surveying, Remote Sensing & Land Information (IVFL), University of Natural Resources and Life Sciences, Vienna (BOKU), Peter Jordan Str. 82, 1190 Vienna, Austria

Heide Spiegel and Taru Sandén
Department for Soil Health and Plant Nutrition, Austrian Agency for Health and Food Safety (AGES), Spargelfeldstrasse 191, 1220 Vienna, Austria

Sajana Maharjan, Faisal Mueen Qamer, Mir Matin, Govinda Joshi and Sanjeev Bhuchar
International Centre for Integrated Mountain Development (ICIMOD), Kathmandu 44700, Nepal

Łukasz Jełowicki
OPEGIEKA Sp. z o.o., 82-300 Elbląg, Poland

Konrad Sosnowicz
Skysnap Sp. z o.o., 02-001 Warsaw, Poland

Wojciech Ostrowski, Katarzyna Osińska-Skotak and Krzysztof Bakuła
Department of Photogrammetry, Remote Sensing and Spatial Information Systems, Faculty of Geodesy and Cartography, Warsaw University of Technology, 00-661 Warsaw, Poland

Isabel Luisa Castillejo-González
Department of Graphic Engineering and Geomatics, University of Cordoba, Campus de Rabanales, 14071 Córdoba, Spain

Anna Dalla Marta, Marco Mancini and Simone Orlandini
Department of Agriculture, Food, Environment and Forestry, University of Florence, Piazzale delle Cascine 18, 50144 Firenze, Italy

Giovanni Battista Chirico and Guido D'Urso
Department of Agricultural Sciences, University of Naples "Federico II", Via Università 100, 80055 Portici (NA), Italy

Salvatore Falanga Bolognesi and Carlo De Michele
Ariespace s.r.l., Spin off Company, University of Napoli "Federico II", Centro Direzionale IS A3, 80143 Napoli, Italy

Filiberto Altobelli
CREA Research Centre for Agricultural Policies and Bioeconomy, Via Po 14, 00198 Roma, Italy

Quanlong Feng
College of Resources and Environmental Sciences, China Agricultural University, Beijing 100193, China

College of Land Science and Technology, China Agricultural University, Beijing 100083, China

Jianyu Yang, Yiming Liu, Cong Ou, Dehai Zhu, Bowen Niu and Baoguo Li
College of Land Science and Technology, China Agricultural University, Beijing 100083, China

Jiantao Liu
School of Surveying and Geo-Informatics, Shandong Jianzhu University, Jinan 250101, China

Isaac Kyere, Thomas Astor, Rüdiger Graß and Michael Wachendorf
Grassland Science and Renewable Plant Resources, Universität Kassel, Steinstraße 19, D-37213 Witzenhausen, Germany

Index